T0134869

# The Anthropocene: Politik—Economics—Society—Science

## Volume 32

**Series Editor**

Hans Günter Brauch, Peace Research and European Security Studies (AFES-PRESS), Mosbach, Baden-Württemberg, Germany

More information about this series at http://www.springer.com/series/15232
http://www.afes-press-books.de/html/APESS.htm
http://www.afes-press-books.de/html/APESS_32.htm

Clarilza Prado de Sousa ·
Serena Eréndira Serrano Oswald
Editors

# Social Representations for the Anthropocene: Latin American Perspectives

 Springer

*Editors*
Clarilza Prado de Sousa
Post Graduate Program in Education
Psychology/Post Graduate Program
in Professional Education: Training of
Trainers
Pontifical Catholic University of São Paulo
(PUC-SP)
São Paulo, Brazil

Serena Eréndira Serrano Oswald
Regional Multidisciplinary Research Centre
(CRIM)
National Autonomous University of Mexico
(UNAM)
Cuernavaca, Mexico

ISSN 2367-4024          ISSN 2367-4032    (electronic)
The Anthropocene: Politik—Economics—Society—Science
ISBN 978-3-030-67777-0          ISBN 978-3-030-67778-7    (eBook)
https://doi.org/10.1007/978-3-030-67778-7

Scientific Committee: Susana Menin – São Paulo State University (UNESP), Brazil
Anamerica P. Marcondes – Pontifical Catholic University of São Paulo (PUC-SP), Brazil
Editor: PD Dr. Hans Günter Brauch, AFES-PRESS e.V., Mosbach, Germany
English Language Editor: Dr. Vanessa Greatorex, Chester, England, UK

This Springer imprint is published by the registered company Springer Nature Switzerland AG
The registered company address is: Gewerbestrasse 11, 6330 Cham, Switzerland

*To Luca Kai*

*Let the beauty of what you love be what you do*
*Rumi*

# Preface

Humanity is experiencing highly complex times. Defining them requires many angles and viewpoints to be considered, as they encompass interactions between a variety of opinions, representations and social practices. This diversity has its advantages; it helps us understand the complexities and the importance of lived experience in diverse situated contexts across the world. What we recognise as common sense knowledge—i.e. the capital that grants us access to the wisdom of the ingenuous, to that understanding that is debated both in the public and private spheres from the mnemic footprint and from historical and everyday-life refer-ents—represents the most democratic meeting point for humanity. No matter what culture, age, sex or class, level of wealth or poverty, all human beings have the capacity to experience and count on common knowledge as a tangible element that ought to be recognised as the most important capital that humanity has devel-oped. We should not forget that "everyday judgements and explanations form a normative interrelation that determines our quotidian thoughts and experiences" (Moscovici, 2011: xi).

Following Moscovici, the art of interpreting the world, social communication and our everyday coexistence is not exempt from judgements or norms that grad-ually become incorporated into our ways of thinking and acting as a result of our interactions. They give sense to our reality, helping us to accommodate and anchor new elements and information, objectifying and emotionally directing novelty and enriching the internal coherence of our representations.

Understanding the mechanisms through which a social representation is constructed is no easy task, precisely because of the cognitive naturalization that takes place in the individual who tends to simplify the socio-cultural process. Examining this is crucial in the field of psychosociology; it unveils the strength of collective thought.

Studying and analysing social representations as systems—meaning a non-hierarchical relational order that structures the elements which make up the corpus of meanings and referents of culture for human beings—implies recognising the role played by norms, beliefs and ideology in a diversity of collectives and social imaginaries.

In the concrete case of Latin America, since its inception in the 1980s, Social Representations Theory (SRT) has focused to a large extent on conflict as a dynamic process, encompassing diverse social objects and cultural phenomena. This requires both a comprehensive analytical lens and, in particular, responses that enable the formulation of strategies that promote resistance and well-being for our cultures. This has also given Social Representations Theory a new role, especially when facing unprecedented new challenges that clearly call for a critical stance in the face of reality: invigorating methodologies that increasingly privileged processes of intervention and impinged on processes of construction and deconstruction. This new role is similar to what Wolfgang Wagner describes as "a kind of practical logic of orientation and communication that has nothing to do with the logic of explanation used in scientific affirmations".

We could say that, following the importation of Social Representations Theory to our region in the 1980s, there have been many generations of social psychologists who have devoted their lives to social representations. They have developed research into social, cultural and political phenomena that have led to scientific explanations based on a more critical and inclusive vision than that of normative models. In addition, principles of community psychology have been incorporated, almost by contextual inheritance, adding a distinctive touch to our research, strengthening the social and political commitments that identify us as a collective. It is not an overstatement to say that in our countries a new way of approaching Social Representations Theory has developed, using it also as a powerful device that introduces guidelines for understanding "subjective space" (González Rey 2002: 22) while simultaneously making advances in the analysis of the social imaginary to which it relates.

First and foremost, it is important to recognize and thank Clarilza del Prado Sousa and Serena Eréndira Serrano Oswald for the originality of their idea that considers the Anthropocene as a societal turning point, an era that needs a critical intergenerational and multidisciplinary dialogue and benefits from the main premises of Social Representations Theory. This book has motivated an interesting intergenerational dialogue, offering a critical perspective that established researchers make of the brilliant intellectual production of younger researchers, providing a repertory of theoretical, conceptual and applied studies and configuring an immensely rich panorama of the continuity and advances of Social Representations Theory in Latin America. Altogether, it also shows a clear political positioning of researchers in the face of the realities and groups they study. This recovers Moscovici's project of social transformation based on a critical analysis of the reasons that are displayed as resistance to change in his own research and, in general terms, refer to social, group and network conflicts resulting from everyday life interaction.

I believe that the originality of this book, besides this intergenerational dialogue, lies in its ability to portray the versatility of Social Representations Theory and its capacity to embrace a vast array of themes of clear relevance at this societal turning point, the Anthropocene: research objects such as public policies, education, the environment, violence, and vulnerability, as well as deep-seated beliefs and research categories such as gender. At the same time, the texts remind us of the processes through which social representations are structured and transformed, incorporating active minorities that have been crucial in this research tradition.

The end of the book is a gem that illustrates how Social Representations Theory 'in motion' is a great ally for all those of us who are committed to social change in Latin America, despite the impact of globalized politics, of that avalanche that has been responsible—among many other things—for annihilating any sidewalk that might lead towards the recognition of the subject of rights, especially for the poor, anonymous and dispossessed members of the masses who make up the majority of the population in our contexts; people and groups who lack political and economic power, who are precarious and deprived, without any social recognition, and are struggling to keep their territory, which—according to this exclusionary logic of globalization—is the only thing of value surrounding them. They are cast off, abject,[1] according to psychoanalysis.

These scraps, abject and implementable, are evidence of the enormous inequality gap and social injustice in our region. They are also the object of the masquerade built by the hegemony of the neoliberal regime, which is detrimental to human freedom.

I must warn you that if you are willing to read this book with a critical eye, you will understand why I have pointed out the importance of focusing on the abject, with a passion that is perhaps inappropriate in the preface of an academic book. This said, I must emphasize that women make up the weakest link in this chain of violence and inequality. Gender discrimination and femicide in our Latin American countries are the overwhelming proof of this oblivion. In order to understand this abjection, we also need to consider the mark of colonialist thought. It has a permanent imprint in the Latin American social imaginary, confounding "the radical heterogeneity of subaltern social subjects" (Chakrabarty, 1997) in a cultural hybrid that lacks definition but is still not defeated.

---

[1]The abject situates us in the face of archaic *fragile states* where humans are driven into animal-like territories (Kristeva, 1989), obeying their drives and instincts without integration, rejecting and being rejected, rebelling against customs and culture.

From its epistemic roots, sociological social psychology, or psychosociology, recovers the importance of the hidden strength of these forgotten, excluded and segregated masses, analysing their potential as triggers for change, despite the fact that they were constituted in colonised social imaginaries. From this perspective, the deconstruction of social systems anchored to a sense of defencelessness and exclusion is a priority. Through analysing and studying the impact of meaning, we can promote novel practices that point towards a dignified life, with effective rights, eradicating impositions and abuses of power, as well as symbolic and structural violence. In order to forge everyday life alternatives and equitable relations of mutual recognition, this should be a key deconstruction aim.

Finally, the editors of this volume should know that we are grateful for the opportunity to get to know more about Latin American research on social representations, and the scope of interests, aims and limitations of this critical scientific thought. However, it is also important to highlight the challenges we face, especially if we recognise that we are reaching the end of an era that has been devastating in many ways, where a new social order is being promoted, but where some of the very fragile liberties that were reached are being put at risk. Covid-19 has provided a good pretext.

Merida, Mexico                                          María de Fátima Flores Palacios[2]
2020

# References

Chakrabarty, Dipesh, 1997: "Poscoloniality and the artifice of history: who speaks for the "Indian" past?", in: Guha, Ranajit (Ed.): *A subaltern studies reader* (Minneapolis, University of Minnesota Press): 263–294.

Gonzalez Rey, Fernando Luis, 2002: "La subjetividad; su significación para la ciencia psicológica", in: Furtado, Odair; González Rey, Fernando Luis (Eds.): *Por una epistemología da subjetividade: um debate entre a teoría socio-histórica e a teoría das representacoes sociais* (São Paulo: Ed. Casa do Psicólogo): 17–42.

Kristeva, Julia, 1989: *Poderes del horror: un ensayo sobre la abyección* (Mexico City: Siglo XXI).

Moscovici, Serge, 2011: "Prólogo", in: Wagner, Wolfgang; Hayes, Nicky; Flores Palacios, Fátima (Ed.): *Discurso de lo cotidiano y el sentido común; la teoría de las representaciones sociales* (Barcelona: Anthropos).

[2]Translated by Serena Eréndira Serrano Oswald.

# Acknowledgements

This volume on *Social Representations for the Anthropocene: Latin American Perspectives* emerged from the collective work and dialogue between highly established scholars and emerging talented consolidating scholars interested in social representations and a deep understanding of the social problems that impact the Latin American region. The co-editors of this book are grateful to all authors who passed a double-blind anonymous peer review process and subsequently revised their papers to take into account the many critical comments and suggestions from these reviewers. Each chapter was assessed by at least three reviewers from different countries.

We would like to thank all reviewers who spent much time reading and commenting on the submitted texts and who made detailed, perceptive and critical remarks and suggestions for improvement—even for texts that could not be included here. The goal of the editors has been to enhance the quality of all texts. The editors were bound by the reviewers' reports.

The following colleagues (in alphabetical order) contributed reviews:

- Antunes Rocha, Maria Izabel, Universidade Federal de Minas Gerais, Brazil
- Arruda, Prof. Angela, Universidade Federal do Rio de Janeiro, UFRJ, Brazil
- Castorina, Prof. José Antonio, University of Buenos Aires (UBA), Argentina
- Camargo, Prof. Brigido Vizeu, Postgraduation Program in Psychology and Psychology Department – Federal University of Santa Catarina, UFSC, Brazil
- Campos, Pedro Humberto Faria, Universidade Salgado Filho, UNIVERSO, Rio de Janeiro
- Chamon, Prof. Edna Maria Querido de Oliveira Chamon, researcher at Taubaté University (UNITAU), São Paulo and Estácio de Sá University, Rio de Janeiro
- Cordeiro, Prof. Maria Helena Baptista Vilares, Universidade Federal da Fronteira Sul
- de Oliveira, Prof. Denize Cristina, State University of Rio de Janeiro, Brazil
- Guareschi, Prof. Pedrinho A., Universidade Federal do Rio Grande do Sul, UFRGS, Brazil
- Guerrero Tapia, Prof. Alfredo, Faculty of Psychology, UNAM, Mexico

- Macedo, Prof. Elizabeth, Universidade do Estado do Rio de Janeiro, Brazil
- Mazzotti, Prof. Alda Judith Alves, Emeritus Professor at Universidade Federal do Rio de Janeiro, Emeritus Professor at Universidade Estácio de Sá
- Mazzotti, Prof. Tarso, Federal University of Rio de Janeiro, UFRJ, Brazil
- Oswald Spring, Prof. Úrsula, UNAM, Mexico
- Ornellas, Prof. Maria de Lourdes S., Universidade do Estado da Bahia, Brazil
- Santos, Dra. Maria de Fátima de Souza, Universidade Federal de Pernambuco, Brazil
- Seidmann, Dra Susana, Universidad de Buenos Aires, Universidad de Belgrano, Argentina
- Serrano Moreno, Prof. Jorge Ramón, UNAM, Mexico
- Trindade, Prof. Zeidi Araujo, Universidade Federal do Espírito Santo, Brazil

We would also like to thank the series editor, Prof. Hans Günter Brauch (Germany); our members of the scientific committee Prof. Susana Menin and Prof. Anamerica P. Marcondes (Brazil); our language editor Dr. Vanessa Greatorex (England); Prof. María de Fátima Flores Palacios for writing the Preface (Mexico); and the Springer team in Heidelberg, Dr. Christian Witschel, Birte Dalia, and Marion Schneider (Germany), as well as the Springer team in Chennai led by Arulmurugan Venkatasalam (India).

São Paulo, Brazil                                              Clarilza Prado de Sousa
Cuernavaca, Mexico                            Serena Eréndira Serrano Oswald
May 2020

# Contents

# Abbreviations

| | |
|---|---|
| AEO | Average Evocation Order |
| AMECIDER | Mexican Regional Science Association |
| ANPEPP | National Association in Postgraduate Programmes and Research in Psychology, Brazil |
| CAPES | Coordination for the Improvement of Higher Education Personnel |
| CCF | Carlos Chagas Foundation, Brazil |
| CEESTEM | Centre for Economic and Social Studies of the Third World, Mexico City, Mexico |
| CEFET | Federal Centre for Technological Education Celso Suckow da Fonseca, Brazil |
| CEMERS | Mexican Centre for the Study of Social Representations |
| CEPHCIS | Peninsular Centre in Humanities and Social Sciences, Mexico |
| CIERS-ed | International Centre of Studies on Social Representations and Subjectivity – Education, Brazil |
| CIS-INAH | Centre of Higher Research of the National Institute of Anthropology and History, Mexico City, Mexico |
| CLAIP | Latin American Council for Peace Research |
| CNPq | National Council of Scientific and Technological Development, Brazil |
| ColMex | The College of Mexico |
| CONACyT | National Council for Science and Technology, Mexico |
| CONICET | National Council for Research and Technology, Argentina |
| CRIM | Regional Multidisciplinary Research Centre of UNAM, Cuernavaca, Mexico |
| EHESS | School of Advanced Studies in Social Sciences (*École des Hautes Études en Sciences Sociales*), France |
| ENTS | National School of Social Work, Mexico |
| FAPERJ | Amparo Research State Foundation in Rio de Janeiro (Fundação de Amparo à Pesquisa do Estado do Rio de Janeiro) |
| FIAP | Ibero-American Federation of Associations of Psychology |
| FLACSO | Latin American Faculty of Social Sciences |
| FM-UFMG | Faculty of Medicine of the Federal University of Minas Gerais |

| | |
|---|---|
| GEPPE-rs | Study Group in Education Research and Social Representations |
| GERES | Social Representations Study Group |
| GISSA | Engineering Geo-solutions and Systems Company, Mexico |
| GPPIN | Research Group in Infant Psychology, Brazil |
| HIV/AIDS | Human immunodeficiency virus infection and acquired immune deficiency syndrome |
| I. Crisol | Crisol Institute, Mexico |
| I. Gestalt | Gestalt Institute, Mexico |
| IICE | Institute of Research in Educational Science, Argentina |
| INAH | National Institute of Anthropology and History, Mexico |
| IRaMuTeQ | R interface for multidimensional analysis of texts and question-naires software package (from French *Interface de R pour Analyses Multidimensionnelles de Textes et de Questionnaires*) |
| ISCTE-IUL | University Institute of Lisbon in Portugal |
| LAC | Latin America and Caribbean |
| LACCOS | Laboratory of Communication and Cognition |
| LARSA | Latin American and Caribbean Regional Science Association |
| LSE | London School of Economics and Political Science, University of London |
| NASA | National Aeronautics and Space Administration, USA |
| NEARS | Centre for International Studies in Social Representations |
| NYU | New York University |
| PIBIC | Scientific Initiation Scholarship Program |
| PLH/A | Person living with HIV/AIDS |
| PPCTE | Science, Technology and Education Post-Graduation Program |
| PPGP | Psychology Graduate Program |
| PROCAMPO | Countryside Education Licentiateship Support Programme (Brazil) |
| PSE | School Health Program, Brazil (*Programa de Saúde na Escola*) |
| PUC-Goiás | Pontifical Catholic University of Goiás, Brazil |
| PUC-Minas | Pontifical Catholic University of Minas, Brazil |
| PUC-PR | Pontifical Catholic University of Paraná, Brazil |
| PUC-RGS | Pontifical Catholic University of Rio Grande do Sul, Brazil |
| PUC-SP | Pontifical Catholic University of Sao Paulo, Brazil |
| RedePso | Coordinator of the Social Psychology Network |
| ReLePe | Latin American Network of Epistemological Studies in Educational Policy |
| REMOSCO | Serge Moscovici World Repository at the École des Hautes Études en Sciences Sociales (EHESS) in Paris, France |
| RENIRS | National Social Representations Research Network, Mexico |
| SADAF | Argentinian Society for Philosophical Analysis |
| SBP | Brazilian Society of Psychology |
| SIERS | State Symposium in Social Representations and Education |
| SIM | System of Mortality Information in Brazil (*Sistema de Informações sobre Mortalidade*) |
| SNI | National Council of Researchers, Mexico |

| | |
|---|---|
| SOAS | School of Oriental and African Studies, University of London |
| SPSS | Statistical Package for the Social Sciences |
| SR | Social Representations |
| SRT | Social Representations Theory |
| UAB | Autonomous University of Barcelona, Spain |
| UAM | Autonomous University of Madrid, Spain |
| UAM | Metropolitan Autonomous University in Mexico City, Mexico |
| UANL | Autonomous State University of Nuevo Leon, Mexico |
| UBA | University of Buenos Aires, Argentina |
| UBC | University of British Columbia, Canada |
| UCLA | University of California, USA |
| UCPel | Catholic University of Pelotas, Brazil |
| UCSAL | Catholic University of Salvador, Brazil |
| UCV | Central University of Venezuela, Venezuela |
| UERGS | State University of Rio Grande do Sul, Brazil |
| UERJ | State University of Rio de Janeiro, Brazil |
| UFCSPA | Federal University of Health Sciences in Porto Alegre, Brazil |
| UFES | Espírito Santo Federal University, Brazil |
| UFF | Fluminense Federal University, Brazil |
| UFG | Federal University of Goiás, Brazil |
| UFMG | Federal University of Minas Gerais, Brazil |
| UFMT | Federal University of Mato Grosso, Brazil |
| UFPE | Federal University of Pernambuco, Brazil |
| UFRGS | Federal University of Rio Grande do Sul, Brazil |
| UFRJ | Federal University of Rio de Janeiro, Brazil |
| UFSC | Federal University of Santa Catarina, Brazil |
| UFSM | Federal University of Santa Maria, Brazil |
| UIS | Industrial University of Santander, Colombia |
| UNAM | National Autonomous University of Mexico, Mexico |
| UNEB | State University of Bahía, Brazil |
| UNESA | University Estácio de Sá, Brazil |
| UNESC | University of the Far South of Santa Catarina, Brazil |
| UNESCO | United Nations Educational, Scientific and Cultural Organization |
| UNESP | Paulista State University, Brazil |
| UNIARA | University of Araraquara, Brazil |
| UNIBO | University of Bologna, Italy |
| UNICACH | University of Science and Arts of Chiapas, Mexico |
| UNICAMP | University of Campinas, Brazil |
| UNICID | City University of Sao Paulo, Brazil |
| UNIFIA | Amparense University Centre, Brazil |
| UNIPD | University of Padua, Italy |
| UNIPE | Pedagogical University of Argentina |
| UNITAU | University of Taubaté, Brazil |
| USA | United States of North America |
| USP | University of Sao Paulo, Brazil |

# Chapter 1
# Building a Sand String: Social Representations for the Anthropocene

Serena Eréndira Serrano Oswald and Clarilza Prado de Sousa

*According to Rabbi Haggai, there was once a wealthy and prosperous country called Kid-Elin. However, the country's youth was filled with spite and grudges against the old. They proclaimed that old age hindered progress; they believed that their beautiful country would be a thousand times stronger, a thousand times happier and more glorious if given to young men. They said, "Let us put an end to all this useless senility!" But not everyone agreed. There was one old man who was saved.*

*A boy named Zarmã held his father in great affection. On the day of the revolt he hid the old man in a subterranean tunnel, thus saving him from falling into the power of the rebels. After exterminating the old, the country of Kid-Elin was governed only by the young men, who were joyful but inexperienced.*

*One day, an embassy of technicians, jurists and economists was sent from the neighbouring kingdom of Beluã. On arriving, the emissaries sought out the young President of the Republic and made him aware of their thorny and delicate mission. The Beluanite government required the immediate delivery of Kid-Elin territory.*

*"We have a treaty to that end with Kid-Elin. Here are the documents."*

*Kid-Elin's President, ministers and magistrates examined the titles and deeds. Everything seemed clear, fluid, unmistakable. The Kid-Elin Republic was forced to surrender all of its land to its powerful neighbour in accordance with an agreement made many years earlier. What could they do? It would be the ruin of the country. It would be a disgrace for young people!*

Serena Eréndira Serrano Oswald is a professor in the Regional Multidisciplinary Research Centre at the National Autonomous University of Mexico (CRIM-UNAM).

Clarilza Prado de Sousa is a professor at the Pontifical Catholic University of São Paulo (PUC-SP) in Brazil.

© Springer Nature Switzerland AG 2021
C. Prado de Sousa and S. E. Serrano Oswald (eds.),
*Social Representations for the Anthropocene: Latin American Perspectives*,
The Anthropocene: Politik—Economics—Society—Science 32,
https://doi.org/10.1007/978-3-030-67778-7_1

*It was then that young Zarmã had an inspiration. He remembered his old father and went to ask him for his advice about the case. On the following day, when he met with the embassy in the presence of the President, ministers, magistrates, generals and high officials, who were all very young, he faithfully repeated the words he had heard from his father: "The Kid-Elin Republic is resolved to return the territory you claim. But it demands that in return the kingdom of Beluã must fulfil its commitments and return the sand string intact!"*

*Upon hearing this sentence, the Beluanite ambassadors were alarmed. That counter-demand, specified in the clause about the return of the "string of sand" fell like a bomb in the middle of them. And after exchanging a few words, in a low voice, with his counsellors, the spokesman conveyed their nervous decision: "We give up our request. You may preserve forever the territory with all its lands and fields!" At this point the spokesman paused, and then, in a tone of irony and rancour, he said: "But it is for me to deny the news spread by your agents and emissaries. I can swear that the Youth Republic does not exist. There are old people still in this country!"*

*And, concluded Rabbi Haggai, the string of sand links the past to the present and links the present to the future.*[1]

**Abstract** In this chapter, we present the main objectives and contents of this book, and the theoretical discussion bridging Social Representations Theory and the Anthropocene as societal era, introducing some of the main research groups and emerging scholars in Latin America. This first volume centres on Brazilian contributions, since they are the oldest and most visible in the continent, although there are also works from Argentina, Venezuela, Mexico and Colombia. The structure of the chapter is as follows: following the 'Building a Sand String' opening quote, we present the objective of the book and a brief outline of the regional research groups represented in this text. The second section discusses the common sense relationship between the Anthropocene as a societal era and social representations. The last sections detail the structure and contents of each chapter of the book.

## 1.1  The Objective of This Book

This book has a purpose and a symbolic meaning. Its creators intend to demonstrate the robustness of a theory that, across generations, resists, improves and expands its heuristic capacity and its ability to engage with other disciplines and theories in the social sciences and the humanities. The potential of Social Representations Theory (RST) to unveil the everyday, bring to light the workings

---

[1] Summary of the story "Strings of Sand" in: Tahan, Malba, 1985 (1941): *Lendos do Povo de Deus*, 11th edn. (Rio de Janeiro: Record). See also: https://www.malbatahan.com.br/ and https://encantadorasdehistorias.wordpress.com/2017/10/27/a-corda-de-areia.

of common sense and foster awareness of the trajectories of lives has made it possible to think about processes of change in practices that involve new methods, but above all it indicates that transformations, however desired, must be constructed while considering the past, its history, its achievements and its mistakes.

Serge Moscovici stated that he purposely proposed an open theory, i.e. a theory that is constructed and established with the support of research and reflection. Knowing historical studies, when proposing the development of a concept, a theory, which considers time, experience, memory, social and cultural context, he understood that a process of constant reconstruction is necessary to guarantee its updating and the communicability of their assumptions. As Oliveira (2000) has already stated, Moscovici's study now tends towards generality, sometimes toward singularity, and indicates that ideas can be a pre-existing social and cultural environment in the life of the subject, who will carry these ideas and translate them into social representations.

As already stated in the Preface, this volume emerged from the collective work and dialogues between established and emerging scholars in the field of SRT in Latin America. These are committed and critical academics who have a special interest in collectively thinking about the usefulness and potential of SRT in order to characterize, explain, analyse, and transform some of the key issues in our territories, collaborate with individual and social agents, and envisage creative and critical alternatives that are rarely considered by more hegemonic types of social scientific endeavour. The value and ethnographic depth of common sense speaks of the coherence and deep cultural roots of everyday knowledge and practices, and the need to consider everyday life knowledge, relations and practices in order to analyse and transform current societal challenges in the Anthropocene.

In this book, the editors began to gather Latin American groups that are developing research to deepen the study of SRT where transgenerational continuity has been established. Obviously not all research groups are represented here. However, this book may be only the first of a series that we organize. Therefore, without seeking to devalue anybody's work, it is not intended to provide a comprehensive register of all the researchers who conduct research in this field and form research groups. This is not a survey of studies in the area. For now, it is just the beginning of a fruitful discussion.

This book presents different perspectives that bridge expert and lay knowledge, as case studies and research trajectories in the Latin American horizon. It does not seek to establish 'a path' or 'a perspective' and is by no means exhaustive. Rather, we present it as the first stage of an open invitation to a multidisciplinary debate that cross-fertilizes the field of study, one that involves roads more and less travelled by. Cycles and processes are social and natural; scientific knowledge and traditional wisdom together form local and global alternatives to an era of collective threat to our Earth System. Therefore, we hope this book will spark discussions and reflections on the Anthropocene in the field of Social Representations Theory. Additionally, this book is intended to suggest that an everyday life knowledge and action approach to important and specific societal problems is a useful and necessary lens that may also be incorporated in the dialogue on the Anthropocene

surrounding issues such as education, health, history, gender, generation, ethnicity, beliefs, justice, representational systems, the environment, traditional groups,
vulnerable groups, active minorities and social change. As Lourdes Arizpe rightly
claims:

> humanity is unique in its capacity to find inspiration and sense in our creations and use
> them to give meaning to the present and the future… We have to move on, guarantee-
> ing 'living', that is to say, the liberty to constantly create new meanings that enable us
> to cooperate beyond divisive ideologies. It is this attribute of peoples, their ability to
> be aware of their capacity of reflection and action, of insisting in an ethics of world
> conviviality that may guarantee a sustainable transformation towards the future (Arizpe
> 2019: 444).

Thus, in the next section of this chapter we want to start a discussion on the
common sense relation between the Anthropocene as a societal era and social representations. The subsequent section presents the researchers who are building
research sand strings regarding Social Representations Theory in Latin America.
The last sections detail the structure and contents of the book, revealing the continuity of SRT among emerging scholars and their dialogue with research elders
and with society around pressing social issues that need to be addressed in the
Anthropocene.

## 1.2   On the Anthropocene as a Societal Era and Its Common Sense Relation to Social Representations Theory

The Anthropocene is a burgeoning field of research and action. 'Officially' stemming from the Earth, geographers and natural sciences, in two decades this
concept has become a field of study extending to the social sciences and the
humanities, and has translated into and recycled everyday discourses, practices,
and vice versa. It is not unproblematic or undisputed, and its validity should not
be limited to scientific discourses and authorship, since reflections on the impact
of anthropogenic activities and the relationship of humanity with the Earth System
have been present for millennia.

The Anthropocene, the 'new' planetary and geological epoch following the
Holocene (12,000 years ago), in which human activity has become the dominant
force shaping the Earth, has been portrayed as a moment of rupture in planetary
history. Linked to anthropogenic global environmental change (GEC), it marks
the period in which human beings and human groups have become a threat which
could potentially lead to the collapse of the Earth System and civilization, while,
at the same, being the potential solution.

This grand narrative of the potential doom of the planet, or at least of human
and many non-human species, and its long-term sustainability vision, is not
exempt from criticism, lack of consensus, politicking, insufficient multilateral
agreements, or even the lobby of business-as-usual deniers. According to Chua

and Fair (2019: 10), the Anthropocene has two main sides to reflect upon. On the one side, it emerges as an awakening narrative based on ignorance: humans suddenly realize the extent and negative impact their actions have produced on the environment and act accordingly. On the other hand, it follows on from a longer-standing ideological battle that is both scientific and social, unitary and diverse, political or depoliticized among humans as unitary subjects or in the context of very varied groups. It is a battle over the ways in which humans have engaged with the non-human world, and the short, medium and longer-term consequences of this.

## 1.2.1 The Anthropocene and the Earth System: A Conceptual Unfolding

Historically, according to Hamilton and Grinevald (2015) and Steffen and co-authors (2011), there were conceptual 'precursors' for the Anthropocene as a humanly distinct geological age, although the understanding of such a concept was very distant from the current understanding of the Anthropocene and caution should be exercised in equating earlier concepts. Precursors distinguished historical epochs and development, emphasizing the uniqueness of modern humankind, which they viewed in a very positive light, unlike the current outlook on humanity as a disturbance, trouble or plague. Additionally, they regarded humans as geological agents, even on a global scale, although the current understanding of the Anthropocene necessarily involves the conceptualization of the Earth as a complex, interlocking and adaptive system with multiple loops that surpasses humanity. This systemic understanding in the occidental scientific tradition is relatively recent, since scientific disciplines were rooted in linear models of science and were distinguished from one another on the basis of their specificity and compartamentalization. Systemic understandings date back to the inter- and post-World War period, for example in the work of Talcott Parsons' Social Systems, Norbert Wiener's Cybernetics, Jay Forrester's system dynamics and Ludwig von Bertalanffy's General Systems Theory. Such a perspective was applied to the Earth as holistic, systemic and interconnected planetary system later, in Lovelock's 1970s Gaia hypothesis.[2]

As for the Anthropocene, most precursors within the scientific occidental tradition recognized a distinct epoch marked by the centrality of humans in the planet, although the impact of humankind was not viewed in a systemic, critical

---

[2]Currently, Lovelock (2019) argues that the Anthropocene, an age in which humans developed technologies on a global scale, is ending after three centuries. Thus, he suggests the utility of the 'Novacene' concept, an era in which artificial intelligence is transforming humanity and will lead to hyper-intelligent beings whose organic relationship with the planet and its conservation is fundamental – instead of the predatory, mechanistic and violent Anthropocene.

or negative light. Almost one hundred and fifty years ago, as early as 1873, Italian geologist Abbot Antonio Stoppani put forward in his three volume *Corso di Geologia* the concept *Anthropozoic Era* to refer to "the introduction of a new element into nature, of a force wholly unknown to earlier periods" relating to civilized Man and the impact 'he' had on landscapes and local climates (Hamilton/ Grinevald 2015: 63). There were other similar concepts such as 'Période Anthropique' by Swiss geologist Eugène Renevier, 'Psychozoic Era' by Joseph LeConte (1877) and Charles Schuchert (1918), and the notion of 'Noösphere' following the work and exchanges between Vladimir Vernadsky, Edouard Le Roy and Teilhard de Chardin (1923–27). The Noösphere was understood as a late stage of progress and evolution, "a psychozoic era of Reason" (according to Russian biogeochemist Verdansky), and an Omega Point (according to French theologian-geologist Teilhard), characterized by collective human reason and consciousness acting on and beyond the biosphere. The Noösphere was the planetary sphere of mind and reason, following two centuries of scientific advancements and enlightenment, the third, last and highest stage of biosphere development following the mineral (geosphere) and organic stages (biosphere): "The 'last of many stages in the evolution of the biosphere'… signalled by the human transformation of its chemistry, including the transmutation of its elements, a task soaked in utopian Promise" (Hamilton/Grinevald 2015: 65).

The current concept of Anthropocene dates back to the turn of the century and mainly credits Paul Crutzen and Eugene Stoermer as pioneers. According to Trischler (2016 cited in Brauch 2020: i.p.), limnologist Eugene F. Stoermer started using the term informally in the 1980s, but it was atmospheric chemist and Nobel laureate Paul J. Crutzen –"Mr. Anthropocene" – who helped most popularize the term. Allegedly, at a meeting in 2000 in Cuernavaca, Mexico, after rounds of discussion on the Holocene, Crutzen had an insight, a 'eureka' moment of revelation and impatiently and spontaneously claimed that the epoch we are currently living in is the Anthropocene. This moment has an aura of 'founding myth' (Trischler 2016: 309). Nevertheless, Crutzen later realized that Stoermer had previously and independently used the term, so together they authored an article claiming that humanity had driven the word into a new geological epoch. It was published in the International Geosphere-Biosphere Programme (IGBP) Global Change internal newsletter 41, under the title "The Anthropocene", crediting them as co-creators of the concept (Crutzen/Stoermer 2000). This work also drew on the ideas and debates generated at the Human Dimensions of Global Environmental Change Programme, and the International Council of Scientific Unions through its 1986 International Geosphere-Biosphere Programme and the International Council of Social Sciences (Arizpe 2019). In 2002, Crutzen published a scientific article in the journal *Nature*, elaborating on the concept. He claimed that the effects of humans on the global environment have significantly affected climatic behaviour. For example, anthropogenic emissions of methane and carbon dioxide have led to the highest concentration levels in 400 millennia, while per capita resource use, population growth and land surface exploitation have resulted in species extinction, freshwater and energy use, nitrogen fixation, toxic waste and global warming.

Thus, "mankind will remain a major environmental force for many millennia", requiring "environmentally sustainable management" and "appropriate human behaviour at all scales" (Crutzen/Stoermer 2000: 23).

The Anthropocene has become first a concept and then an epoch that is unparalleled; the impact of humanity on Earth is so profound and enduring that making it visible challenges previous conceptions of planetary history altogether. For the first time, the story of Earth and humanity are co-creating each other side by side. In addition, the context has changed; the understanding of Earth is now holistic, systemic and complex, just as the human *uni*verse has translated into a *multi*verse of change, plurality and diversity. Nevertheless, so far the Anthropocene remains a predominantly scientific domain with limited cross-disciplinary dialogue and no consensus. However, the challenges and opportunities in the Anthropocene are such that scientific knowledge is not enough, as we shall discuss. The 'Age of Humans' (as it was called by Crutzen – and this anthropocentrism has been heavily criticized by Haraway) already has a career as a geological and cultural term, according to Trischler (2016); it is part of popular culture. From the perspective of social representations, we all co-construct the Anthropocene; it is an emerging relational field of knowledge with action guidelines. This should be the bottom line, instead of obsessing and quarrelling about its footprints and exact meaning, as we shall see next.

In interdisciplinary science there is a dual movement between the assimilation of and adaptation to new knowledge about systems theory and theories of complexity (Piaget 1950, 1972). Both processes establish a dynamic balance between integration and differentiation. Therefore, when the subject approaches the 'object' this requires new approaches as additional questions and problems appear. Research objects become more complex, better structured and allow for empirical research (Piaget/García 1982). This process of gradual integration and differentiation allows reflective abstractions and meta-systemic reflections, slowly involving thematization and integration of external and border conditions that influence ongoing processes (for example, socio-economic growth, public policies, energy use and greenhouse gas emissions, crisis, violence). Beck, IPCC, Steffen and co-authors (2011) and many other scientists have noted that the processes of change or even potential tipping points of the Earth System are no longer linear and deterministic, but chaotic, unpredictable, unstable, dynamic and non-linear with deep unexpected feedbacks, thus a different epistemological approach is required to analyse them. SRT, as the theory of chaos, although in different scenarios and with different objectives, analyses this dual process between assimilation and adaptation.

Within science, the Anthropocene emerged as a field initially linked to the natural sciences, with evidence and debates centred on the importance of its recognition as a unit of the Geological Time Scale (Zalasiewicz et al. 2019) by the International Commission on Stratigraphy (ICS). In 2009, an Anthropocene Working Group (AWG) was established in order to provide evidence for the International Union of Geological Sciences (IUGS) to ratify it as an epoch, a subdivision of the geologic time scale. It would mark the transition from the Holocene, a period that dates back 11,650 calendar years before the present. Although the literature already refers to an Anthropocene 'stratigraphic turn', there is no consensus as to when this epoch commenced. Looking at air trapped in polar ice layers, the original proposal of Crutzen (2002) was that the Anthropocene started in the latter part of the eighteenth century, coinciding with James Watt's design of the steam engine. His initial argument was that since the start of the Industrial Revolution, the planet has endured changes that have a distinct stratigraphic signature. However, there is mounting controversy among geologists and non-geologists, and Crutzen recently asserted that Anthropocene and the alteration of the Earth started in the late 1950s.

Other dates and arguments that have been put forward as the start of the Anthropocene are: i) the origins of patriarchy (understood as the dominance of men over women, or the sexual division of labour, as well as the origins of the family and private property, among other things); ii) the agricultural revolution; iii) the 'Thin Anthropocene'; iv) Capitalism; v) the Columbian Exchange, colonialism, the transatlantic slave trade and the rise of the global capitalist economy; vi) the Industrial Revolution; vii) the use of coal; viii) the Trinity tests and the start of the Atomic or Nuclear Age; the New World Order; ix) the 'Great Acceleration' with cheap fossil fuel energy sources.

Besides the start date, there have been other political and scientific debates and critiques (see Arizpe 2019; Schulz 2017; Merchant 2020; Brauch 2021). Some alternative concepts are science writer Andrew Revkin's *Anthrocene* describing a geological era significantly influenced by humans in his 1992 book *Global Warming: Understanding the Forecast*, that was echoed in popular culture through Australian rock band Nick Cave & The Bad Seeds' song 'Anthrocene' in their 2016 music album *Skeleton Tree*.[3] In 1999 South African entomologist Michael Samways published an article entitled "Translocating fauna to foreign lands: here comes the Homogenocene" in the *Journal of Insect Conservation*. Since this article, there has been debate surrounding this 'biological era'. Journalist and writer Charles C. Mann further developed the concept in his 2011 book *Uncovering the New World Columbus Created*, also entitled *1493: How the Ecological Collision of Europe and the Americas Gave Rise to the Modern World*. Following the Columbian Exchange and waves of globalization and modernization on a planetary scale, the *Homogenocene* has come to mean the process of mixing formerly isolated people, cultures, tools, diseases, flora and fauna around the world that has made ecosystems and peoples more alike. Although this 'homogenizing' conception is problematic – since globalization has led to greater cultural diversity and complexity rather than homogenization, as can be seen in UNESCO's world

---

[3]"The dark force that shifts at the edge of the tree
  It's alright, it's alright
  When you turn so long and lovely, it's hard to believe
  That we're falling now in the name of the Anthrocene."
    (Extract, Nick Cave & The Bad Seeds' 2016 song, *Anthrocene*.)

culture reports – it is interesting to note that the key agents in this account are not only human animals (Sollund 2017), going beyond the conception of the Anthropocene as the 'age of Man', and that although humans are the original driving force and still have an important role to play, this process is not only in human hands.

Many authors believe that 'Anthropocene' is an inadequate or insufficient term. This discussion is most visible in the field of the *Capitalocene*, 'the Age of Capital', where proponents argue that the main enemy and the fundamental turning point is not humanity but capitalism, braiding imperialism, neocolonialism, gender inequality, class struggle, racial order and environmental collapse together with a thoughtlessness that is a hallmark of the 'business as usual' mentality. Capitalism, as an economic, social and natural system – a capitalist world ecology outlook – has modes of production and social relations that are so deeply engrained that it is easier for people to imagine the end of history than the end of capitalism. The book *Anthropocene or Capitalocene? Nature, history and the crisis of Capitalism*, edited by Jason W. Moore in 2016, summarizes the main arguments, criticizing the Anthropocene as 'substantialist', 'bourgeois', erasing capitalism, of 'arithmetic character', and thus 'descriptively powerful, but analytically anaemic' (Moore 2016: 88). Instead, the rise of capitalist civilization is seen as the pivot of the environmental and social crises, crossing boundaries at systemic and planetary level, built upon terrible structural inequalities and malaise, pressuring biospheric stability as well as potentially leading to civilizational collapse. This ecologically orientated history of capitalism contemplates reformulating social relations as well as creatively addressing environmental degradation (Moore 2016).

The *Chthulucene* as part of the creative and alternative thinking within the framework of the Capitalocene means 'tentacular thinking in staying with the trouble'. That is to say, using intra and inter-action, forging networks and 'making kin' in order to overcome individualist thinking about human exceptionalism and transform human waste and planetary anthropogenic effects, favouring instead a multispecies, multi-layered and multi-systemic Gaia (Haraway 2016). Concepts that encompass similar elements are *Econocene* (Norgaard 2013), *Manthroposcene* (Raworth 2014; di Chiro 2017), *Misanthropocene* (Patel 2013), *Technocene* (Hornborg 2015), *Plasticene* (Corcoran et al. 2014), *Novacene* (Lovelock 2019), *Necrocene* (Mc Brien 2016), *Myxocene* (Pauly 2010), *Anthrobscene* (Parikka 2014; Swyngedouw/Ernstson 2018), "Anthropocene Noir" (Rose 2013), and "Gore Capitalism" (Valencia 2010), among others (Moore 2016: 6).

*Patriarchalocene*, *Phallocene* and *Androcene* are concepts used to link women, gender, local and indigenous thinking to the Anthropocene. It is an umbrella discussion linked with very diverse forms of political activism and grass-roots action. The main argument is that gender inequality, environmental degradation, Eurocentrism, capitalism-inequality and patriarchy are interrelated and can be reversed. Its multiple forms of violence can be deconstructed by fair treatment, diversity, decolonialism, justice, ethics, restorative and local alternatives, such as ecofeminism or the umbrella term *Gynocene* (Merchant 2020). It is sometimes linked to an anti-colonialist reading (as in the work by Latin American anthropologist Rita Segato), and also to the concept of *Plantationocene* that makes visible the use of slave labour in the colonized world. The fundamental argument is that while "capitalism is a 600-year old social and economic system, while the Anthropocene is a 60-year old Earth System epoch" (Angus 2016: 232), patriarchy is the root system of dualism, the inequality in social relations on which capitalism, imperialism and anthropocentrism build, and it dates back at least 6,000 years (Lerner 1986).

As can be seen, the concept of the Anthropocene has become useful albeit controverted, an umbrella term that encompasses important reflections and actions, with increasing popularity in scientific discourses and political decision-making. As Arizpe (2019) notes, although environmental concerns have been present for centuries, there was no confirmed knowledge among natural and social scientists about anthropogenic environmental and climate change until the 1980s and 1990s, and policy-makers were both unaware of and totally ignored environmental factors in their political planning, strategies and policies. It is interesting to note that the Anthropocene could boost collaboration between the social, natural and human sciences, as well as pave the way to address the divorce between expert knowledge and traditional wisdom. As Arizpe (2019: 378) has pointed out as meta-reflections following consultations with scientists at the International Science Council (ISC) and the International Council for Scientific Unions (ICSU), it is surprising that the lenses and domains through which global environmental change phenomena have been predominantly addressed are the 'natural sciences', although 80% of these phenomena can be attributed to anthropogenic actions. The Anthropocene concept has been helpful in this respect, since before the year 2000 multidisciplinary reflections and collaborations were scarce.

According to Angus (2016), "from 1988 to 2015, the International Geosphere-Biosphere Program (IGBP) coordinated the research of thousands of scientists from dozens of countries in a concerted effort to learn how human activity has changed and is changing the world – not just local environments and ecosystems, but 'the planetary life support system as a whole'". Brauch (2021) notes that since the concept of the Anthropocene was officially coined at the turn of the millennium, scientific publications on the Anthropocene have boomed within only two decades, resulting in over 1,000 printed books and more than 4,000 peer-reviewed articles in scientific journals, most produced in the last five years. Debates on the subject have spread across all disciplines, although the field remains underdeveloped within the social sciences, making multi-, inter- and transdisciplinary research scarce. Nevertheless, given the implications the Anthropocene has for human groups, living species, ecosystems and the planet, it cannot remain a matter which is exclusively subject to multidisciplinary scientific debate.

Even among critics, the Anthropocene remains useful as a concept when speaking to a broader audience (Moore 2016: xii). In the human domain, the Anthropocene calls for serious public debate, political will, but especially for much deeper reflection among diverse cultures. As Schulz rightly highlights, critics of the Anthropocene have been quick to note that the 'New Human Age' is not only hard to grasp from a geological point of view, but also greatly contested from a cultural perspective (2017: 127). Others insist on the political realm, world governance and a societal sustainability transition (Brauch et al. 2016). Jürgen Renn argues that humanity has entered a new stage of evolution, one of epistemic evolution. He goes for a different option and rejects the attempts to relabel the Anthropocene, as doing so would "tear down an important bridge between the natural sciences and the social sciences and humanities". In his view, we simultaneously inhabit the Capitalocene, the Anthropocene and the Ergosphere, a sphere of human work characterized by "the transformative power of human labour both with regard to global environment and humanity itself"; it is "still open in its evolutionary logic to different ways of shaping the relationship between humanity and its planetary home in terms of the cumulative effects of human interventions embodied in their 'works'" (Renn 2020: 382). Today we also inhabit the Corona anthropocene.

Without reinforcing the anthropocentric bias, a linear outlook of *his*tory or claiming that all human beings are homogenous, we are aware that as a human species our footprints and actions at a meta-systemic planetary level are decisive, part of the problem and potentially part of the solution. We believe that recognizing the stratigraphic markers of the 'common era' and agreeing on the most comprehensive concept is important, although it is only a minor step in the face of the potential societal and planetary consequences of our time. Science and politics at the service of highly specialized interest groups are not enough. This is the main reason for starting a reflection that links the Anthropocene with Social Representations Theory, with its emphasis on cultural diversity and centrality of common sense knowledge and action, lay people and everyday life. Agreeing with Arizpe (2019: 16), "culture is made up of meanings that are activated through human relations that enable people to transform their lives and their ecosystemic and planetary environment".

## *1.2.2   Establishing the Common Sense Relation Between the Anthropocene and Social Representations Theory*

After introducing the concept, its possible time frames and some of the main arguments surrounding it in the previous section, we want to go back to the opening metaphor and title of this chapter. The idea of opening the book by recounting the allegory of the sand string and its story relates to the importance of linking transgenerational dialogues and sustainability with a systemic outlook. We firmly believe and have witnessed that there are limits to science, technology and goodwill if they are not integrated as a coherent part of cultural heritage, diversity, social groups and relations, as well as everyday life agency. This book aims to start a conversation, and perhaps signal a few routes of research and action that highlight the centrality of social representations as a relevant framework in which individual and social subjects with very different cultural heritages have the freedom to create and transform their lives, their relations, their production and political systems, and their ecosystemic and planetary environment.

We identify the Anthropocene as a reflexive, systemic and agentic period in planetary and human *their* story. We are not interested in legitimizing or securitizing the Anthropocene as a purely chronostratigraphic, conceptual, technological or disciplinary debate. Its scientific validity, popularity, authority and solutions are important for sure, although that is just the tip of the iceberg; it needs to be part of the dialogue within governance systems and become enriched by the perspective of peoples at grass-roots level. Maybe some authors believe single-discipline scientists and denialist politicians need to cross the bridge to the side of multidisciplinary scientific knowledge and technological responses to understand the implications of the Anthropocene. This would no doubt be useful, but we have all witnessed the limitations of the Paris Agreement, the Millennium Development Goals (MDGs) and the Sustainable Development Goals (SDGs) when faced with the economic interests of elite groups and emerging nations. According to the UN, the SDGs belong to the entire global community, but they have not been transversally gestated or appropriated, they lack funding, remain a matter of political polarization, and overall they remain far from delivery. Thus, crossing the bridge is not enough; what we need is a quantum leap, to go beyond the reified domain of science and make the Anthropocene accessible to people in everyday life, and to provide feedback on the Anthropocene in a cross-cultural dialogue from the perspective of lay people.

So, how to devise this quantum leap? Our take is that we need to root the Anthropocene culturally as a meta-dialogue with diverse viewpoints and specific actions. Unless we regard the freedom to create as one of the most profound essential values of human beings and groups (Arizpe 2019), and we consider its transformative, proactive, polysemic, intersubjective, and relational dimensions in the face of the challenges – some that are known and yet others still to be unveiled – we will miss a great opportunity, worldwide: the chance to let human individual and collective subjects from different generations and contexts reflect and act, linking themselves, their environment, their relations and activities in the face of very complex phenomena that threaten the survival of the planet and humankind, providing a sense of ethics, purpose, sensibility, responsibility, exchanges, cooperation, heritage, decision-taking and action. This is the reason we believe it is relevant to consider the work of Serge Moscovici. A French social scientist of Romanian origin, a migrant and Jew during a convoluted time in European story, Moscovici had a deep understanding of the complexities of our times, which is reflected in his *oeuvre*. In a time of extremes, with structuralism and functionalism on the one hand, and social constructionism and linguistic postmodernism on the other, Moscovici stressed the importance of the dialectic relationship between self-alter-object and context, understanding knowledge and action as deeply, culturally rooted socio-genesis. His contribution to academia and to society become relevant through his work devoted to Social Representations Theory and everyday life, as well as to the role of active minorities and the relationship between humans and the environment. Nevertheless, beyond his personal contributions, an important field of critical research and situated intervention has developed around the world. In Latin America, it has become a challenging and stimulating arena across generations that is increasingly linked to other endogenous critical perspectives of knowledge, action and the environment, such as de-colonial indigenous feminisms, the perspective of coloniality, and the theology of liberation, among others.

In his 2002 *Nature* article, Crutzen argues that "unless there is a global catastrophe – a meteorite impact, a world war or a pandemic – mankind will remain a major environmental force for many millennia". Today we are in a critical pandemic, but it has only exacerbated our actions as anthropogenic driving force. The situation calls for "appropriate human behaviour at all scales, and may well involve internationally accepted, large-scale geo-engineering projects, for instance to 'optimize' climate", although at this stage "we are still largely treading on *terra incognita*" (Crutzen 2002). Although this quote is illustrative of the current predicament, it also shows that the mainstream outlook on the Anthropocene is Euro, ethno, techno and androcentric, it is linear, totalizing and it dismisses power relations, relegating agency to the imposition of multilateral agreements. Clearly most societies in the world today have been greatly influenced by occidental science, although the universality, objectivity, validity and neutrality of science has been profoundly questioned, as has its stance on nature as a commodity, an object to be appropriated, dominated and bent to human will (Schultz 2017). Although our main argument is in favour of taking the Anthropocene beyond the scientific realm and enriching it in exchanges with lay knowledge and multiple cultural knowledge systems, for a start it would be useful for Anthropocene dialogues to recognize their own blind spots. Thus, Dalby (2016) distinguishes between "the good, the bad and the ugly" Anthropocene. Justice – social and otherwise – is not only a matter of enforcement but also of exchanges and respect (Reardon 1996). The Anthropocene is not only about managing a crisis, but also about the very way it is constructed, validated, appropriated and acted upon. There are significant human paradoxes in the era of the Anthropocene, for example: i) specific people are mainly responsible for certain anthropogenic actions but they do not have to be accountable for them, whereas other people will have to pay for the former's actions, perhaps even with their lives, thus it is an ethical problem; ii) some people might not even realize they are being affected, whereas others might not be affected but perceive the effects. Furthermore, the Anthropocene has to go beyond *anthropos* (Sollund 2020); it has to be achieved by linking the natural and social contracts.

Reflecting on 'threats' in our late modern society, Denise Jodelet has developed an analytical comparative framework that considers important dimensions we wish to adopt with regard to the Anthropocene. The aim is to define specificity, relevance, comparability and intervention, avoiding generalizations, confrontations, and polarizations (Jodelet 2020: 23):

i.   it is important to locate issues in the physical, material and natural world as well as in the discursive, symbolic and ideological one;
ii.  we ought to engage human responsibility and consider pernicious intentionality;
iii. responsibility can be direct or indirect, individual or collective, even if it is attributed or contracted in the name of protection of the common good;

iv. attention should be paid to the intentions underlying perpetration of threatening actions coming directly from a human source, referring to concrete or ideological objectives that can range from intimidation, destabilization or change of opinion to nuisance and destruction;

v. it is necessary to look at conditions and effects: uncertainty and imprecision of its occurrence, the extent of damage caused; and the dimensions involved and their link to the imaginary;

vi. the state of the threat targets: resilience or vulnerability; the effects at the level of experience and cognitive processes; emotional dimensions;

vii. collective processes inducing the threat or induced by it: the role of political and media outlooks in crystalizing antagonistic social positions; massive mobilizations; vulnerable populations likely to become socially threatening.

The recent coronavirus SarsCov2 poses an interesting and relevant example. The transmission of viruses from wild animals to humans is not new or rare, but the impact of anthropogenic activities on ecosystems and wildlife traffic has had exponential effects, as can be seen in recent pandemics such as AIDS, MERS, SARS and Ebola. Nevertheless, with COVID-19 we have seen the unprecedented scale, speed and fragility of health at planetary level. As a comparison, according to WHO figures, during the H1N1 influenza pandemic in 2009–2010, there were around 61 million people infected and 12,500 deaths (0.02% mortality rate). Since the 1976 registers, Ebola has killed approximately 12,950 people (out of 31,094) and it has one of the highest fatality rates at around 42%. However, regional containment has been a key factor in reducing the spread of Ebola. Since 1981, AIDS has affected 40 million people and killed 750,000 (about 1.9%) whereas in the space of nine months COVID-19 has affected almost 40 million people, killing 1,100,000 people (roughly 2.75%), which is why COVID-19 provides a paradigmatic example and has been called 'the disease of the Anthropocene' since it:

> follows a complex sequence involving disruption of the natural, social, economic and governance systems. The destruction of natural habitats and the extinction of species, the poorly regulated capture, marketing and consumption of non-human animals, the influence of lobbies to nullify or delay measures to protect natural and social systems, the limitation of current scientific knowledge and the contempt by governments and companies of the available evidence, have all worked in an orchestrated sequence to facilitate the current COVID-19 pandemic. This sequence of distal causes is closely related to the global climate crisis and the rest of environmental disruptions of the Anthropocene (O'Callaghan-Gordo/Antó 2020: 2).

Looking at COVID-19 from a more psychosocial outlook, besides wildlife traffic and the alteration of ecosystems, the anthropogenic component of the disease becomes more evident as different individuals and groups of people across these nine months have made sense of emerging knowledge regarding COVID-19. Two examples. First, the UN has called gender violence "the pandemic of the pandemic of Coronavirus", given the way in which it has impacted and escalated gender-related forms of violence and inequalities across the globe. Second, taking a local example, as of mid-July 2020, in Mexico one out of ten people still thought COVID-19 was not real, reporting it as mere government propaganda and distraction from more pressing societal issues; another 5% of the population was not sure, despite 50% of the sample knowing somebody directly affected by it.

The examples linked to coronavirus help us see that moves from a geological era to an anthropogenic one are centred on situated reflexivity and systemic connection. This should make us wary of falling into the anthropocentric trap of Cartesian thought, of once again thinking that we are the centre of the universe, the planet and all life forms. It should also warns us against the grand narratives of development and democracy that coerce people into 'freedom'. Humanity is at a reflexive corner, questioning the hierarchies and totalizing forces of modernity and scientific hegemonic thought, of a deeply centralized development model that would require the Global South to use ten times the energy and 200 times the resources of the Earth in order to 'pair up' with the Global North: a model that dismisses the strategies of innovation, resistance, resilience and adaptive continuity that have been developed over centuries in the Global South (Arizpe 2019: 95, 132).

Caution is also advisable with regard to the mystification of technologies and 'green' technologies as the single way forward. Technologies are multiple and have been developed over millennia; they go well beyond patents. In addition, some current technological capabilities surrounding the reproduction of life forms and artificial intelligence call for 'civilizatory' dialogue and caution. However, despite all advances in the era of the communications revolution, access to technologies remains highly unequal. Of course, it is of paramount importance to encourage technologies that enable a sustainable coexistence between humanity and nature, but it is important to remember that neither humanity nor nature is as homogenous or separate as is often assumed. There are multiple outlooks, as well as historical and structural gaps. Environmental footprints are not even, and highly developed nations and elite groups have not been historically accountable for environmental damage; nor are some emerging nations. There is a private-public divide in terms of gains and responsibilities, although disasters and socio-environmental consequences are public and collective. Furthermore, innovations are costly. Integrating technological innovations into productive and life processes takes time, and this generates further inequalities.

Green technologies are a means, one of many, not an end. The *oikos* calls for a more profound reflection. According to the United Nations Framework Convention on Climate Change (UNFCCC), the green economy is worth as much as the fossil fuel sector, accounting for 6% of the global stock market, and if it maintains its current course, it will represent as much as 10% of the global market value by 2030.[4] It currently seems that the 'green' market is booming because it 'sells' and translates into large 'profits', not because there has been an ontological reflection surrounding the art of living (Fromm 1956) or social aesthetics (Fernández

---

[4]United Nations Framework Convention on Climate Change (UNFCCC 08.06, 2018), "Green Economy Overtaking Fossil Fuel Industry – FTSE Russel [sic] Report", at: https://unfccc.int/news/green-economy-overtaking-fossil-fuel-industry-ftse-russel-report#:~:text=As%20such%2C%20it%20is%20central,water%2C%20waste%20and%20pollution%20services.

Chriestlieb 2003), for example surrounding the value of being, creativity, cultural heritage and relationships. The dynamism of green technologies is heavily reliant on economic gains, the net worth of accounts, products and consumption, not on the intrinsic value of life and diversity. The underlying preoccupation is that betting on a profit-orientated version of science and technology, even if it implies a collaboration between academia, society and the private sector as a solution to the challenges of the Anthropocene, remains simplistic at best. It is a partial, troublesome, colonialist and unfair gamble, deeply disconnected from everyday life dynamics and cultural diversity. At the end of the day, there is no clarity that technologies and green technologies are enough to revert already existing anthropogenic damage, much less to create alternatives. Believing that all will be solved by science and technology instead of taking the opportunity to critically self-reflect as humanity and in differentiated cultures and sectors might lead to greater damage and partial solutions. Filipe Duarte Santos (2020) formulates it as a conceptual, strategic and technical question: "Is the anthropogenic climate change still controllable or not? The possible positive scientific and technological answers to this question seem to pave the way to new threats." The prospects remain dim down that road. This does not mean we should disregard green technologies; it is simply a word of caution as well as an invitation to provide technological solutions and green innovations from the lab down and from the grass roots up, taking into account the wisdom of different cultures and everyday life.

This invites reflection regarding the intertwined socio-psychological dimension of the Anthropocene. On the one hand, we find a societal dimension, affecting social relations, group identities and cultural heritages; on the other hand, we find emotions that are interwoven. Thus, a holistic approach to the Anthropocene cannot be based solely on impersonal science; it needs to consider and account for individual, interpersonal, societal and affective aspects. Challenges and solutions are constructed and evaluated interpersonally from the micro to the macro level and vice versa. Knowledge and behaviour are not automatic; they are intersubjective and negotiated. This also applies to social polarization, animosity, exclusion and violence in myriad forms. When individual, group and planetary identities are at stake, identities and representations have both an adaptive and a defensive component. This does not mean that knowledge or behaviour is justified, just that its logic of coherence has to be comprehensively addressed. No individual or collective subject is too trivial; all are agents.

To summarize, the Anthropocene is a moment when human beings and human cultures become the driving environmental force of the Earth System. It translates into a complex meta-systemic relationship between human beings, their diverse cultural systems and the Earth System with all its living beings. The Anthropocene is an era of threats and opportunities in which human beings are the main menace and, paradoxically, may become the solution. It calls for a sustainable revolution, although this is far from automatic or mechanic. Scientific knowledge is relevant, but so is culture, politics and everyday life. If we consider interests, the process, the context, identities and exchanges, "the main obstacles to reaching sustainability come from the ways in which human beings envisage their relationship with bioecosystemic and geoatmospheric systems" (Arizpe 2019: 22). This includes postmodern, modern, indigenous and alternative knowledge systems, without excluding or romanticizing any of them.

The perspective of social representations is useful since it centres on the coherence of everyday life knowledge and action that is both changing and deeply embedded in culture. Culture is seen as dynamic and complex, with defensive and evolving processes. We are reflexive beings with multiple scripts and structural determinants, gestating dialogic and participative arenas to reconfigure culture and agency. In political science terms, it includes the threefold context: the field of *policy*, the process of *politics*, and the norms, laws and institutions that make up the *polity*. It also considers a second-order cybernetics, a meta-reflexive position that is far from neutral, where we are situated observers and part of our observations as well as agents. The term 'culture' etymologically stems from 'growing' and 'cultivating'; it is thus that through culture, with our creative freedom as human beings, we are at a crossroads, asking ourselves what it is we want to cultivate. Do we want the commodification, reification and mercantilization of our being, our planet, our *oikos* (home) and life ways? Do we want a sustainable present and future? Is sustainability a luxury? Do we all want the same? 'Can we' in the same ways? How do we mitigate the systemic impact of the choices of different groups that affect the planet and other people? Is technology enough? How should we deal with geoengineering and its associated global risks? Schultz poses a spot-on reflection in this regard: the "fundamental shift in the Earth System, which is at the heart of the Anthropocene concept, arguably requires an equally fundamental shift in our understanding of the human condition and its symbiotic intersections with nature, society and technology" (2017: 127).

Perhaps we need to reinvent ourselves as well as to recuperate deeply embedded subaltern cultures, forms of knowledge and action that are more in line with the understanding of the human condition in its symbiotic intersection with nature, society, technology and cultural diversity. In doing so, it is important to recognize that human beings and human groups are diverse and have divergent cultural forms, competing interests and power structures, although we all share the same planet with each other and with other life forms and we are all interconnected and our life is *interdependent*. This understanding and connection is present in many cultures across history and geography. It speaks of the inter-linkages we have had, we still have and we will have in the future. Related to this, it is important to become explicitly accountable for our decisions regarding not only ourselves and our interest groups, but also other groups and species that are currently co-residents of the planet, and to take sustainability seriously. We are connected with future residents of the planet, both human and non-human. Humans have already transformed between a third and a half of the land surface on Earth, disrupted nitrogen fixation, used more than half the world's readily accessible freshwater run-off, changed the composition of the atmosphere, heated, contaminated and altered all oceans, driven at least 680 vertebrate species to extinction, and currently threaten the extinction of around a million species including our own. The point is that in the past four centuries, but especially since the industrial revolution and the Great Acceleration, we are having impacts at a speed that formerly took millions of years. Therefore, in order to avoid the sixth mass extinction (Kolbert 2014), with a *sand-string perspective* that considers the evolution of knowledge, future generations of humans and other species, we have to reflect on long-term pathways. We are the process. Lastly, following Arizpe (2015, 2019), *there can be no sustainability without conviviability*. If the emphasis is only on conflict and imposition, instead of negotiation, dialogue and compromise, individuals, peoples and governments will have a hard time in developing their cultural freedom and creativity. Human solidarity and reciprocity feature exchanges, dynamic equilibriums and accountability that is culturally coherent and rooted in the deepest elements of social representations. This makes sense; it is at the heart of both sustainability and creative, systemically coherent and situated well-being.

## 1.3   A Brief Regional Outline of the Social Representations Research Groups Presented in This Book

This book assembles key contributions by Latin American scholars working with social representations in the social sciences that are of conceptual relevance to the study of our complex societal era called the Anthropocene. It presents research with thematic, theoretical and methodological innovations and case studies that have rarely been available in English. It is important to keep in sight – as will become evident in the chapters of this book – that the general framework of our academic endeavours relates directly to the most pressing issues and the everyday realities of Latin America, although most are not exclusive to our territories. Given that SRT has experienced very significant growth in Brazil, this volume will primarily look at the Brazilian production and contributions, and its emerging trends, although it also considers a dialogue with researchers from Colombia, Argentina, Venezuela and Mexico.

Latin American research has made an important contribution to the SRT field, identifying novel possibilities of applying the theory in diverse contexts. For example, key areas have been education, health, minorities, sustainability, gender, collective identities, social conflict and emotions. It is essential to highlight the way in which Latin America creativity has revitalized SRT. For example, an innovative feature of SR research in Latin America is the convergence between social phenomena, academia and activism, establishing a close relationship between the making of scientifically sound research and the generation of alternatives rooted in the situated and temporal co-existence of subjects. In this respect, social psychologist Mireya Lozada from the Central University of Venezuela stands out for devoting the past three decades to the analysis of social violence, polarization in the public realm, reparation processes, democracy and everyday life.

Another characteristic of the Latin American SR field is multidisciplinary works. An emblematic example is the work by philosopher and pedagogue José Antonio Castorina at the University of Buenos Aires (IRES/UNIPE/CONICET), whose research centres on genetic epistemology. Currently, he focuses his investigations on exploring the relationship between knowledge, authority, conceptual change, social representations and school learning. In his research he discusses the problem of "conceptual change", that is, the problem of the transformations of "previous ideas" by students and teachers towards socially established knowledge; and discusses constructivism as a "position against science", that is, as an epistemology. Following his findings regarding everyday life pre-conceptions by students and teachers, we can question many of Piaget's findings and interpretations regarding educational processes.

An important feature of SR studies in Latin America refers to the construction of research networks. This is very salient in Brazil and Mexico. In Mexico, the National Social Representations Research Network (RENIRS) and the Mexican Centre for the Study of Social Representations (CEMERS) based at the Autonomous University of Nuevo León (UANL) have, since 2010, created groups composed of various researchers and students interested in SRT from the most important public and private universities. RENIRS and CEMERS have organised five national conferences, which have been central in deepening the epistemological, theoretical and practical application of SRT linked to some of the most pressing social issues. The founding members of RENIRS-CEMERS are María Estela Ortega Rubí from UANL in the north as Coordinator; Silvia Valencia Abundiz from the University of Guadalajara in the west, who organised the International Conference on Social Representations in 2004; and Eulogio Romero Rodríguez from BUAP in Puebla, in the country's central region. In Mexico City, we find Silvia Gutiérrez Vidrio from UAM-Xochimilco; Martha de Alba González and the late Javier Uribe Patiño, both from UAM-Iztapalapa; and Alfredo Guerrero Tapia from the Faculty of Psychology at UNAM, who had the opportunity to respond to the call for chapters in this first book, and participates by establishing a profound dialogue with the contribution by Brazilian Lúcia Villas Bôas.

Other research networks can be found, mainly established under the great umbrella of the National Autonomous University of Mexico (UNAM) in its different centres and campuses. In the south-eastern region of Mexico in the city of Merida, at the Peninsular Centre for Humanities and Social Sciences (CEPHCIS), we find the Permanent Seminar in Social Representations, Gender and Vulnerability chaired by María de Fátima Flores Palacios, who generously prefaced this volume. The initiative by Gilberto Giménez at UNAM's Institute of Social Research (IIS), with the assistance by Guillermo Peimbert (CRIM), is noteworthy. They established the electronic journal *Cultura y Representaciones Sociales* (Culture and Social Representations), which has gathered scientific articles with a transdisciplinary perspective since 2013, many based on SRT. Also from CRIM in Cuernavaca, Morelos, we find the contributions to gender and social representations by Serena Eréndira Serrano Oswald, who has extended the debate on SRT in the fields of anthropology, political and peace studies and regional sciences. Retired from CRIM, we find philosopher and sociologist Jorge Ramón Serrano Moreno, who generously reviews the chapter by Colombian Deysi Jerez.

Deysi Ofelmina Jerez-Ramírez is Colombian social worker from Santander who transitioned into political science at UNAM in Mexico, in order to research the social construction of disaster risk. Her doctoral dissertation was awarded the 2019 Serge Moscovici prize. Her multidisciplinary research based on SRT led to her being invited to NASA in order to discuss the colonization of Mars. She currently leads a research and training team at the University of Science and Arts (UNICACH) in the southern Mexican state of Chiapas.

Brazil is the country in Latin America with the largest output on Social Representations Theory. It has active, fruitful and prominent research groups and networks. Amongst the most salient we find Serge Moscovici's Network formed by Angela Almeida in Brasilia, currently coordinated by Maria de Fatima de Souza Santos at UFPE. The web of researchers who make up the International Centre for Studies in Social Representations and Subjectivity – Education (CIERS-Ed) links more than forty research groups that have already published 600 articles and ten books in the area. It was created and initially led by Clarilza Prado de Sousa and the Research Department of the Carlos Chagas Foundation. She received all sorts of support from Serge Moscovici and Denise Jodelet and from the Foundation Maison Science of L'Homme, guaranteeing its development to date. The CIERS-Ed was subsequently coordinated by Lúcia Villas Bôas and is currently led by Adelina Novaes; it represents a space for enriching the study of SRT in Latin America.

In addition, it is necessary to highlight active research groups that work with researchers supporting the training of Master's and doctoral students at these centres and research networks in Brazil, from the southern to the northern tip.

Looking at the regional Brazilian output on Social Representations Theory, in Rio Grande do Sul, at UFRGS, Professor Pedrinho Guareschi has been developing extraordinary work for years, mainly on the theoretical basis of SRT. Guareschi – 'Pedrinho' to friends and colleagues – has offered a profound contribution to the studies of SRT, having had a solid background in philosophy, theology, literature, social psychology and social sciences.

In Santa Catarina, at UFSC, one of the pioneers in the study of SRT in Brazil, Brígido Vizeu Camargo, founded the Laboratory of Communication and Cognition (LACCOS). With a background in psychology, he became interested in European sociological social psychology from an early age, completing doctoral and postdoctoral degrees in France. His performances at the Brazilian Society of Psychology (SBP) and the Ibero-American Federation of Associations of Psychology (FIAP) show his concern with the field of social psychology in Brazil, France and Italy and his commitment to SRT. His work has been greatly enriched by his collaboration with Clélia Nascimento. Currently, he seeks to analyse the diffusion of the theorization of social representations in Brazil, the construction of ideas and new ideologies as systems of representations. Still in southern Santa

Catarina, Maria Helena Cordeiro specialized in the training of teachers of early childhood education and methodology with significant contributions to SRT and education.

This book lacks contributions from Brasilia, where the International Center for Studies in Social Representations 'Serge Moscovici' is located. It was there that Angela Almeida started her interesting work and networks. Nevertheless, with the transfer of the 'Serge Moscovici' Centre to Pernambuco, currently directed by Maria de Fátima de Souza Santos, we aim to present the strings of continuity in their SRT work.

In São Paulo, the research groups focus mainly on the area of education and SRT. These studies investigate the construction of a field of study, rather than just the application of the theory. In this rich and dynamic research area, the following groups stand out: CIERS-Ed, founded by Clarilza Prado de Sousa and currently coordinated by Adelina Novaes; the UNESCO Chair in Teaching Professionalization coordinated by Lúcia Villas Bôas; the Franco-Brazilian Serge Moscovici Chair, also coordinated by Lúcia Villas Bôas; the Centre for International Studies in Social Representations (NEARS) coordinated by Clarilza Prado de Sousa at the Pontifical Catholic University of São Paulo.

Still in São Paulo, we find the multidisciplinary contribution by Edna Maria Querido de Oliveira Chamon is a professor and researcher at the University of Taubaté (UNITAU), in Taubaté, São Paulo, and a collaborating researcher at the University of Campinas (UNICAMP) in the Department of Architecture and Construction at the College of Civil Engineering, Architecture and Urbanism. She has been guiding students and developing research that articulates SRT and education. Her contribution to research on field education with the support of SRT has been an indispensable reference in this field of studies.

In Minas Gerais in Brazil, the group led by Maria Isabel Antunes-Rocha at UFMG at the Social Representations Study Group (GERES) focuses on professional training in basic education, contemplating education and active minorities and developing fascinating research surrounding social representations in movement.

In Rio de Janeiro, SRT has received contributions from very distinguished authors, such as Celso Sá and Angela Arruda, who practically started research in SRT in Brazil, following from which many research groups were formed. Celso Pereira de Sá (1941–2016), whose life was dedicated to social psychology and substantial contributions to SRT, has a well-established international reputation. The constitution of his "string of sand" is not restricted to the works he conducted at the State University of Rio de Janeiro (UERJ). His books and his research echo through most SRT research in Latin America and Brazil. Angela Arruda, based at the Rio de Janeiro Federal University (UFRJ), and associate professor at the University of Évora, is one of the indisputable pioneers of SRT in Brazil and Latin America, and one of the most prominent researchers still active. Her extensive experience of social psychology, with its emphasis on SRT and qualitative methodologies, and her studies in gender, health, social movements, social imaginary,

Brazil, Brazilian cultures and thought, political and social polarization are essential references to all who study SRT in Latin America.

Denize Cristina de Oliveira, at the State University of Rio de Janeiro (UERJ), has developed important research in the fields of Public Health and Nursing, linking SRT to the field of health. Her contributions to social imaginaries and professional practices in health, to the symbolic incorporation of health systems, to health promotion among children and adolescents, as well as to work and professional practices related to HIV/AIDS and their symbols are also a must-read, and have contributed to both theory and public health policies.

Alda Mazzotti, emeritus professor at both the Federal University of Rio de Janeiro (UFRJ) and the Estácio de Sá University, focuses on the area of educational psychology. Her academic output that focuses on the topics related to the research methodology is a standard reference in the area of education. With methodological rigour she has developed countless studies on social representations. She leads a research group devoted to analysing teacher knowledge, teacher training and work, teacher identity, school failure, public school students, and child labour.

Tarso Mazzotti is associate researcher at the Carlos Chagas Foundation and associate professor at Estácio de Sá University. He has made significant theoretical contributions, which are multidisciplinary, mainly in the field of the philosophy of education, working on the link between SRT and rhetoric, philosophy of education, and epistemology.

Maria Elizabeth Macedo is based at the State University of Rio de Janeiro (UERJ) and is a visiting scholar at the University of British Columbia (UBC). She is currently the President of the International Association for the Advancement of Curriculum Studies, and undertakes important international editorial work. Her involvement in SRT relates to the research undertaken by CIERS-Ed, in which she is developing a political analysis of teacher training and teacher training curricula from a SRT standpoint.

In Espírito Santo it is necessary to emphasize Zeidi Araujo Trindade of the Federal University of Espírito Santo. She has experience in the area of social psychology and is the coordinator of the Network of Studies and Research in Social Psychology (RedePso). Her research in social representations describes her interest in social practices and culture, gender, youth, parenthood and reproductive health.

In Bahia, Maria de Lourdes Ornellas, a professor at UNEB, has been trying to articulate psychoanalysis with SRT. As a researcher, she is also linked to the CIERS-Ed and the Carlos Chagas Foundation and contributes to the field of SRT and education.

Fátima de Souza Santos, at the Federal University of Pernambuco, conducts research in the areas of social psychology and developmental psychology linked to SRT. Her contributions weave the topics of SRT with violence, adolescence, old age, health and social practices. She currently coordinates the International Centre for Studies in Social Representations 'Serge Moscovici'.

## 1.4   About the Chapters in the Book

As has been already explained, this book gathers innovative and solid research from a generation of young mid-career academics, whose trajectories are blooming, and who are going to enrich the field of social representations in Latin America and abroad in the coming years. As editors, we asked distinguished professors in the field of SRT to select and orientate us towards proposals of younger high-quality academics whose contributions could be useful not only for our endogenous field of studies, but would have the reflexive and action-orientated potential to face this era of civilizatory crisis at a time in which working with common sense and everyday life in different fields can help make science more engaged and highlight useful paths and ways forward.[5] Thus, it seemed important to us that each of the authors, their research trajectories and pieces should be presented and commented on by the senior scholar, engaging in a cross-generational, meta-situated and systemic dialogue. As shall be seen, the authors also dialogue with other theoretical, disciplinary, epistemological, methodological and thematic referents, in such a way that they provide feedback to the field of social representations and at the same time produce interesting research proposals that enhance other academic areas and fields of knowledge.

In this way, we consider that it is also possible to see how second and third generation academics in social representations have re-appropriated the theory while simultaneously contributing to it. This is because representatives of the younger generations have the classical European training in SRT derived from their teachers and their own travels, but also benefit from the mature reflections of over forty years of reflection on SRT in the Latin American context, with its concrete problematic and territories. This is the reason that led us as editors to embrace the metaphor of the 'sand string'. This book tries to weave time and space in an era that calls for complex thinking and solutions with the shared experience of individual and collective research subjects, who have systemically reconfigured a field that is vigorous and critical, innovative and current, despite of – or maybe precisely because of – the passing of the years. This mechanism enables it to project itself into the future. Therefore, in this section we shall briefly outline the contents of each chapter.

Chapter 2, written by Lúcia Villas Bôas (Carlos Chagas Foundation, Brazil), deals with social representations and history from a theoretical perspective. In its presentation, it establishes a transgenerational, transdiciplinary and transpacific dialogue, going from the south to the north of the American continent, as it is introduced by a deep and critical reflection written by Alfredo Guerrero Tapia (UNAM, Mexico). The chapter invites us to consider the relationship between social psychology and history, disciplines that are traditionally mapped

---

[5]From the start, we are dealing with social representations in Latin America as a broad research field that is characterised by diversity rather than unified currents or research traditions.

as distant. It discusses the idea of representation as both affirmation and negation of historiographic discourse, based on what have been called the 'modern' and 'postmodern' historiographic conditions. The chapter begins by discussing how memory has become an epistemological object within the discipline of history. It thus offers reflections on the relationship between history and memory, understood as two forms of managing the past, while exploring their implications for social representations theory. Building upon some considerations of a current in historiography called 'history of the present time', the chapter presents the concepts of 'regime of historicity' and 'presentism', developed in particular by François Hartog and Reinhart Koselleck, and discusses some of the challenges they present in relation to current processes surrounding the intelligibility of memory. Finally, it emphasizes that the possibility of developing common ground between history and Social Representations Theory depends on building a post-disciplinary and epistemo-political agenda.

Chapter 3, authored by Colombian Deysi Ofelmina Jerez-Ramírez (University of Science and Arts of Chiapas, Mexico), with an introductory comment by Jorge Ramón Serrano Moreno (UNAM, Mexico), looks at social representations in the study of disaster risk based on a case study in Colombia. The study develops around the characterization of the processes of the social construction of risk and the daily relationship "to know-to do". It presents part of the final results of the research titled: "Social constructions of disaster risk from social representations in the municipality of Piedecuesta, Santander". Disaster is viewed as the confluence of environmental factors and social-cultural components, and risk as a dynamic, polysemic and questionable conceptual entity – non-inert, unambiguous, unequivocal. According to this logic, the thematic is examined as a social construction closely related to the experiences, knowledge and social actions that are generated from daily interaction with conditions of vulnerability and threat. The perception of risk recreated in intersubjectivity connects with other consciences and dialogues, circulates between speeches and social praxis, is anchored into the human psyche, and returns to the context whence it was generated. It is, in any light, a representational phenomenon that has been approached from the Moscovician theory, within the framework of a social study of disasters and the conceptual supports of identity and territory. Social representations (SR) of the Piedecuestana population are analysed in the development of three dimensions: Social Cognitive, Social Structural and Social Territorial.

Chapter 4, written by Aline Reis Calvo Hernandez (Federal University of Rio Grande do Sul, Brazil), is introduced by Pedrinho Guareschi (Federal University of Rio Grande do Sul/ Pontifical Catholic University of Rio Grande do Sul, Brazil). It addresses the confluences between Social Representations Theory and the Psychology of Active Minorities, looking at the production in Brazil, using qualitative methodology, through an exploratory bibliographical, descriptive and interpretative study. First, it presents the output of research groups registered in the Directory of Research Groups and Lattes Platform of the National Council for Scientific and Technological Development (CNPq) in order to highlight some descriptive statistics and map the scenario of the groups and production areas in the country. Subsequently, it analyses the relationship between Social Representations Theory and the Psychology of Active Minorities, the two fundamental theoretical proposals in Moscovici's work, in order to deepen the epistemological discussion of social representations and their importance in sociological social psychology. Finally, it discusses some possible horizons and challenges faced by social representations researchers in Brazil.

Chapter 5, introduced by Maria de Fátima de Souza Santos (Federal University of Pernambuco, Brazil) and authored by Renata Lira dos Santos Aléssio (Federal University of Pernambuco, Brazil), reflects theoretically on the relationship between beliefs and social representations. In studies about social representations it is common to think of beliefs as a class of elements which exist in representational contents of the most diverse objects. This chapter will provide, in an abridged form, Moscovici's conceptual ideas regarding the relationship between beliefs and representations. The next part presents two different theoretical perspectives that encompass that relationship. The conclusion suggests avenues to be explored in the context of the development of Social Representations Theory in Latin America.

Chapter 6, by Alicia Barreiro (University of Buenos Aires, Argentina), explores social representations of justice as developing structures of sociogenesis and ontogenesis. It is introduced by José Antonio Castorina (University of Buenos Aires, Argentina). Social injustice has been present in Argentina, as in other Latin American countries. Retributive justice has become a daily matter of debate since the fear of crime and the degradation of political and social participation have led to the weakening of social bonds and fractures in the sense of community. Reflections and discussions on diverse – even opposite – ways to understand and promote justice have a long tradition in social sciences that can be traced to Ancient Greece. The justice-injustice opposition could be considered a theme that permeates the entire history of Western culture, leading, within the same society, to the coexistence of different ways of understanding justice based on different ideological groundings. Social representations (SR) are the product of everyday exchanges and are constructed as a way to understand social objects that challenge the available cultural meanings and require a symbolic coping process. The ontogenesis of SR is the process through which individuals reconstruct SR while appropriating them and, in so doing, develop diverse social identities. The chapter presents a set of studies aimed at understanding the sociogenesis and ontogenesis of SR of justice, as well as the relationship between the two processes.

In Chapter 7, Suzanna Alice Lima Almeida (State University of Bahía, Brazil), addresses common sense in Gramsci's and Moscovici's writings. Introduced by Maria de Lourdes Soares Ornellas (State University of Bahía, Brazil), it investigates convergences between Gramsci's and Moscovici's theories and, through these convergences, puts forward an educational proposal for praxis in order to help train people to become proactive, politically-minded members of their community who take action to improve living conditions for all. Theoretical research is undertaken using both authors' key books on the development of the theory, such as *The Prison Notebooks* (Gramsci 1975) and *Psychoanalysis: its Image and its Public* (Moscovici 1979), besides other sources by scholars who have studied the work of these authors. Results suggest that the main theoretical convergence between the authors is related to their approach to common sense, from which emerged subcategories that give better visibility to such approximations between them.

Claudomilson Fernandes Braga (Federal University of Goiás, Brazil) in Chapter 8 looks at the genesis of research into social representations and communication. Introduced by Pedro Humberto Faria Campos (Salgado Filho University, Brazil), the chapter seeks to discuss the actuality of social communication and its three core constructs: diffusion, propaganda and propagation. Throughout his work, Moscovici gave a preponderant role to social communication. Communication has a fundamental role in the maintenance and dissemination of social representations. Moscovici characterizes communication systems, and Vala speaks of a typology of communicative acts of social representations: diffusion, propagation and advertising. Five decades after the seminal work of Serge Moscovici was released, there is an exciting theoretical legacy. However, the legacy left by the author in relation to the role of communication in the process of disseminating social representations and, above all, communication systems has been neglected over the years.

Chapter 9, by Rita Cássia Pereira Lima (University Estácio de Sá, Brazil), with an introductory comment by Tarso Mazzotti (Federal University of Rio de Janeiro, Brazil), proposes reflections on the 'figurative core' of social representations by highlighting the roles of figures of thought. The starting point is the expression 'figurative model' in the seminal work of Serge Moscovici, *Psychoanalysis: Its Image and Its Public,* and later theoretical proposals of the 'core', such as Jean Claude Abric's Central Core Theory and, more recently, Pascal Moliner's Matrix Nucleus Theory. The notion of 'image' runs through these reflections. Starting from these referential concepts, Tarzo Mazzotti's studies on the relationship between 'figurative core' and figures of thought are prioritized, especially when the author proposes an 'argumentative core' for analysing speeches. An example of some research in which analysis uses rhetorical devices is presented. Overall, the work highlights the presence of a core in social representations which approximates to the approach of rhetorical social representations, and seeks ways to both identify and analyse it.

Chapter 10 is co-authored by Mariana Bonomo (Espírito Santo Federal University, Brazil) and Sabrine Mantuan dos Santos Coutinho (Fluminense Federal University, Brazil). Introduced by Zeidi Araujo Trinidade (Espírito Santo Federal University, Brazil), it investigates representations and social practices related to ethnic and gender identities among members of a gypsy community in a Brazilian territory. Analysis is undertaken in the light of the theoretical-conceptual contribution of Social Representations Theory. In the gypsy community where the study was conducted, there were few women who were born in gypsy tents, or were daughters of gypsy parents. Many of the women who lived there did not have the same origin and only became part of the group because they had married gypsies. Among the children present in the visited territory, the boys were not only the majority, but also almost the entire child population. These initial observations, resulting from an exploratory and ethnographic study, led to questions regarding the absence of girls in the territory and, consequently, the configuration of gender relations within the group, a scenario that implied the need to consider gender and ethnicity as interdependent dimensions in the analysis of the identity phenomenon in this sociocultural context. It is important to emphasize, however, that this phenomenon cannot be interpreted as a characteristic of the gypsy culture, but rather was verified in the dynamics of social relations established within the specific group investigated.

Chapter 11, authored by Adriane Roso (Federal University of Santa Maria, Brazil), with an introductory comment by Susana Seidman (University of Buenos Aires, Argentina), looks at social representations, symbolic violence and the gendered medicalized body. It presents two theoretical approaches that can shed light on the phenomenon of the medicalization of women's bodies – Social Representations Theory and Social Dominance Theory. Two key hypotheses sustain this reflection: (a) different kinds of social representations interfere in the decision-making process regarding contraceptive use; and (b) social dominance benefits from hegemonic social representations of masculinity and femininity when the legitimized beauty myth serves as a tool used by medical professionals to dominate women and limit their sexual and reproductive autonomy. The reflections on medicalization are underpinned by the experiences of Brazilian women regarding contraceptive methods.

Chapter 12, by Antonio Marcos Tosoli (State University of Rio de Janeiro, Brazil), addresses Social Representations Theory in the field of Nursing, looking at professional autonomy, vulnerability and spirituality/religiosity as representational objects. It is introduced by Denize Cristina de Oliveira (Rio de Janeiro State University, Brazil). 'Social representations' is a polysemic term with specific interpretations in the context of different knowledge fields, which tends to make the situation of first contact even more complex. In Brazil, the theory has been intensively used in areas such as health and education in order to aid understanding of their main objects of study, and also with a view to supporting interventions in different areas of practice. In health, nursing has gained prominence over the last few decades, given the amount of production in the area and the involvement of a large number of researchers in its study and application. From the explanation of this scenario follows the approach of three representational objects that mark, at different moments, the trajectory of a researcher in the context of the postgraduate programme in nursing at the State University of Rio de Janeiro in the city of Rio de Janeiro, Brazil.

In Chapter 13 Daniela B.S. Freire Andrade (Federal University of Mato Grosso, Brazil), introduced by Angela Arruda (Federal University of Rio de Janeiro/University of Évora, Brazil/Portugal), establishes a dialogue between social representations and the multiple ordinations of reality in the case of children, underpinned by the work of Lévy-Bruhl. The expression 'universes of socialization' proposed by De Lauwers (1991) reveals the importance of social representations without delineating children's learning or coming close to what

Brougére (2004) describes as perspectives – possibilities available to children and their capabilities of perceiving. Social representations, beliefs and values related to childhood are presented and linked to the contexts in which a child lives daily, through either their speech or organically, through social practice and space organizations, which end up setting universes of socialization, filled with structures of opportunities for human learning and development. The multiple ordinations of reality, based on Lévy-Bruhl's work and social representations, are the matter of this chapter.

Chapter 14, written by Alcina Maria Testa Braz da Silva (Federal Centre for Technological Education Celso Suckow da Fonseca, Brazil), with an introductory comment by Edna Maria Querido de Oliveira Chamon (University Estácio de Sá and the University of Taubaté, Brazil), focuses on the contribution of Social Representations Theory to science education. The chapter initially contextualizes the current scenario of science education in Brazil and the various assessment systems, focusing on the PISA (Programme for International Student Assessment) evaluation in science. Then, the contribution of Social Representations Theory to science education is discussed, in respect of knowledge, science and schools. It presents some research results on the projects developed by the research group Education in Sciences and Social Representations (EDUCIRS), in the Science, Technology and Education Postgraduate Programme (PPCTE) of the Federal Center for Technological Education Celso Suckow da Fonseca (CEFET), Rio de Janeiro. Finally, it reflects on future developments in the interface of investigation between scientific education and social representations.

In Chapter 15 Romilda Teodora Ens (Carlos Chagas Foundation, Brazil), introduced by Elizabeth Fernandes de Macedo (State University of Rio de Janeiro, Brazil), addresses the possible dialogues between social representations and educational policies and the dilemma of data analysis. Based on Social Representations Theory (SRT) and educational policies, it aims to demonstrate the fertility of SRT for the analysis of policy texts and how these anchor the representations of teachers and undergraduate students on school and on being a teacher. It adopts a combined qualitative, theoretical-bibliographic and exploratory interpretative approach, which yielded reflections that indicate ways to analyse texts about teaching policies through the lens of SRT. From this perspective, it explains the contribution of SRT to the analysis of these texts. It presumes the SR of the teachers on educational policies to be one of the macro-regulators of their craft which allows them to give meaning, signification and resignification to carry out these policies, because they are intrinsically related to ideas, representations, translations and simulacra situated according to the historical, political and economic moment in which the institution and its actors are inscribed.

Chapter 16 looks at social representations of violence among public school students. Introduced by Brigido Vizeu Camargo (Federal University of Santa Catarina, Brazil), it is co-authored by Andréia Isabel Giacomozzi (Federal University of Santa Catarina, Brazil), Amanda Castro (University of the Far South of Santa Catarina, Brazil), Andrea Barbará da Silva Bousfield (Federal University of Santa Catarina, Brazil), Priscila Pereira Nunes (Federal University of Santa

Catarina, Brazil), and Marlon Xavier (Federal University of Santa Catarina, Brazil). Violence is a social phenomenon, for it exists in specific contexts and is effected in relationships with others, due to socio-economic, political and cultural factors. This study, of a quantitative, qualitative and descriptive nature, aimed to investigate the social representations of violence among 349 students from Florianópolis, Brazil, through a self-administered questionnaire. The participants' average age was sixteen years and three months old. It found that violence is part of the students' everyday life. Their elaboration and dissemination of social representations of violence focused on three main aspects: physical violence; verbal violence; and internal causes (perpetrator characteristics) and external causes (social inequality, drugs, and games) of violence. It is important to know the social representations of violence and their associated factors in the context of adolescents' daily lives in order to contribute to preventive works and to promote a culture of peace in schools.

In Chapter 17 Sandra Lúcia Ferreira (City University of São Paulo, Brazil), introduced by Alda Judith Alves Mazzotti (Federal University of Rio de Janeiro/ Estácio de Sá University, Brazil), reflects on quality school education from the perspective of young students, asking the question 'what is the future?' The chapter is the result of an information-centric survey of 227 pre-adolescents – thirteen to fourteen years old – from elementary school. Its objective is to reflect on the teaching of these young people, whose different faculties, potentials and skills equip them to be part of the decision-making process about teaching methods. The discussion is supported by Social Representations Theory because it reflects the interdisciplinary potential, combining students, teachers and managers in a one-dimensional perspective. For this purpose, it is a precondition to involve these educators in the process of continuing training, making them active researchers of the educational phenomenon, capable of improving daily practice by means of scientific methods. The methodology is defined by the application of questionnaires to focus groups in order to reveal their opinions about the daily routine at school. The results indicate that Quality in School Education is associated with respect and trust between students and educators, activities that make students feel valued, and the participation of students in processes which affect the school dynamics. There is also evidence that substantiates the view that Quality in School Education is based on a fictional social space, a school which does not actually exist. In this sense, Quality is designed for an imaginative future far removed from the present situation in which the students find themselves.

Cristene Adriana Silva Carvalho (Municipality of Belo Horizonte, Brazil) and Luiz Paulo Ribeiro (Federal University of Minas Gerais, Brazil) author Chapter 18. With an introductory comment by Maria Isabel Antunes-Rocha (Federal University of Minas Gerais, Brazil), the text presents the construction process and first delimitations of the concept of social representations in motion (SRM), formulated from the work of the Social Representations Study Group (GERES – Federal University of Minas Gerais). GERES is organized by professionals with different fields of knowledge who work in educational contexts. The group is interested in the way in which social representations, as part of the production of knowledge, are constructed by subjects who are experiencing situations that demand changes to their ways of thinking, feeling and acting. According to Moscovici, social representations are created to make the unfamiliar familiar. The authors realise that there is the possibility of change – and a limit to the extent of change – in the representational universe of the subject when facing the unfamiliar. In the face of what is "strange", subjects tend to have one of three reactions, or 'motions': (a) refusing to experience the new, rejecting the opportunity to change and strengthening knowledge that is already instituted; (b) fully adhering to what is strange, breaking with the past; or (c) initiating a process of familiar re-elaboration, integrating novelty progressively. Thus, between 2004 and 2020 the GERES study group, under the coordination of Maria Isabel Antunes-Rocha, conducted nineteen studies, including dissertations, theses, postdoctoral reports and professional research. All the works contributed in some way to the construction of the concept of RSM. In the text, to build the concept of social representations in motion, the authors elaborate its conceptual, analytical and methodological contours. The results of this research answered the question as to why we create representations.

In order to close the book, Mireya Lozada (Central University of Venezuela, Venezuela) and Adelina Novaes (Carlos Chagas Foundation, Brazil) masterfully discuss social representations as a bet on social change in Chapter 19. Unlike all other chapters in this book, in this closing chapter two researchers from different generations and national contexts share their reflections on the importance of promoting social change in the Latin American context based on the outlook of social representations. In this sense, the adoption of Social Representations Theory by all the studies presented in this book is linked to its potential to promote creativity and change, since SRT is radically different from both social determinism, which explains human beings as products of society, and pure voluntarism, which regards the human being as a free agent. Thus, the studies that bring us together share the same dissatisfaction that feeds Latin American social psychology, namely: its constant preoccupation with immediate social reality, its emancipatory vocation and its political-reflective character. This chapter shows exactly what we would like to highlight as a working model of present and future psycho-social perspectives for Latin America in the Anthropocene based on the potential of Social Representations Theory to promote ethico-political change.

# References

Angus, Ian, 2016: "Anthropocene or capitalocene? Misses the point", in: *International Socialist Review*, 103 (July).

Arizpe, Lourdes, 2019: *Cultura, transacciones internacionales y el Antropoceno* (Mexico City: CRIM-Miguel Ángel Porrúa).

Arizpe, Lourdes, 2015: *Vivir para crear historia: antología de estudios sobre desarrollo, migración, género e indígenas* (Mexico City: CRIM-Miguel Ángel Porrúa).

Brauch, Hans Günter, 2021: "Peace ecology in the Anthropocene", in: Oswald Spring, Úrsula; Brauch, Hans Günter (Eds.): *Decolonising conflicts, security, peace, gender, environment and development in the Anthropocence* (Cham: Springer Nature Switzerland AG): 51–185.

Brauch, Hans Günter; Oswald Spring, Úrsula; Grin, John; Scheffran, Jürgen (Eds.), 2016: *Handbook on sustainability transition and sustainable peace* (Cham: Springer).

Chua, Liana; Fair, Hannah, 2019: "Anthropocene", in: Stein, Felix; Lazar, Sian; Candea, Matei; Diemberger, Hildegard; Robbins, Joel; Sanchez, Andrew; Stasch, Rupert (Eds.): *The Cambridge encyclopedia of anthropology* (Cambridge: Cambridge University Press).

Corcoran, Patricia L.; Moore, Charles J., Jazvac, Kelly, 2014: "An anthropogenic marker horizon in the future rock record", in: *GSA Today*, 24,6 (June): 4–8.

Crutzen, Paul, 2002: "Geology of mankind", in: *Nature*, 415,23 (January).

Crutzen, Paul; Stoermer, Eugene, 2000: "The Anthropocene", in: *Global change newsletter— International geosphere-biosphere programme*, 41: 17–18.

Dalby, Simon, 2016: "Framing the Anthropocene: The good, the bad and the ugly", in: *Anthropocene Review*, 3,1: 33–51.

di Chiro, Giovanna, 2017: "Welcome to the white (M)Anthropocene? A feminist-environmental critique", in: MacGregor, Sherilyn (Ed.): *Routledge handbook of gender and environment* (London: Routledge): 487–505.

Fernández Chriestlieb, Pablo, 2003: "La psicología política como estética social", in: *Interamerican Journal of Psychology*, 37,2: 253–266.

Fromm, Eric, 1956: *The art of living* (New York: Harper & Brothers).

Hamilton, Clive; Grinevald, Jacques, 2015: "Was the Anthropocene anticipated?", in: *The Anthropocene Review*, 2,1 (January): 59–72.

Haraway, Donna J., 2016: *Staying with trouble: making kin in the Chthulucene* (Durham, NC: Duke University Press).

Hornborg, Alf, 2015: "The political ecology of the Technocene: Uncovering ecologically unequal exchange in the world-system", in: Hamilton, Clive; Gemenne, Francois, Bonneuil, Christophe (Eds.): *The Anthropocene and the global environmental crisis: Rethinking modernity in a new epoch* (London: Routledge).

Jodelet, Denise, 2020: "Uses and misuses of threats in the public sphere", in: Jodelet, Denise; Vala, Jorge; Drozda-Senkowska, Ewa (Eds.): *Societies under threat* (Cham: Springer): 13–26.

Kolbert, Elizabeth, 2014: *The sixth extinction: An unnatural history* (New York: Henry Holt & Company).

Lerner, Gerda, 1986: *The creation of patriarchy* (Oxford: Oxford University Press).

Lovelock, James, 2019: *Novacene: The coming age of hyperintelligence* (London: Penguin).

Mann, Charles C., 2011: *Uncovering the New World Columbus created, also entitled 1493: How the ecological collision of Europe and the Americas gave rise to the modern world* (London: Granta Books).

McBrien, Justine, 2016: "Accumulating extinction: Planetary catastrophism in the Necrocene", in: Moore, Jason M. (Ed.): *Anthropocene or capitalocene? Nature, history, and the crisis of capitalism* (Oakland, CA: PM Press): 78–15.

Merchant, Carolyn, 2020: *The Anthropocene and the humanities: From climate change to a new age of sustainability* (New Haven, CT: Yale University Press).

Moore, Jason M. (Ed.), 2016: *Anthropocene or capitalocene? Nature, history, and the crisis of capitalism* (Oakland, CA: PM Press).

Norgaard, Richard B., 2013: "The econocene and the delta", in: *San Francisco Estuary and Watershed Science*, 11,3: 1–5.

O'Callaghan-Gordo, Cristina; Antó, Joseph M., 2020: "Covid-19: The disease of the Anthropocene", *Environmental Research*, 187, 109683: https://doi.org/10.1016/j.envres.2020.109683.

Oliveira, Denize Cristina de (Ed.), 2000: *Estudos interdisciplinares em representação social* (Goiânia: AB).

Palsson, Gisli; Szerszynski, Bronislaw; Sverker, Sörolin; Avril, Bernard; Crumley, Carole; Hackmann, Heide; Holm, Poul; Ingram, John; Kirman, Alan; Pardo Buendía, Mercedes; Weehuizen, Rifka, 2013: "Reconceptualizing the 'Anthropos' in the Anthropocene: integrating the social sciences and humanities in global environmental change research, in: *Environmental Science & Policy*, 28: 3–13, https://doi.org/10.1016/j.envsci.2012.11.004.

Parikka, Jussi, 2014: *The Anthrobscene* (Minneapolis: University of Minnesota Press).
Patel, Raj, 2013: "Misanthropocene", in: *Earth Island Journal*, 28, 1 (Spring): 21.
Pauly, Daniel, 2010: *5 easy pieces: The impact of fisheries on marine systems* (Washington: Island Press).
Piaget, Jean, 1950: *Introduction à l'épistémologie génétique* (Paris: PUF).
Piaget, Jean, 1972: *Psychology and epistemology: Towards a theory of knowledge* (Harmondsworth: Penguin Press).
Piaget, Jean; García, Rolando, 1982: *Psicogenesis e historia de la ciencia* (Mexico City: Siglo XXI).
Raworth, Kate, 2014: "Must the Anthropocene be a manthropocene?", in: *The Guardian* (20 October).
Reardon, Betty A., 1996: *Sexism and the war system* (New York: Syracuse University Press).
Renn, Jürgen, 2020: *The evolution of knowledge: Rethinking science for the Anthropocene* (Oxford: Princeton University Press).
Revkin, Andrew, 1992: *Global warming: Understanding the forecast* (New York: Abbeville Press).
Rose, Deborah Bird, 2013: "Anthropocene noir", in: *Arena Journal*, 41, 42: 206–219.
Santos, Filipe Duarte, 2020: "Climate change in the 21st and following centuries: A risk or a threat?", in: Jodelet, Denise; Vala, Jorge; Drozda-Senkowska, Ewa (Eds.): *Societies under threat* (Cham: Springer): 143–156.
Schulz, Karsten Alexander, 2017: "Decolonising the Anthropocene: The mytho-politics of human mastery", in: Woons, Marc; Weier, Sebastian (Eds.): *Critical epistemologies of global politics* (Bristol: E-International Relations Publishing).
Sollund, Ragnhild, 2017: "Going green: Critical criminology with an auto-ethnographic, feminist approach", in: *Critical Criminology: An International Journal*, 25: 245–260.
Sollund, Ragnhild, 2020: "Wildlife crime: A crime of hegemonic masculinity?", in: *Social Sciences*, 9, 93: 2–16.
Steffen, Will; Persson, Åsa; Deutsch, Lisa; Zalasiewicz, Jan; Williams, Mark; Richardson, Katherine; Crumley, Carole; Crutzen, Paul; Folke, Carl; Gordon, Line; Molina, Mario; Ramanathan, Veerabhadran; Rockström, Johan; Scheffer, Marten; Schellnhuber, Hans Joachim, Svedin, Uno, 2011: "The Anthropocene: From global change to planetary stewardswhip", in: *Ambio*, 40,7 (Noviember): 739–761.
Steffen, Will; Rockström, Johan; Richardson, Katherine; Lenton, Timothy M.; Folke, Carl; Liverman, Diana; Summerhayes, Colin P.; Barnosky, Anthony D., Cornell, Sarah E.; Crucifix, Michel; Donges, Jonathan F.; Fetzer, Ingo; Lade, Steven J.; Scheffer, Marten; Winkelmann, Ricarda; Schellnhuber, Hans Joachim, 1998: "Trajectories of the earth system in the Anthropocene", in: *Proceedings of the National Academy of Sciences of the USA- PNAS*, 115,33 (August): 8,252–258,259.
Swyngedouw, Erik; Ernstson, Henrik, 2018: "Interrupting the Anthropo-obScene: Immuno-biopolitics and depoliticizing ontologies in the Anthropocene", in: *Theory, Culture & Society*, 35,6: 3–30.
Trischler, Helmut, 2016: "The Anthropocene: A challenge for the history of science, technology and the environment", in: *NTM Zeitschrift für Geschichte der Wissenschaften, Technik und Medizin*, 24: 309–335.
Valencia, Sayak, 2010: *Capitalismo gore* (Barcelona: Melusina).
Zalasiewicz, Jan; Waters, Colin N.; Williams, Mark; Summerhayes, Colin P. (Eds.), 2019: *The Anthropocene as a geological time unit: A guide to the scientific evidence and current debate* (Cambridge: Cambridge University Press).

# Chapter 2
# Social Representations and History: Theoretical Problems

Lúcia Villas Bôas

## Introductory Comment

Alfredo Guerrero Tapia

*The analysis offered in this chapter by Lúcia Villas Bôas lays out a genuine problematic field[1] regarding the relationship between social representations and history. The main role of a problematic field is to open up a set of questions that demand answers, and in the process drives the movement of the theory. She does so by offering a concise review of the visions and positions of diverse thinkers in the fields of both history and social representations; identifying the links between these two fields, highlighting their differences, distinguishing the reach of their perspectives; highlighting the polysemy that often attends the use of the notions of both social representation and history; and also by referring to the schools of*

Dr Lúcia Villas Bôas is Coordinator of the UNESCO Chair in Teaching Professionalization at the *Fundação Carlos Chagas (FCC)*, scientific lead of the Franco-Brazilian Serge Moscovici Chair (FCC/French Consulate in São Paulo) and Professor on the Academic and Professional Master's Programme at the City University of São Paulo (UNICID). Email: lboas@fcc.org.br.

[1]Dr Alfredo Guerrero Tapia is Senior Researcher at the Faculty of Psychology, National Autonomous University of Mexico and has a PhD in Social-Environmental Psychology from UNAM. He has worked in the field of social representations, theoretically and empirically, since 1989, and is Co-Editor, with Denise Jodelet, of the book *Develando la cultura* (México: Facultad de Psicología/UNAM, 2000). He belonged to the International Research Group on Latin American Imaginaries and Social Representations, supported by the Laboratoire Européen de Psychologie Sociale, of the Fondation Maison des Sciences de l'Homme, whose research project "Images of Latin America and Mexico through mental maps" was published in the book *Espacios imaginarios y representaciones sociales* (Barcelona, Anthropos/UAMI, 2007). Email: alfredog@unam.mx.

© Springer Nature Switzerland AG 2021
C. Prado de Sousa and S. E. Serrano Oswald (eds.),
*Social Representations for the Anthropocene: Latin American Perspectives*,
The Anthropocene: Politik—Economics—Society—Science 32,
https://doi.org/10.1007/978-3-030-67778-7_2

*thought to which they belong or within which the different authors are situated. Together, these elements generate a series of complex problematizations that are expressed in distinct conceptual axes that, according to Lúcia, intersect with one another in two "historiographical conditions" – the modern and the postmodern – and their respective underlying epistemological problems. Thus, the **problematic field** fulfils its purpose by raising a wide range of questions in the fields of both social representations and history, both under contexts that are shared and particular to each, as well as, of course, in their connections. As such, this response is made through the formulation of some questions that are separate from those stated, and drawn from other areas of study that are assembled around a central theme: human consciousness. Due to space restrictions, these questions and the contextualization of them are, of course, only stated and not analysed in depth.*

*Before stating some of these questions (which do not exhaust those that arise from the chapter) I would like to make clear that, in thinking about the relationship between social representations theory and history, Lúcia Villas Bôas's approach to the site where the theory is located – at the crossroads between modernity and postmodernity – is absolutely pertinent to the contemporary debate in science, including the philosophy and epistemology of science. The approach correctly identifies the gravitational centre of the problematic field, since it defines the perspective, the method and the epistemology through which the relationship can be investigated and analysed. In effect, thinking about this within the parameters of modernity means doing so within a fragmented matrix of disciplines, their territories delimited by their respective objects of study and theories. It is to be imprisoned within the boundaries that delimit their knowledge, which may be narrow or broad depending on their multidisciplinary, or even interdisciplinary, scope. Taking this path opens up a host of problems, as it becomes necessary to assess what is common and what is different in each field and theory, while facing obstacles that are more a product of the theoretical construction than of the nature of the actual phenomena. For its part, thinking about the relationship between social representations and history from a postmodern perspective is to travel along the road that does not stop at any disciplinary frontier, and whose perspectives transcend the concern to preserve some kind of identity or commitment to the discipline. It means thinking with greater complexity, allowing for movement between categories and a greater approximation to the phenomenal aspects of social representations and history. That is to say, I try here to make "critical use of the theory" by means of a "postmodern epistemology", as Lúcia's invitation suggests.*

*A first question raised by the problematic field that Lúcia presents concerns the problem of time and space. While in the discipline of history time and space appear consubstantial to the very idea of history, in the theory of social representations the presence within the phenomenology of social thought and the subjectivity of groups within societies is not so clear. In studies of social objects represented by social groups and collectivities, where phenomena such as perception, memory, reminiscence, imagination, expectation etc. come together, these representations are understood as expressions of socio-cognitive processes, because they are particular to the psychology of these groups or collectivities. As such, they are manifestations of a present time. Nonetheless, one of the traditions within studies of social representations, known as the 'genetic' approach, has offered interesting observations about the genesis and transformation of representations. Permanence and change, as Lúcia Villas Bôas indicates, is a paradoxical property within social representations theory. Both may be observed, however, if one chooses interpretations that go beyond collected data, whether quantitative or qualitative.*

*The past, however, can only be recovered as testimony, narratives of events that have occurred, or as living memory, images, reminiscences, translated into written language or oral communication. In all cases, no type of recovery is true to the fact; it's not isomorphic. The more types of recovery that exist for the events that occurred (whether in history or any other discipline), the more the past will be present and the more comprehensive the recovery of it will be. Common sense, which has been associated more directly with social representation, being a type of epistemology of social groups, is present in oral tradition, and part of oral tradition, rather than written texts. Thus, the past lives in the present as a permanent recreation. However, in studies of social representations that emphasize the notional structure of the object of representation, seeing it as a sociocognitive phenomenon, the historical contents do not appear or disappear just because they are usually at the periphery of the structure. Genetic studies, by their own method, identify the historical elements of the object of representation. Hence the importance of Lúcia Villas Bôas' recognition of the importance of history within the thought of Denise Jodelet in her work on the body and madness. In addition, we could ask ourselves: how do time and space appear within the mental schemes of groups?*

*The question of time and space necessarily leads us to two of the most thorny and complex, but crucial themes-problems-phenomena of modernity: human consciousness and the problem of origins. The problem of consciousness continues to be one of the greatest and most fundamental problems confronting all scientific, philosophical, and artistic knowledge, in the face of the prolonged and profound crisis of civilization in which humanity finds itself. What can the knowledge provided by history and social representations theory contribute to our understanding of human consciousness? In the evolution of civilizations and the broad stages through which their evolution passes, ideas, knowledge, wisdom, about themselves and their moment of existence, are produced. In these civilizations, representations made of reality are always relative to a substratum. These have been identified and named using various concepts and ideas: "reflections of reality", religious beliefs, ideologies, fantasies, myths, etc., in whose contents consciousness is manifested. All of these are anchored to a past. This has always been an object of focus for both the historian and the philosopher, but it can be also for the social psychologist who is interested in unravelling different cultural products and identifying the causes that led to their creation, and in exploring the mysteries that contain the symbolisms of consciousness, their meanings, and all those elements from which consciousness is nourished.*

*Consciousness and origin are amalgamated in such a way that first Nietzsche and later Foucault both warned that there are no beginnings but only genealogies. Later, in his profound research on the transitions and mutations of human consciousness over the course of the different epochs through which civilizations have passed, and which provide clues for understanding our current context, with the bifurcation between the course of technological development and its contradictory consequences and the evolution of human consciousness, Jean Gebser (2011: 15–16) reminds us that 'the origin' has always been present. He states:*

> The origin is always present. It is not a beginning, since every beginning is linked to time. And the present is not merely the now, the today or the instant. It is not a part of time, but an integral result and, consequently, always original. Whoever is able to make the origin a reality and the present integral, whoever is capable of making them concrete, overcomes the beginning and end, and mere present time. What we experience today is not only a European crisis. Nor is it a crisis of morality, of the economy, of ideologies, of politics, of religion. It doesn't only affect Europe and America, but also Russia and the Far East. It is a world crisis and a crisis of humanity, only comparable to those that occurred during conjunctures of great transcendence, which were definitive and decisive for the life of the earth and of humanity […].

*The question of human consciousness is, in my view, the nuclear aspect in which a set of connections between forms of knowledge and understanding, coming from a diversity of scientific, philosophical, humanistic, artistic etc. disciplines, converge, and among which can be included history and social representations theory. It is a kind of crucible that needs, under the premises of postmodernism, to integrate the broadest and most varied range of conceptual elements.*

*An ever-present element in human consciousness is the collective memory of society and social groups. Lúcia Villas Bôas' careful review of the relationship between memory and history – highlighting the contributions of Durkheim, Bloch, Halbwachs, Ricœur, and the historians Hartog, Koselleck, and Nora, among others, as well as much research within the field of social representations and memory, including by Licata, Prager, Castorina and Carretero, Celso Sá, and the author herself – sheds light on the controversies that emanate from the ways these authors approach memory and their relationships with other forms of disciplinary knowledge, but also on the property of memory as a socio-cognitive (sometimes traumatic) phenomenon of human groups. The extensive literature published in recent years on the subject of memory, on the other hand, leads us to believe that the great interest aroused within the academy by memory and the growing tendency of societies to look towards their past is driven by the great uncertainties surrounding how we live in the present and look towards the future. When certainties in societies are diluted, it is common for them to look to their past. But in this response I am interested in situating the theme of memory and its relationship to history and social representations within human consciousness, because we have argued that in this era consciousness takes on a relevant and transcendental place in the course of societal and civilizational development. Human consciousness is not national consciousness or the consciousness of any group; it is consciousness housed in many individuals who inhabit different latitudes of the planet; it is planetary consciousness as recognized by Erving Laszlo. This is the dimension of consciousness in which the knowledge constructed by social representations theory and the knowledge produced by history can converge.*

*Human consciousness is not only nourished by perceptions of the past, or the historiographic construction of facts; it also moves with imagined futures, with utopian thinking, or with the "daydreams" described by Ernst Bloch. History also works with this dimension, although its constituents are part of the past events, because it accounts for and testifies the way societies and social groups have thought about and imagined their futures, how near or far they were from such events, and how they happened. Social representations theory, on the other hand, as a theory about the ways in which social groups translate the knowledge of science into common sense knowledge, did not propose to discuss the future of social representation. Moscovici's foundational study on psychoanalysis, carried out in France, was at no time concerned with knowing the future of psychoanalytic theory, nor its transformation as a dominant theory within French culture. The body of research and theorization on social representations, in its more-than-fifty-year history, has not generated knowledge of the future representations of the social objects studied, nor of the future dimension of social thought. This does not mean that it does not exist as a phenomenon of social subjectivity or of the social production of knowledge. It is an unexplored terrain that calls out to be investigated theoretically and empirically, especially in moments like the current one in which young people speak of the future like a void and there appears to be an absence of utopias in our globalized societies.*

*Within the problematization that Lúcia Villas Bôas offers regarding the notion of a "duty of memory", we are led to a series of questions about the role that social representations play in so-called "historical truth", "transitional justice", and other concepts highly relevant to different Latin American countries that have gone through periods of extreme violence or polarization that have left their populations in a state of permanent conflict; concepts such as "forgiveness", "anger", "revenge", "reconciliation", "compensation for damage ", and others. What place do these concepts, together with their respective representations, have in the construction of social consciousness? What is their nature? Are they obstacles or catalysts for the flow of these processes? To what extent are they anchored in the historical peculiarities of the development of different societies and culture? Without a doubt, answering these questions requires consideration not only of legal frameworks and the facts of a given conflict, but also of the psycho-emotional, subjective, representational, present and past contents of each of the social actors. Lúcia Villas Bôas's observation is correct when she says that there is not one but a plurality of memories in these contexts, and each of them has a place within the reference systems generated by the conflicts.*

*Finally, I would like to close this comment by reiterating the value of Lúcia Villas Bôas' work in an area that is understudied in the fields of both social representations and history. The problematic  field that it provides us with reflects the power of her insights into the relationships, movements and controversies of history, as well as the gaps and problems half hidden in social representations theory. It constitutes an outstanding contribution to the task of understanding thought in societies in today's world. Pursuing the image of the crossroads, I have chosen the path of postmodernity, following it towards the phenomenon of human consciousness.*

## References

Gebser, Jean, 2011: *Origen y presente* (Girona, Spain: Atalanta).

**Abstract** Addressing the relationship between social representations and history, this chapter discusses the idea of representation as both affirmation and negation of historiographic discourse, based on what have been called the 'modern' and 'postmodern' historiographic conditions. It begins by discussing how memory has become an epistemological object within the discipline of history. It thus offers reflections on the relationship between history and memory, understood as two forms of managing the past, while exploring their implications for social representations theory. Building upon some considerations of a current in historiography called 'history of the present time', the chapter presents the concepts of 'regime of historicity' and 'presentism', developed in particular by François Hartog and Reinhart Koselleck, and discusses some of the challenges they present in relation to current processes surrounding the intelligibility of memory. Finally, it emphasizes that the possibility of developing common ground between history and Social Representations Theory depends on building a post-disciplinary and epistemo-political agenda.

**Keywords** History · Social representations · Memory

## 2.1   Introduction

This chapter calls for us to think about the articulation between history and social psychology, disciplines usually thought of as estranged from one another. Seemingly estranged, but, on deeper inspection, not so much, especially if we consider a paradoxical aspect of social representations: of simultaneously articulating permanence and change, insofar as they remain dependent on the past, which anchors them, while, nonetheless, being able to evolve within different contexts.

Looking in particular at research that has been carried out to date in Brazil, we can see that despite its importance for understanding the genesis and transformation of the contents of social representations, the study of their historicity has been little explored.

Against this backdrop, this chapter offers some reflections on the relationship between social representations and history, since it is mainly via memory and anchoring that the articulations between these two areas become clearer.

However, these links are broader. In fact, already in 1989, Jodelet showed the urgent importance of adopting a historicist perspective on social representations, a position clearly shaped by her studies on madness and the body in the 1970s.

Since then, especially since the 2000s, several studies have emerged which highlight the relationship between social representations and history, identifying the latter as either the object of social representations, or the source of their content. Among these are the works of Darío Páez in Spain, James Liu in New Zealand, Laurent Licata in Belgium, Elena Zubieta in Argentina, and Celso Sá in Brazil.

In spite of the importance of these studies, adopting a historicist approach to social representations, as Jodelet (1989) called for, has not yet been explored from this perspective, which still constrains the possibilities of creating a genuinely common field.

In this context, this chapter explores the relationship between social representations and history, discussing the idea of representation as both affirmation and negation of historiographic discourse. The point of departure for this discussion is the distinction between the 'modern' and 'postmodern' historiographic conditions, although unfortunately, due to lack of space, it is not possible here to problematize these qualifiers.

The chapter then goes on to discuss how memory becomes an epistemological object within the discipline of history. It thus offers reflections on the relationship between history and memory, understood as two forms of managing the past, while exploring their implications for social representations theory. Building upon some considerations of a current in historiography called 'history of the present time', it discusses the concept of 'regime of historicity', developed mainly by Hartog (2001, 2003, 2010). The concept is used to discuss the idea of presentism and some of the challenges related to current processes surrounding the intelligibility of memory.

Finally, the chapter stresses that the possibility of developing common ground between history and Social Representations Theory depends on building a post-disciplinary and epistemo-political agenda (Ferraroti 2013).

## 2.2  Representation: At the Crossroads of Two Historiographic Paths

In broad terms, 'representation' is recognised as a key concept across the social sciences. The field of psychology is no exception. Several authors, for example Jodelet (1989, 2003), have devoted themselves to discussing use of the concept in different subdisciplinary fields within psychology itself.

The field of history is no different: Carlo Ginzburg, Roger Chartier, Gérard Noiriel and many others have shown how different historiographic schools

understand the term 'representation', and highlight its ambivalence and even contradictions, as it varies according to the frames of reference adopted.

There is no space here to offer a detailed examination of these discussions, but
it should be emphasized that in the Brazilian context it was the subfield of cultural
history in particular that pioneered the diffusion and popularization of the term
'representation' among historians, a process that did not necessarily translate into
a more effective discussion of its meanings.

In the Brazilian context, Ciro Flamarion Cardoso deserves special mention as
the historian who has come closest to engaging with social representations theory.
According to Cardoso, in a text published in the 2000s, it was social psychology,
through the studies of Serge Moscovici, that most effectively explored the meaning of representation, a concept which had remained disappointingly vague in
other areas of the social sciences (Cardoso 2000).

The well-known Italian historian Ginzburg (1991) highlights the importance of the semantic ambiguities surrounding use of the term 'représentation'
in French and 'representation' in English. According to him, some researchers
have included the study of symbolic expressions of mental or affective content
within the field of the history of representations. Noiriel (1998) points out that in
German these dimensions do not mix. Whereas *Vorstellung* designates mental content, *Darstellung* refers to symbolic expression. By contrast, the polysemy of the
French or English usage produces considerable confusion.

Ginzburg (1991) also points out that it is necessary to highlight the semantic
ambivalence between the mental and visual domains. This causes confusion in
research since these terms, which, according to him, refer to the mental and the
visual, have been often used in research on material images with fruitful results,
for example, between the history of the representations and the anthropology of
images in the field of medieval studies.

For Cardoso (2000), one of the reasons for the lack of a clear definition of what
constitutes 'representation' in the discipline of history is that its use is relatively
recent, initially with reference to the so-called history of mentalities, despite being
more effectively mobilized by Chartier (1990) in the cultural history approach.
According to these authors, the theoretical categories of the history of ideas centre on
conscious/unconscious, time/duration, which has given rise to concepts such as 'collective representations', 'visions of the world', and 'spirit of the age'.

Regardless of these difficulties, related to the specific perspectives that each
field and sub-field have brought to bear on this notion, the fact is that, within the
scope of history,

> the debates about representation actually involve the conditions of possibility of historical
> discourse as the discourse of a specific disciplinary practice, committed to the production
> of a certain type of knowledge, which we usually call 'history'and/or 'historiography.'
> (Falcon 2000: 42–43)

Broadly speaking, the problematic of 'history and representation' can be situated at the crossroads of two historiographical paths – the modern and the
postmodern. It is as if we had before us the two 'faces of Janus', in a beautiful
metaphor used by Falcon (2000: 42):

one looks towards representation as a category inherent to historical knowledge; the other looks in the opposite direction, and sees representation as the denial of the possibility of this same knowledge. The first gaze focuses on the intellectual environment of 'modernity', while the second focuses on the horizon of 'postmodernity'.

That is to say, this doesn't merely express a different vision of historical discourse, but of the very notion of historical reality.[2] Over the course of its constitution as a discipline, history has always claimed, at least in principle, that its purpose is to say something true about past events. Even if we are aware of the limits imposed on the historian, we do not question the historical reality of the past (Falcon 2000: 59). Postmodern historiography, by contrast, proposes an identification between historical and narrative discourse, reinforcing the separation between discourse and reality.

In this case, representation is always a text equivalent to many others, historical truth being a self-referential product of language. Thus, history-discipline becomes epistemologically fragile since it is always a personal construction of the historian as narrator.

From this perspective, which may be called 'postmodern' and which would include historians such as Hayden White and Keith Jenkins, the past is always contingent and there is no extra-discursive historical reality. As a kind of

> prisoner of language, the historian produces interpretations, but, strictly speaking, her discourse can no longer be put forward as the representation of a presumed historical reality, only if we admit that this reality is also a set of discursive practices. (Falcon 2000: 63)

Added to these two different historiographic perspectives is the very polysemy of the term 'representation'. However, it was within this modern perspective of the historiographical tradition that the history of mentalities emerged, which has perhaps engaged more than any other with the perspective of social representations. It is no coincidence that, to some extent, it has ended up being the progenitor of what is now, in the historiographic field, called the 'history of representations' or 'cultural history'. Moscovici himself (2001: 45) asserts that the Durkheimian notion of 'collective representation', which prefigures the concept of social representations, "would have fallen into disuse had it not been for a school of historians who preserved its insights, through research on mentalities".

Jodelet (1989) also points to the reciprocity between social representations theory and the 'history of mentalities' with respect to the definition of objects, the collective character of phenomena studied and awareness of the affective dimension, as well as their concern with historical time frames (long, medium and short).

---

[2]Here it is important to open a parenthesis because, in general, modern historiography distinguishes between two levels: History, with a capital 'H', referring to historical reality; and history, with a small 'h', referring to historical knowledge. Thus, History is the substance or referent of the discourse that the historian produces as history, or historical knowledge. It is to this distinction that Vilar's famous phrase refers (1980: 17): "history speaks of History", that is, history-discipline speaks of History-Substance (Falcon 2000).

Offering a similar interpretation, Castorina (2007: 76) points out the following convergences between social representations theory and the history of mentalities:

> Both have played a critical, remarkably similar role in the recent history of each discipline; the notes which characterize the respective definitions of these categories are equally nebulous; for this reason, their relations with ideology are debatable, but illustrative of its most important features; both are the result of processes of the imaginary of intellectual production; moreover, each decisively influences the practical life of individuals; finally, the understanding each has involves the articulation between society and individual.

Emiliani and Palmonari (2001) point out that the approach taken in the history of everyday life, related to the history of mentalities, also engages with the concept of social representations, not only because their object of study focused on the realm of ideas, but because it also concerns common sense.

In fact, Moscovici (1991: 77) never denied a link between social representations theory and history as a disciplinary field. According to him, history and psychology have different, though complementary, questions to answer. As such, the historians' perspective differs from that of social psychologists insofar as they

> [...] highlight the production of ideologies and ask where the ideas that we have about society and politics come from. Are these ideas socially determined? What validity can they claim? However, these are not the questions that I fall in love with and which, as a social psychologist, I try to answer. The questions of my discipline are different: How are ideas transmitted from generation to generation and communicated from one individual to another? Why do they change people's way of thinking and acting until it becomes an integral part of their lives?

Even if these differences are offset, it is possible to conclude that work carried out in both history and social representations theory take into account the existence of border areas. And the history of mentalities has contributed much to this in that it widened the field of historical research.

On the other hand, this expansion led to the emergence of important epistemological and methodological problems, which led the perspective of the history of mentalities to be gradually replaced by new perspectives (Dosse 2000).

Thus, during the 1980s and 1990s, in the context of a historiographical landscape marked by eclecticism, what would later be called cultural history, or history of representations, emerged in France, with a somewhat nebulous perspective, that bit by bit gradually absorbed different historiographical currents.

Even in the 1970s, discussions of popular culture contributed to a definition of cultural history that was gradually drifting away from the history of mentalities. The reception given to the works of Carlo Ginzburg and Robert Darnton also contributed to this debate.

It was in this context that, in the early 1980s, Chartier (1990) proposed, contrary to the history of mentalities approach, going from cultural facts to social configurations, abandoning the idea of a social and professional framework. That is, we no longer go from the social to the cultural, as in the history of mentalities, but from the cultural to the social.

With this shift, the notion of representation is distinguished from the notion of mentality. Now, the notion of representation attempts to account for the symbolic constructions of social groups, practices that manifest a social identity, and institutions that attest, in a stable way, to the existence of these groups.

The fact is that cultural history is the heir to the history of mentalities and is fundamentally considered to be a history of representations and social practices.

Another heir of the evolution and diversification of the history of mentalities is the so-called history of memory, which became, in the 1970s, one of the facets of the history of representations. Nora (1978: 400) notes that

> L'histoire s'écrit désormais sous la pression des mémoires collectives tentant de compenser le déracinement historique du social et l'angoisse de l'avenir par la valorisation d'un passé qui n'était pas jusque-là vécu comme tel.[3]

## 2.3  Memory and History

In the relationship between memory and history, a long road had to be travelled before memory became a central epistemological question within the discipline of history. Several factors contributed to this absence. Among these, we can cite: the conviction that historical construction should be carried out by observing a certain regression in time; a distrust of first-hand accounts, considered to be unreliable and partial; and the priority given to archives and documents. It is no coincidence, therefore, that the process that led memory to be considered as a specific object of the social sciences has not progressed in the same way in the historiographic field.

Working from a Durkheimian perspective, it was in the 1920s that Maurice Halbwachs began to investigate how memory is rooted in social communities, thus removing memory from a purely individual sphere. There is no doubt that his contribution is still highly relevant. However, what should be emphasized here is the striking opposition he establishes between memory and history.

For Halbwachs, history was characterized by being critical, conceptual and problem-orientated, whereas memory was fluctuating, concrete, lived and multiple. This radical opposition, based on the Durkheimian perspective, makes history appear to be a space of absolute objectivity incarnate, and therefore an abstract theoretical form of knowledge that retained a past beyond the lived dimension, which was the territory of memory (Halbwachs 1994).

Although Halbwachs' discussions did not go completely unnoticed at the time by historians – Bloch (1925), for example, adopted the idea of collective memory in his analysis of peasant customs in seventeenth-century France – memory remained a residual concept in the field of historiography.

It was with the epistemological renewal that took place in the late 1970s, driven above all, in the European context, by the emergence of local identities and

---

[3]"History is now written under the pressure of collective memories that attempt to compensate for the historical uprooting of the social and anxiety about the future through the valorization of a past that was not previously lived as such."

minority differences, that memory became the subject of debate within at least two historiographical currents during the period: one called the 'New History' – which had Pierre Nora as one of its leading representatives – and the other called 'history of present time', whose engagements with the question of memory became fundamental, since it placed the historian on the terrain of the contemporary, which until then had been dominated by psychologists, social scientists and journalists.

It is at this moment that historians became interested in memory in two ways – as a historical source and as a historical phenomenon – and began, from then on, to discuss aspects related to the principles of the selection of memory, of variation from place to place, or from one group to another, how it changes over time, its modes of transmission and the uses of forgetting. In short, they began to explore problems that required memory to be discussed as an object of historical analysis.

This discussion was driven, especially in the 1980s, by the publication in France of *Places of Memory* (1984), edited by Pierre Nora, which, based on the studies of Halbwachs, once again presented the relationship between memory and history as an opposition, bringing old dichotomies back to the discussion: oral *versus* written, nature *versus* culture, lived *versus* analytical, affective *versus* rational.

More recent studies have tended to argue that while this opposition need not be completely rejected, it requires greater nuance. For example, in *Memory, History and Forgetting*, Ricoeur (2000) presents memory as the very matrix of history, arguing that the interplay between history and memory is inevitable and that there is therefore no possible separation: history is a tributary of memory, and memory is a tributary of history.

As two models of managing the past, their differences can be seen in their different claims as ideal models. Thus memory claims to be faithful to the past, while history, founded on the pursuit of knowledge, has the pretence of being true.

There is insufficient space here to lay out Ricoeur's entire philosophical argument. But it is important to point out that in his work the discussion shifts from the image of a cleft between history and memory to the idea of circularity, to reciprocal constitution from one sphere to another, insofar as that which is remembered from the past is inseparable from what is known about it.

So, within this perspective of circularity, what questions about memory are historians confronting today and what interest do they hold for studies of social representations?

This is an important question if we look, for example, at the categories of analysis mobilized by Hartog (2001, 2003): 'regime of historicity' and 'presentism'.

The concept of historicity has been used to discuss the relationship that a society has with its past, its present and its future, especially in moments of crisis. Clearly, this is a broad characterization designed to operationalize an epistemic category.

In the words of Hartog (2010), a regime of historicity is not a metaphysical entity, but a framework of long-term thinking that enables and interdicts certain forms and contents of thought.

Thus, the regime of historicity, that is to say, the experience of time established, for example, by European societies until the mid-eighteenth century, would be characterized by an idea of the future directly linked to the past and understood as a model to be followed, to be imitated. We would thus, in this context, be under a regime of historicity based on the idea of 'history as the master of life'.

Between the late eighteenth and mid-twentieth centuries, there was a change in this relationship, as the future came to be seen as no longer determined by the past, and instead to be interpreted as a completely different project from the past.

For Hartog (2001), we are now living in a new regime of historicity. We see the past neither as a model, as had been the case in the eighteenth century, nor as the differentiation of the future relative to the past, as prevailed until the middle of the twentieth century. For him, what characterizes our current condition is the omnipresence of the present, which he calls presentism, whose symbolic origin refers to the fall of the Berlin Wall. On this point, the conclusion of François Furet (1995: 572), in *The Past of an Illusion*, is quite emblematic: "the idea of another society has become almost impossible to think… We are condemned to live in the world in which we live".

This discussion takes different forms in the social sciences. For example, engaging with an altogether different body of literature, Lyotard (2005) proposes the concept of 'post-modernity' in analysing the crisis of modernity, characterized, among other symptoms, by the collapse of 'grand narratives' which precludes the very idea of a project and, therefore, of any form of intelligibility guided by the future.

Without entering this vast terrain, which ultimately shows that the social sciences have a set of common concerns, it is undeniable that the idea of a change of regime of historicity is a disturbing proposition that brings with it a challenge to studies in the field of social representations. That is to say, do the most common concepts of memory that we work with – many of which are constructed for thinking about modernity based on other temporal experiences – respond to this new form of intelligibility demanded by a contemporary context that is defined by presentism?

Of course, this is an open question that, from a historiographic point of view, forces a set of concepts, such as ideology, identity and great narratives, to leave the scene and give way to others, such as present, memory, responsibility and recognition. This establishes the conditions for the reappearance of the figure of the 'witness' as emblematic of historical knowledge, a figure that had been side-lined since the nineteenth century due to the emphasis given to archives as a privileged source for the study of history.

This return of the witness also facilitates the return of a dialogue between history and memory. Reflecting primarily on the post-War European context, Hartog (2001: 21) states that "a witness today is, above all, a voice, a face of a victim, a survivor that we listen to, who we record and film". It is within the scope of this problem that Pierre Nora popularized the notion of the 'duty of memory', especially in the French context, which was associated with not only an obligation to preserve the past, but, above all, the duty to recognize the suffering of others.

Thinking about this notion, of the 'duty of memory', it is worth mentioning that in the Brazilian context it is not part of the lexicon of social movements nor of public policies aimed at managing the past. In this sense, we see a much greater use of the expressions 'places of memory', 'recovery of memory' or even 'recognition of memory' for dealing with the events of our recent past.

Among these events of our recent past, I would like to briefly point to one in particular, which has directly used this semantic repertoire linked to memory: present, responsibility, redress, recognition, etc. This is the military coup that established a dictatorship in Brazil fifty-four years ago (at the time of writing in 2018), and discussions of which, between the State, the armed forces and civil society have revolved around expressions like 'recovery of memory', 'historical truth' and 'recognition'.

As an integral part of this discussion, the Brazilian Government recently created the National Truth Commission, which in its first article identifies as one of its primary objectives the realization of the right to memory and historical truth and the promotion of national reconciliation, "…in the recovery of memory and the truth about the grave violations of human rights that occurred during the period of the dictatorship in order to prevent these facts once again becoming part of the history of our country" (Brazil 2011).

In this context, there is a plurality of memories that claim – from a kind of policy of redress, via the recognition of what had been concealed, to the idea of the recovery of memory, especially by the sectors related to the Armed Forces – to represent revanchism, by repressing specific demands for justice.

There is not enough space here to explore this plurality of memories, which will only reaffirm the political and conflictual dimensions of evoking the past. What I would like to mention, however, is that it is still too soon to see whether, as in post-war France, we have in Brazil a fertile social ground for creating a system of references in which memory is transformed into a virtue, in which the recognition of abuses committed on behalf of the State is viewed positively, and the promotion of forgetting and amnesty is socially unacceptable. This system of references explains, at least in the French context, the popularity of the notion of 'duty of memory', which, though criticized for some of the uses and abuses it engendered, has become associated with social and political obligations on the part of the nation towards parts of its population.

Regardless of context, the recourse to memory and history does not take place on neutral ground, outside the political. On the contrary, in serving as a political instrument of recognition, memory and history are constituted not only as explanatory categories, but also as ways of acting in the social field, which allow different sectors of society to construct their demands for recognition. It is up to us to understand how they act and what strategies are used to historically legitimize discourses of memory in the field of practice.

In these spaces of conflict, it is above all by means of anchoring that the contribution of social representations theory to the study of history and memory becomes visible, since, as we know, it identifies the integration of novelty into already established systems of representation.

Collective memories are the product of the collective work of reinventing the past. They constitute part of the representational system of a group, serving to anchor it. They are always the product of a process of objectification through which the group symbolized its experience in the past. Collective memories, therefore, teach us about this collective experience (Licata 2016: 553).

In a study about the idea of diversity as one of the organizing axes of social representations in Brazil, I showed that the processes of anchoring and objectification make us understand the dual character of historicity presented by social representations: its variation across eras and social, cultural and political contexts, and therefore their sensitivity to the evolution of time and dependence in relation to past traditions or innovative repetitions (Villas Bôas 2010).

As tributaries of pre-existing forms of thought, that originated in distinct chronological eras, and from the pragmatic imperatives of daily life, social representations are, as we know, a "particular mode of knowledge that has as its function the elaboration of behaviours and communication between individuals" (Moscovici 1978: 26), nourished both by the knowledge derived from everyday experience and by reappropriations of historically consolidated meanings.

This reappropriation of the past by the contemporary occurs through a dynamic process in which each generation changes the meanings, and understandings, of pre-existing knowledge. That is to say, every current context selects content from the past and updates it by delineation and interpretion, based, ultimately, on the meaning that a given group attributes to its 'space of experience' and 'horizon of expectation' (Koselleck 2006).

According to Koselleck (2006: 306), 'space of experience' and 'horizon of expectation' are formal categories of knowledge that ground the possibility of a history, without, however, conveying an a priori historical reality, as "all histories have been constituted from the lived experiences and expectations of people who act or who suffer. With this, however, we have not yet said anything about a concrete history – past, present or future".

It is from this space of experience, constructed from historical knowledge produced, or experienced, that a future will be projected. In this future a horizon of expectation is established, a horizon that "…is realized in the today, it is a present future, looking to the still-not, to the not-yet-tried, to what can only be predicted. Hope and fear, desire and will, restlessness, but also rational analysis, received visions or curiosity are part of this expectation and constitute it" (Koselleck 2006: 310). Historical time is constructed from tension between experience and expectation, constituted through the interweaving of what is understood by the past with what is perceived as the future.

This presupposes that the historicity of social representations can only be grasped through its effect[4] in the present (Gadamer 2002) insofar as access to the past is mediated by its connection with the contemporary.

---

[4]The perspective of 'actual history', theorized by Gadamer in *Truth and Method* (2002: 449), is used here, although its use is specific. According to this author, "When we try to understand a historical phenomenon from the historical distance that determines our hermeneutical situation as a whole, we always find ourselves under the effects of this actual history. It determines beforehand what appears to us as questionable and as an object of inquiry, and we soon forget half of what it really is, more than that, we forget the whole truth of this phenomenon, every time we take the immediate phenomenon as the whole truth."

In the studies developed by Licata (2016) in Belgium, and by Castorina (2007) with Carretero, Gonzalez and Alberto (2007) in Argentina, these authors have pointed to a fundamental question about the relationship between history and social representations, concerning the influence between the past and the present.

Roughly, this question can be formulated as follows: Are the ideas that members of a social group have about a given object influenced by the ways they represent their past? Or, on the contrary, do the ideas a given group has influence their representations of the past?

Licata (2016), based on the studies of Prager (2001), distinguishes two approaches that he describes as neo-Freudian and neo-Durkheimian. In the neo-Freudian approach, the tensions within a group are understood to result from the inevitable intrusions of a traumatic past into the present. By contrast, in the neo-Durkheimian approach, the past is interpreted as a symbolic source that the members of a given group mobilize in order to reduce tensions in the present.

Collective memory is thus understood as a complex process of social construction linked to the current identity of the group, which means that different versions of the past can be elaborated at different moments in the life of a group, or from one social group to another.

The origins of social representations theory are linked to the Durkheimian tradition. However, it would be reductive to limit its contributions to the study of collective memories to the neo-Durkheimian approach, described above.

As Moscovici (1984: 54) has written: "Donc nos représentations rendent le non-familier, ce qui est une autre manière de dire qu'elles dépendent de la mémoire."[5] However, even though the role of objectification in the dynamics of collective memories is less visible, it is here that social representations theory more clearly reveals these processes, due to the pre-requisite of a common repertoire of representations.

As Moscovici (1976) has proposed elsewhere, figurative structures must be understood as significant. In describing a historical event, the subject expresses present feelings, takes a stand in the current debate. Here it is necessary to distinguish historical truth and social representations from history. A representation need not be true to provide anchoring. But it can reveal another order of truth. In any case, the past, represented, can in this way "weigh" upon the present.

It is not a matter of confronting what people say with the work of historians in order to point out inconsistencies. Discourses may seem to describe past events, but in fact they denounce a current situation. Thus, collective memories are chosen as a function of social actors' present motivations. They are used to interpret current situations.

Disguised with the illusion of historical truth that lends iconic qualities, they legitimize the taking of positions in the present. They draw attention to the

---

[5]"So, our representations render the unfamiliar, which is another way of saying that they depend on memory."

continuity of the outgroup: they were evil then and remain evil now. When a representation of the past is no longer relevant, it disappears.

## 2.4 Final Considerations

I would like to conclude by suggesting that analysis of the relationship between history and representations (Jodelet 1989; Moliner 2001; Bertrand 2002; Castorina 2007; Villas Bôas 2010; Licata 2016; Páez/Bobowik/De Guissme/Liu/Licata 2016), and also the discussion of the historicity of social representations (Villas Bôas 2014), remain underdeveloped. Although we use many of the same terms, they are mobilized with very different meanings and from different perspectives. The result is that instead of leading to much-desired interdisciplinarity, a kind of cacophony is produced.

Thus, the construction of a common field of study requires the establishment of some shared principles. Here I would like to propose two: one that I will call an epistemo-political principle and another that I will call post-disciplinary, based on the discussions developed in the second half of the twentieth century by the Italian sociologist Franco Ferrarotti.

Regarding the epistemo-political principle, one must not forget that individual memories are updated by a broader memory that is inscribed in a public space. It thus subject to collective modes of thinking and also influenced by paradigms of representations of the past. It is in this sense that, notwithstanding historical inaccuracies, they carry a certain truth. Thus, collective memories are chosen as a function of the present motivations of social actors. They are used to interpret current situations and legitimize positions taken in the present (Licata 2016).

To those working with, and in, the contemporary social world (historians, sociologists, journalists, psychologists), it is necessary to discuss not only the right to interpret the past, but also the ethical imperative of such an interpretation, hence the need for an epistemo-political stance (Ferrarotti 2013).

This is because the relations between history and social representations escape disciplinary boundaries. That is to say, it is not a multi-threaded discussion, neither inter- and transdisciplinary, but post-disciplinary, in the terms of Ferrarotti (2013). In other words, it is necessary to look for heuristic and methodological tools wherever they may be found: in history, in psychology, anthropology, ethnography, psychology and psychoanalysis, but also in literature and poetry, ultimately putting into practice what Marc Bloch, in the 1920s, called the "common market of social sciences".

Against this backdrop, one of the basic problems that must be faced is precisely the discussion about aspects of the field of history that are eminently useful to the social psychologist, showing the place that historiography reserves for representations, and the emergence, among researchers of social representations, of greater consciousness of the historicity of these representations.

This is a challenging discussion that requires humility, or the will to not only drink the same water, but also to drink it from the same glass.

# References

Bertrand, Valérie, 2002: "Dimension historique des représentations sociales: l'exemple du champ sémantique de la notion d'exclusion", in: *Bulletin de Psychologie*, 55,5 (Paris): 497–502.
Bloch, Marc, 1925: "Mémoire collective, tradition et coutume", in: *Revue de Synthèse Historique*, 40 (Paris): 73–83.
Brasil (Government of), 2011: *Lei n° 12.528, de 18 de novembro de 2011: Cria a Comissão Nacional da Verdade no âmbito da Casa Civil da Presidência da República* (Brasília-DF: Governo Federal).
Cardoso, Ciro Flamarion, 2000: "Introdução: uma opinião sobre as representações sociais", in: Cardoso, Ciro Flamarion; Malerba, Jurandir (Eds.): *Representações: contribuições para um debate transdisciplinar* (Campinas: Papirus): 9–39.
Carretero, Mario; Gonzales, Maria Fernanda; Alberto, Rosa, 2007: *Ensino da história e memória coletiva* (Porto Alegre: Artmed).
Castorina, José Antonio, 2007: "Um encontro de disciplinas: a história das mentalidades e a psicologia das representações sociais", in: Carretero, Mario; Rosa, Alberto; Gonzales, Maria Fernanda (Eds.): *Ensino da história e memória coletiva* (Porto Alegre: Artmed): 75–88.
Chartier, Roger, 1990: *A História Cultural – entre práticas e representações* (Lisbon: Difel).
Dosse, François, 2000: *L'Histoire* (Paris: Armand Colin).
Emiliani, Francesca; Palmonari, Augusto, 2001: "Psychologie sociale et question naturelle", in: Buschini, Fabrice; Kalampalikis, Nikos (Eds.): *Penser la vie, le social, la nature: mélanges en l'honneur de Serge Moscovici* (Paris: Maison des Sciences de l'Homme): 129–144.
Falcon, Francisco José Calazans, 2000: "História e representação", in: Cardoso, Ciro Flamarion; Malerba, Jurandir (Eds.): *Representações: contribuições para um debate transdisciplinar* (Campinas: Papirus): 41–79.
Ferrarotti, Franco, 2013: "Partager les savoirs, socialiser les pouvoirs", in: *Le sujet dans la Cité Revue Internationale de recherche biographique*, 4,1 (Paris): 18–27.
Furet, François, 1995: *O passado de uma ilusão: ensaio sobre a idéia comunista no século XX* (São Paulo: Siciliano).
Gadamer, Hans-Georg, 2002: *Verdade e método: traços fundamentais de uma hermenêutica filosófica* (Rio de Janeiro: Vozes).
Ginzburg, Carlo, 1991: *História Noturna* (São Paulo: Cia das Letras).
Halbwachs, Maurice, 1994: *Les cadres sociaux de la mémoire* [First edition 1925] (Paris: Albin Michel).
Hartog, François, 2001: "Note de conjoncture historiographique", in: Hartog, François; Revel, Jacques (Eds.): *Les usages politiques du passé* (Paris: EHESS): 13–24.
Hartog, François, 2003: *Régimes d'historicité: présentisme et expérience du temps* (Paris: Seuil).
Hartog, François, 2010: "Historicité/régimes d'historicité", in: Delacroix, Christian; Dosse, François; Garcia, Patrick; Offenstadt, Nicolas (Eds.): *Historiographies II: concepts et débats* (Paris: Gallimard): 766–771.
Jodelet, Denise, 1989: "Pensée sociale et historicité", in: *Technologies, idéologies, pratiques,* (special issue), 1,4: 395–405.
Jodelet, Denise, 2003: "Représentations sociales: un domaine en expansion", in: Jodelet, Denise (Ed.): *Les représentations sociales* [First edition 1989] (Paris: PUF): 45–78.
Koselleck, Reinhart, 2006: *Futuro passado: contribuição à semântica dos tempos históricos* (Rio de Janeiro: PUC-Rio-Contratempo).
Licata, Laurent, 2016: "La mémoire collective: passé objectivé ou présent ancré dans le passé?", in: Lo Monaco, Gregory; Delouvée, Sylvain; Rateau, Patrick (Eds.): *Les représentations sociales: théories, méthodes et applications* (Louvain-La Neuve, Belgium: De Boeck Supérieur): 553–556.
Lyotard, Jean-François, 2005: *Le postmoderne explique aux enfants* (Paris: Galilée).

Moliner, Pascal, 2001: "Une approche chronologique des représentations sociales", in: Moliner, Pascal (Ed.): *La dynamique des répresentations sociales: pourquoi et comment les représentations se transforment-elles?* (Grenoble: Presses Universitaires de Grenoble): 245–268.

Moscovici, Serge, 1976: *La psychanalyse, son image et son public* (Paris: PUF).

Moscovici, Serge, 1978: *A representação social da psicanálise* (Rio de Janeiro: Zahar).

Moscovici, Serge, 1984: "Introduction, le domaine de la psychologie sociale", in: Moscovici, Serge (Ed.): *Psychologie sociale* (Paris: Presses Universitaires de France): 5–22.

Moscovici, Serge, 1991: "La fin des représentations sociales? (débattant Jean Maisonneuve)", in: Aebischer, Verena; Dechonchy, Jean-Pierre; Lipiansky, Edmond Marc (Eds.): *Idéologies et representations sociales* (Geneva: Delval): 65–84.

Moscovici, Serge, 2001: "Das representações coletivas às representações sociais: elementos para uma história", in: Jodelet, Denise (Ed.): *As representações sociais* (Rio de Janeiro: Eduerj): 45–66.

Noiriel, Gérard, 1998: *Qu'est-ce que l'histoire contemporaine?* (Paris: Hachette).

Nora, Pierre, 1978: "Mémoire collective", in: Le Goff, Jacques; Chartier, Roger; Revel, Jacques (Eds.): *La nouvelle histoire* (Paris: Retz): 398–401.

Nora, Pierre (Ed.), 1984: *Les lieux de mémoire: la République* (Paris: Gallimard).

Páez, Dario; Bobowik, Magdalena; De Guissme, Laura, Liu, James H.; Licata, Laurent, 2016: "Mémoire collective et représentations sociales de l'Histoire", in: Lo Monaco, Gregory; Delouvée, Sylvain; Rateau, Patrick (Eds.): *Les représentations sociales: théories, méthodes et applications* (Louvain-La Neuve, Belgium: De Boeck Supérieur): 539–552.

Prager, Jeffrey, 2001: "Collective memory, psychology of", in: Neil J. Smelser; Paul B. Baltes (Eds.): *International encyclopedia of the social & behavioral sciences* (Amsterdam: Elsevier): 2,223–2,227.

Ricoeur, Paul, 2000: *La mémoire, l'histoire et l'oubli* (Paris: Le Seuil).

Vilar, Pierre, 1980: *Iniciación al vocabulario del análisis histórico* (Barcelona: Ed. Crítica).

Villas Bôas, Lúcia Pintor Santiso, 2010: *Brasil: ideia de diversidade e representações sociais* (São Paulo: Annablume).

Villas Bôas, Lúcia Pintor Santiso, 2014: "Representações sociais: a historicidade do psicossocial", in: *Diálogo Educacional*, 14,42 (Curitiba): 585–603.

# Chapter 3
# Social Representations in the Study of Disaster Risk in the Municipality of Piedecuesta, Santander (Colombia): The Social Cognitive Dimension

Deysi Ofelmina Jerez-Ramírez

## Introductory Comment

Jorge Ramón Serrano Moreno

*The word[1] 'proemium'[2] gets right to the point: to lay out here, in simple and concise words, everything that the chapter will offer the reader, with the technical-scientific rigour appropriate to the case in question. For this purpose, on the one hand, I will demonstrate the stark contrast between the important advantages – or proficiencies – shown throughout the course of the chapter when broaching the heated question of disasters and their risks, which have been scientifically researched and are explained there in detail; and, on the other hand, stress that its method is to confront and overcome the most significant deficiencies that have frequently been covered in numerous texts on the topic. If it were not for the terrible consequences that populations suffer during these disasters, such an approach would not be so severe and cruel. But given such circumstances, it is indispensable to overcome them, the sooner the better – that is, from the moment that risk*

---

Deysi Ofelmina Jerez-Ramírez is a professor and researcher at the University of Sciences and Arts of Chiapas (UNICACH). She completed her Master's degree in social work and her PhD in political and social sciences at the National Autonomous University of Mexico. Email: deysi.jerez@unicach.mx.

---

[1]Prof. Jorge Ramón Serrano Moreno was Senior Researcher (now retired) at the Regional Multidisciplinary Research Centre, UNAM; Email: jrsmhi@gmail.com.

[2]The Latin *proemium* comes from the Greek 'pro' (before, in favour of) and '*míon o mínein*' (I walk, to walk), such that throughout this chapter *proemium* refers to that which is favoured or is 'in favour of walking'.

© Springer Nature Switzerland AG 2021
C. Prado de Sousa and S. E. Serrano Oswald (eds.),
*Social Representations for the Anthropocene: Latin American Perspectives*,
The Anthropocene: Politik—Economics—Society—Science 32,
https://doi.org/10.1007/978-3-030-67778-7_3

indicators are perceived. This is even more the case when we are living in a world ever more threatened by disasters 'so disastrous' that they seem immeasurable.

One important merit of the chapter is that it establishes a solid theoretical base while also developing a rigorous method, both of which are applicable for many types of disaster – not just the one studied in the chapter – given that it is based on a micro-scale study but, with its methodology and 'pedagogy', builds from a precise foundation to become fully usable – with the obvious and appropriate adaptations for each circumstance, i.e. 'factual polychromes' – for a wide range of disasters that can be treated even at macro-scale. It was therefore essential to develop a methodology which is firmly based on highly rigorous theory and conceptualizations and which is able to incorporate, rather than leave out, an integral approach to the combination of enormous and sometimes tragic factors that are involved. This is what Dr Jerez-Ramírez proposes here.

But how does she achieve this? And further, how can we demonstrate that the aforementioned contrast will simplify the process of understanding the basis of her argument? It is precisely for this reason that this Proemium has been written. Offering as its 'leitmotiv' (conducive motive) the framework presented above of 'Proficiency vs. Deficiencies', it applies this to each of the four sections of the chapter in order to make the content more accessible. It is like undertaking the journey with a map in hand to enable users to gain the most from the tour. It means walking on solid ground until landing upon the vitalizing roots that give life to such a formidable, sought-after object, worthy of strengthening ourselves successfully before the great obstacle which is to be avoided: disasters seen from their risks. Let us turn now to the first of the four sections.

## Theoretical-conceptual overview

To start with, I will outline some deficiencies, and, as a way to overcome them, I will then look at the proficiencies. Deficiencies: 1) Traditionally, disasters have been interpreted as chance events produced by natural phenomena. This has even included passivating the victims as a way to understand them and offer appropriate responses. 2) Sometimes the idea of risk has been increased, given the uncertainty of its occurrence and the insecurity or unease that it entails. 3) As one consequence among others, public policies have been established that do not offer responses which address the urgent needs of the victims.

However, in light of these insufficiencies: 1) A social approach to the topic has emerged in order to empower and not passivize society in the face of events that are supposedly natural. 2) To this end, there is a differentiation between what nature delivers – which is the threat of natural phenomena, with possible types being geological (earthquakes, volcanic eruptions), biological (epidemics), meteorological (rain, drought, etc.) – and the risk that is the effect of the threat when society is exposed to it, and with which the idea of vulnerability is introduced because of its association with risk. 3) However, vulnerability per se is not seen as social vulnerability, that is, within society there resides a vulnerability which *must be reduced in order to manage risk; this means that risk is linked to social vulnerability: the higher the society's vulnerability, the more imperilled it is by risks, and vice versa. 4) In addition, a variety of risk types are identified: social and economic, physical and structural, cultural and political. 5) Following this is the topic of social construction, which depends on the perception of risk, which in turn is supported by culture, education, etc. 6) Yet the connection with perception, which is (inter)subjective, along with the social construction, can be explained by Social Representations Theory (SRT). This combination of elements underlies and is the basis of the entire chapter.*

## Methodology

*Given the particularities that are evident in the study of risk from the perspective of social representations, there is a fundamental methodological triangulation of tools and techniques, which develop in two stages:*

*The first stage seeks to analyse in depth the representational phenomenon in terms of content and as a process, and approaches the lines of 'anchoring' and 'objectification'. To do this, interrogatory and associative methods are implemented (such as the in-depth interview and the associative chart). The second stage is realized using techniques such as social cartography (usually based in workshops), and the community diagnosis of vulnerability-threat, the motive of which is to achieve social construction. The objective of this phase is to identify how these dimensions are inserted into the meanings system that operates and mobilizes the social construction of the object of study. This is simultaneously influenced by the generation of experiences that hold affective, rational, and emotional weight, and are very important: those value –aspects which are supported using tools suggested by SRT.*

*It is worth mentioning that the identity dimensions, in terms of identity periods and territory, function as transversal paths for the methodological strategy. In terms of the sample selected for the application of this methodology, the key groups identified during the participation (period) must be considered. (The chapter, which observes a micro-scale setting, focuses on local groups of 'municipal risk management councils' that comprise three sectors – institutional, business, and community – in residential areas, periphery, rural areas, and urban settings, with social actors in each area.)*

*All of the methodological points are demonstrated to be proficient. Some may lack the deficiencies of any other methodology provided that its key characteristic of integral methodology is abandoned.*

## Results

*The third section provides the results. It explains the 'advantages' of using the approach proposed at the core of the chapter both in theoretical-conceptual and methodological terms, but these are seen from the perspective of their results, which are the 'findings' obtained from the in-situ research (in this case of a community on a micro scale). This is the proficiency element. Among the most important findings that demonstrate proficiency – i.e. what turned out to be proficient – are 1) corroboration of the complementarity between the areas of focus that have been explored; and 2) identification of the principal analytical categories which orientate themselves towards community actions and public policies that are much more effective. Meanwhile, among the findings of deficiencies, what stands out above all else is the presence of plausible contradictions on a theoretical-conceptual level, and also on a methodological one. Using the above steps (1 and 2), these types of deficiency can ultimately be overcome.*

*These three points are the most obvious results, but there is one more. Theoretical-conceptual interaction with a social focus is an unavoidable necessity, both for this type of analysis and to avoid the risks associated with disasters. Consequently, it is also necessary to draw on Social Representations Theory. Accordingly, we are introduced to the necessity of making use of conceptual tools, which are essential in situations of risk, as are concepts like 'threat' and 'vulnerability' (referred to above), and are valid at micro scale and any other social conglomerate, even macro scale. On the other hand, this makes evident the analytical poverty – deficiency – of reducing a disaster to the status of a natural phenomenon only. The role of Social Representations Theory thus becomes an indispensable cornerstone. The author therefore follows the Moscovician processual perspective, which is another aspect of proficiency. However, it is not be necessary to explain here the various facets of the said theory, as widespread as it is essential, given that the author herself does this with much rigour, technicality, and finesse in the analytical details.*

## Conclusions

*In an apparent contrast with the previous results section, and using the findings as a basis, this section is quite short and easy to understand in its almost linear simplicity.*

*The generation of institutional strategies was appropriate for risk management but should never disregard common sense, which gathers and organizes social know-how, much less the contextual conditions, both material and symbolic, that precede and/or accompany the course of the event.*

*When common sense, which gathers and organizes the know-how of the community or social conglomerate, as the case may be, is not taken into consideration, institutional strategies will be inadequate.*

*Therefore, if the information and what it generates cannot be articulated with the complex links of everyday actions without entering into conflict, it will be very difficult for behavioural changes to transcend time, a fact which explains the contradictions that are seen in social behaviour. Further, this is very difficult to resolve from a purely traditional intervention by the physical sciences. For this reason, the perception of disaster risk is a social construction, and not purely a consequence of natural phenomena. This is neatly expressed by the wise observation that the author derives from a text by Amartya Sen, which she presents over the course of the chapter: "if an event that could be avoided occurs, it becomes an injustice; accordingly, if an injustice exists, casualties and responsibilities coexist – just like authorities – that cannot be assumed, in their totality, by the social unit". Which could not, take note, be managed within a natural phenomenon approach.*

**Abstract** This study has been developed around the characterization of the processes of the social construction of risk and the daily relationship "to know-to do". It presents part of the final results of the research titled: "Social constructions of disaster risk from social representations in the municipality of Piedecuesta, Santander (Colombia)". Social representations (SR) of the Piedecuestana population have been analysed in the development of three dimensions: Social Cognitive, Social Structural and Social Territorial. The document has been structured in six sections: introduction, theoretical-conceptual review, the methodology and sample of the study, the presentation of results, analytical categories, and final conclusions.

**Keywords** risk · Colombia · social representations · territory

## 3.1 On the Analysis Offered in this Chapter

For a long time, risk has been reduced to a simple "uncalculated" deviation of the processes of modernity. Its origin has sought to situate itself in the externalities of the environment – natural or built – that operate independently of the social actor and his or her intentionality. The study of disaster risk has initially been targeted at monitoring threats and, more recently, at examining vulnerability scenarios.

In this case, the analysis has focused on the different meanings that the notion of risk can acquire in each context and time, trying to trace the social origin of its nature and the role it fulfils in the configuration of resilient knowledge and practices.

I begin by viewing disaster as the confluence of environmental factors and social-cultural components, and risk as a dynamic, polysemic and questionable conceptual entity – non-inert, unambiguous, unequivocal. According to this logic, the thematic is examined as a social construction closely related to the experiences, knowledge and social actions that are generated from daily interaction with conditions of vulnerability and threat. The perception of risk recreated in inter-subjectivity connects with other consciences and dialogues, circulates between speeches and social praxis, is anchored into the human psyche, and returns to the context whence it was generated. It is, in any light, a representational phenomenon that has been approached from the Moscovician theory, within the framework of a social study of disasters and the conceptual supports of identity and territory.

The work exhibited here presents part of the final results of the research entitled: "Social constructions of disaster risk from social representations in the municipality of Piedecuesta, Santander (Colombia)". This study has been developed around the characterization of the processes of the social construction of risk and the daily relationship "to know-to do".

The social representations (SR) of the Piedecuestana population have been analysed in the development of the dimensions Social Cognitive, Social Structural and Social Territorial – analytical categories. However, for the purposes of this chapter, the data corresponding to the category Social Cognitive and Social Structural are summarized for three main reasons:

1) To present research findings regarding psychosocial processes that integrate SR of risk with common sense, incorporating the elements of threat and vulnerability from the way its representation is expressed.
2) To develop a new research approach based on the exercise of theoretical-conceptual integration between the social study of disasters and Social Representations Theory (SRT).
3) To reflect on the innovative element that arises from the peculiarity acquired by SR during the investigation of risks and disasters.

The document has been structured in five sections, including this introduction (3.1), the theoretical-conceptual review (3.2), the methodology and sample of the study (3.3), the presentation of results (3.4), and the exposure of final conclusions (3.5).

## 3.2   Theoretical-Conceptual Review: Risk and Social Representations

Disasters, when interpreted historically as incidental events dependent on natural causes, have been related to risk conditions precisely because of the uncertainty and insecurity surrounding the time and location of the possible event. However, a social approach to the understanding of disasters has been proposed to revitalize the notion of social and system agency arising from the risk-vulnerability link, without leaving aside the physical event itself (the threat).

In the study of disasters, the notion of social vulnerability was used by Levell (1996, 2004) to account for the previous context of needs and shortcomings that particularize a social structure. The categories of vulnerability developed by Anderson and Woodrow (1989) make a fundamental contribution to the thematic. These categories are formulated as compound types of vulnerability: (i) social and economic; (ii) physical and structural; and (iii) cultural and political. The notion of threat, meanwhile, corresponds to the physical event that can be biological, geological, hydro-meteorological, technological or socio-natural, temporal and/or of fixed spatiality. The variables that shape the evaluation of threats or dangers are the phenomenon, the proximity, the level of exhibition and the frequency of the event.

Finally, the concept of risk has arisen from synthesis of the interaction between these two dimensions – vulnerability and threat – although the dynamic and subjective character of the concept has continued to evolve. Consequently, studies on the perception and social construction of risk are relevant, with precedents in French literature.

The notion of social construction is synonymous in the work of French researchers with risk perception (RP), a term incorporated into scientific language by British anthropologist Mary Douglas (1982) in her work *Risk and Culture*. The author defines RP as: "A product of the cultural construction of societies in their historical becoming" (García 2005: 15, in relation to the concept given by Douglas). This concept questions the probabilistic postulates of the risk-benefit method of Starr (1969) and the Bayesian approach.

The perception of risk as a subdiscipline has broadened its horizons to the fields of anthropology, history, sociology and psychology – individual and social. The theoretical and methodological frameworks of studies on disasters did not take long to adapt to these proposals, which is why the approach of social perception is one of the concepts most used in research into this trend. However, although RP arises from the recognition of both mental and sociological and cultural elements, it fails to find the intersubjective bridge between cognitive activity and social construct.

According to Jodelet, the "to know–to do" relationship (bridge) is the process by which the systems of ideas are formed, transformed and circulated among the social agents (Serrano 2010), aspects that can be studied more deeply using Social Representations Theory (SRT), and more specifically the contributions of the processual approach.

### 3.2.1   Social Representations Theory and the Processual Approach

The concept of social representations stems from the premise that we are social subjects who require communicative interaction so that each object or phenomenon is incorporated into our commonsense structure. SR are contextualized systems (Nuño, 2004) that are expressed in shared codes using a linguistic practice that reduces the potential ambiguity in our interpretation of the world.

Social representations have been studied from three specific approaches: the structural, the sociological – from which the systemic approach has emerged – and the processual.

The processual approach maintains the complexity of contents, techniques, instruments and multidisciplinary approaches proposed by Moscovici. It is complemented by the theoretical systematization of Jodelet (1986), developed from the following questions based on the classic interrogatives 'why?' and 'for what?'

(Jovchelovitch 2007): Who knows and from where does it know this? What and how is it known? What effect is it known for in an integrative way?

The processual school recognizes both the cognitive component and the socio-relational component of the representational phenomena. In the words of Jodelet: "We must consider on the one hand the cognitive functioning and that of the psychic apparatus, on the other the functioning of the social system, of the groups and the interactions to the extent that they affect the genesis, the structure and the evolution of the representations" (Jodelet 1989 quoted by Banchs 2000: 3.3). This statement covers two fundamental issues: first, the origin of social representations is dual and this condition is attributed to the interactive frontiers that are shared between mind and reality; and second, SR cannot, consequently, be approached only as products located in a specific place (Banchs 2000; Serrano 2010), psyche or context; they are actually processes that ensure the permanent circulation of the symbolic elements that compose the knowledge and common praxis.

This theoretical orientation considers the thematic in the context of its global character, the plurality of the elements-processes that compose it, and the heuristic approach that it is responsible for. It emphasizes the relationship between object-subject-cognition and reality (Serrano 2010), identifying the processes of objectification and anchorage that originate from the representational phenomenon.

Interesting environmental studies have been developed since the inception of SRT (Calixto 2008; González/Valdez 2012; Molfi 2000), as the study of disaster risk, being an environmental issue, focuses on the relationship between society and the environment. Nevertheless, this research has tried to go beyond the typologies of SR traditionally managed from this field – naturalist, globalizing and anthropocentric (Reigota 1990) – incorporating aspects of temporality and identity that reveal a new approach.

## 3.3 Methodology and Area of Study

Given the particularities manifested by the study of risk from social representations, the methodological work has been based on a triangulation of instruments and techniques carried out in two stages:

- The first stage seeks to analyse in depth the content and process of the representational phenomenon, covering the lines of anchorage and objectification. In this first stage I have implemented interrogative and associative methods, such as the interview in depth (ten) and the associative letter (twenty), tools developed in the context of a contextualization exercise.
- The second stage includes elements of the social study of disaster risk through techniques typical of this approach, such as social cartography (three workshops) and vulnerability-threat diagnosis. The objective of this phase is to

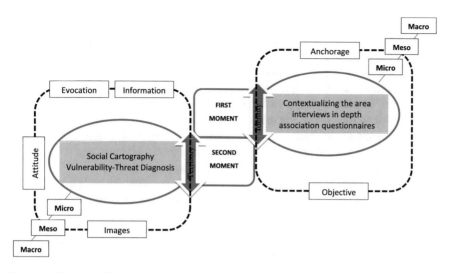

**Fig. 3.1** Methodological moments (*Source* Author's own elaboration)

identify how these dimensions are inserted into the sense system that activates and mobilizes the social construction of the object of study, simultaneously influencing the generation of experiences that contain affective, emotional, rational and value burdens – aspects that have also been addressed via the instruments suggested by SRT.

The dimensions of identity and territory serve as transversal axes of the methodological strategy, which operates on three levels of time-space: the micro, the meso and the macro (Fig. 3.1).

It should be noted that, in order to apply this methodology, the selected sample includes the key groups identified during the participation in the municipal risk management councils. These groups represent three sectors in particular: institutional, business[3] and community. For the last of these sectors it was pertinent to develop subcategories in relation to the place of residence – periphery, rural or urban area. In accordance with the objectives of the study, the representative sample focuses mainly on the social actors of the three areas already identified. It includes information about the institutional scene both to complement the findings and to point out the particularities of the representational content according to the context.

Finally, the vulnerability-threat diagnosis provides relevant information about risk-related issues. The municipality of Piedecuesta, with a territorial extent of 481

---

[3]The business sector was identified as an area of interest in the development of the vulnerability-threat diagnosis, as well as in the risk management process itself. However, it was not possible to access the comments and opinions of this population during the development of the investigation, so further analysis relating to this sector is not included.

square kilometres and a population of 156,167 inhabitants (according to the census projections of 2005; see DANE 2018), exhibits the following conditions: low levels of rurality; concentration of goods and services in the urban area; deficiencies in the availability of basic services (education, health, sport, culture, housing, public services and security); precarious settlements in areas of high threat; the presence of displaced people and the absence of effective care policies, with the peripheral locations being the areas of greatest reception; threats associated with geological activity and hydro-meteorological events; high vulnerability of the base sector of the municipal economy to extreme changes in climate; and bad agro-industrial practices in the use of soil and water resources.

However, this vulnerability-threat box presents variants according to the area of study –periphery, rural or urban area – which will not be enlarged upon in this section, but will be indicated in the presentation of the results.

## 3.4 Results

The findings presented as research results are not only related to the information generated by the development of the main categories. The theoretical-methodological process by which the identification of these categories and their different analytical levels have been identified also make a valuable contribution to studies on social representations and disaster risk.

Although both the process and the content can be located within the work that has been done on the environment since the inception of Moscovician theory, using an innovative analytical approach adds an extra dimension to the research, bearing in mind that disaster risk analysis is a subject that already has well-defined theoretical, conceptual and methodological elements – aspects that reiterate the disaster as a social phenomenon and, therefore, as an area ripe for study from the perspective of SRT. This entails theoretical dialogue, an exercise which results in a scheme of theoretical-conceptual integration – a kind of analytical tissue between the elements that make up each of the selected approaches (social study of disasters and SRT). This makes it possible to: (i) corroborate the complementarity of the approaches; (ii) rule out possible contradictions at the theoretical-conceptual level; and (iii) determine the main categories of analysis.

### 3.4.1 Theoretical-conceptual Interaction of the Social Approach and the Moscovician Perspective

The social approach has already chalked up the analytical poverty of reducing a disaster to the occurrence of a natural phenomenon. Accordingly, and without ignoring the event itself, the elements of social susceptibility and risk have been investigated to determine whether such conditions result from the dynamics of collective perception. Specifically, the concept of risk involves a social construction that depends on both cognitive and socio-cultural processes, which mediate in the appropriation of reality.

The incorporation of SRT makes it possible to go beyond the cognitive framework to explore socially shared patterns, which avoids the danger of studying the topic of interest (disaster risk) as a fragmented phenomenon. Following this approach, I have worked with the notions of micro, meso and macro systems (in interrelation).

It seeks to emphasize the continuity and synergy between the different scenarios in which social representations circulate, breed and transform, as well as between the types of relationships that are established from 'nearby environments' (microsystems[4] and mesosystems[5]) and the larger contexts (macrosystems[6]) in which the environments are included (Colás 2006: 32). This systemic vision is retrieved from the sociological school of SR and can be equipped with the ecological model of Bronfenbrenner (1987). These aspects recognize the importance of both the subject – as a starting point – and the immediate contexts and macro-contexts in which key relationships are developed (interpersonal, family, group and structural).

The physical characteristics and symbolic representation of reality, configured in a micro, meso and macro context, directly influence the way in which the risk of disasters is perceived. For that reason it is essential to make use of conceptual tools such as threat and vulnerability.

The concept of 'threat' (physical phenomenon) incorporates the notions of proximity, level of exposure and frequency of the event. Regarding the conditions of vulnerability, the contexts that together represent the social structure (the psycho-social and the cultural, the economic and the political, the physical and structural) come into debate. This dimension constitutes the socio-structural scope, which: a) contains the most notable contextual variables to which the representations of the figurative nucleus are adapted; b) defines the type of relationship gestated between the social group and its physical-subjective environment.

When studying risk it is essential to identify the elements of common knowledge that are repeated and that constitute the traits that predominate as social archetypes; to do this, in addition to the ideological factors (macrosystems), it is necessary to analyse the meso level within a specific group, since this is where the shared knowledge and the generalized actions at risk can be explained more easily, – always within the social understanding, in other words, as common know-how.

---

[4]The microsystem involves the type of close, face-to-face interactions in which the person can participate actively, including interpersonal roles and activity.

[5]The mesosystem is configured by the interconnection of microsystems and is expanded as the person integrates into new environments.

[6]The macrosystem is determined as the pattern of beliefs, values and ideologies that commands and sustains the broad sociocultural base.

The notion of agency also becomes important at this point. Although the structure exerts a decisive force in the formation of the system of thought, the subject-group has the potential to modify the information received through processes of anchorage and objectification in order to make it familiar and intelligible while the 'information' is re-built and re-created.

However, the specific role of social representations is to adapt abstract concepts and ideas – such as those produced by science – to society and, through the formation of behaviours and the orientation of social interactions, to ensure the adaptation of the society to new sets of categories and information. This is achieved through two processes that show the interdependence between psychological activity and its social conditions: objectification and anchorage (Villarroel 2007: 436-437).

We are not talking about mental mechanisms or cultural impositions, or of simple automatic reproductions; these are sociocognitive processes that enable the circulation and daily exchange of 'the represented' at the immediate intra-personal and interpersonal levels (microsystem), on a broad social basis. The functionality of the sociocognitive resources is varies according to the difficulties faced by the subject when encountering new, generally abstract, decontextualized data which is significantly charged (i.e. has an excess of meanings). In response, the objectification process materializes and synthesizes. At the same time, the anchorage provides the modelling schemes that serve as points of reference in the decoding of the newly incorporated element (Hollisch 2014).

We start from the recognition of risk as a collective construction where, in this specific case, identity operates as an element of articulation. Following Serrano (2010), identity recreates the point that connects psychological action and social action, configuring a dialogic instrument between the immaterial and material dimensions that make up reality. Consequently, the sociocognitive and socialstructural dimensions (Categories 1 and 2) are oblique aspects of what is termed here the identity of time – the narrative of the present that goes backwards (past) and forwards (future) between the certainty that is defined by a personal-group history and the uncertainty that represents the risk per se.

Theoretical-conceptual integration also offers a third range of analysis related to territory, which, like disaster risk, is conceived as a product of society: the socioterritorial dimension (Category 3). Although systemic vision points to the macro level as the scenario that contains the set of psycho-socio-cultural interactions, in this work it will be socioterritorial, as a category and transversal dimension, which fulfils the functions of scope-totalizing unifier of the different levels (micro, meso and macro).

Disaster risk is categorized according to processes and forms of knowledge-practice that create and maintain a shared vision of the world. In this sequence of ideas, the notion of 'socioterritorial' allows key aspects of risk to be defined according to a basic spatial interpretation, such as the material and subjectivities that express actions and knowledge, the set of relationships that characterize the representational phenomenon (cognitive, social, cultural, political, environmental) and the spatial agency of urban, rural and peripheral groups. It should be noted that, as with the temporal scope, the territory is not one-dimensional, much less

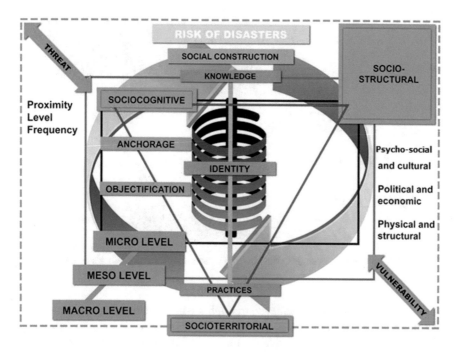

**Fig. 3.2**  Theoretical-conceptual integration (*Source* Author's own elaboration)

fixed. The spatial attributes of the social processes have a defined existence in the present, as well as an identity memory composed of the spatiality losses and the expectations of permanence and spatial control prior to conditions of risk. Thus, space becomes time and time in space (or territory, also present in the first and second categories of analysis), in such a way that the micro, the meso and the macro, being transversed by the axes of identity and territory, must be addressed as a temporary space system.

All the above leads to the generation of a complete image of the aspects and processes that characterize the topic of study. Figure 3.2 outlines the relationship between the different factors that were reviewed above and that, according to the theoretical-conceptual considerations, configure the social construction of disaster risk. It emphasizes the synergies and interactions that are generated between the sociocognitive, sociostructural and socioterritorial processes that are presented as inputs of the main analytical category.

## 3.4.2    Categories, Integrative Elements and Basic Content

Regarding the multifactor that must integrate the social study of risk, it is advisable to generate an analytical framework that establishes meeting points and visible synergies and generates unity in the possible dispersion of the elements. The

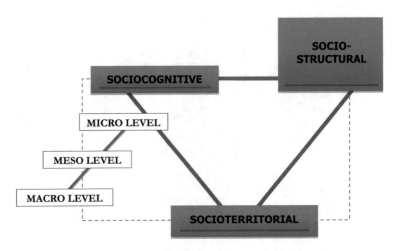

**Fig. 3.3** Intersections between analytics and major categories (*Source* Author's own elaboration)

analytical framing has been achieved through the triangulation of three dimensions that jump into view in the scheme of theoretical-conceptual integration: sociocognitive, sociostructural and socioterritorial. The dimensions mentioned have two characteristics in common: firstly, they are formed of the various elements of interest; secondly, they operate at one of the three levels – micro, meso and macro – in which the SR are manifested.

Using this analytical framework, the main categories of analysis have been identified, as well as each of their components, which have been called integrative elements (IE). Thus, for example, we know that the sociocognitive dimension relates to the mechanisms by which the information obtained from the medium is objectified and incorporated into the existing baggage, serving as a frame of reference not only for interpretation but also for action in that environment. The components that make up this first category are the processes of objectification and anchorage, closely linked to behaviours and practices.

The sociostructural category, on the other hand, contains the elements of vulnerability-threat, more specifically the way in which the population of the municipality perceives and represents the conditions of susceptibility and exposure to risk. Vulnerability relates to social, cultural, economic, physical, environmental, and political conditions that reduce the ability of the community to cope with dangerous events. With regarding to threat, it is essential to investigate the way in which the inhabitants relate to their environment and the emotional, symbolic and value implications that are configured from the perception of proximity, level of exposure and frequency that lead to typify a natural element as dangerous (Fig. 3.3).

The last category, i.e. socioterritorial, is explored from the reading of the inhabited space; the SR that the population has built of their environment's exposure to risk and how they have acted in the place they occupy as a result of that representation. In this case the integrative elements include:

- The materials and subjectivities that express knowledge and practices in the face of disaster risks (common know-how);
- The set of relations that characterizes the representational phenomenon (psychosocial, economic, cultural and political);
- The socio-spatial agency; and
- The socio-environmental dynamics (territorializing, de-territorializing and reterritorialization).

Thus, the socioterritorial dimension is the totalizing-integrative dimension that contains the elements of the temporal space system, but in its relational expression: the intra-contextual (i.e. relational, social relations and broad social base), bearing in mind that the territory operates, together with the identity, as a transverse axis of the different categories. Although the integrative elements (IE) are more specific aspects than the analytical categories, they still have an abstract character that can hinder the intelligibility of the data. For this reason, a third item has been incorporated which identifies, via questions, the essential aspects into which the IE are translated, called basic contents.

Triangulation of the information gathered in the fieldwork has been developed within the framework of the categories and subcategories that have emerged from the exercise of theoretical-conceptual integration, in such a way that the first two levels (analytical categories and integrative elements) have guided the search and review of field information, while the development of such basic content [the answers to the questions] yields a fuller picture.

The correlation between the analytical categories, the integrative elements and the basic contents – in other words, the link between the theoretical aspects and the research findings – is recorded in a data matrix configured as a tripartite scheme (see Table 3.1). The functionality of the matrix, in terms of organization and the correlation of components, has facilitated a more complex analysis which seeks to weave the contents (analysis categories) into a critical and argumentative form.

## 3.5  Analytical Categories

### 3.5.1  Socio-Cognitive Dimension

The socio-cognitive dimension contains the mechanisms that define the psychological and social activity from which the representations of disaster risk in the shared thought system are generated, built and integrated. It involves aspects classified as knowledge and practices at risk, integrating elements of this first category that have been developed with the guidance of the questions posed from the

**Table 3.1** Data matrix

| Analytical Categories | Integrative Elements | Basic contents |
|---|---|---|
| Sociocognitive | Anchoring/ objectification | Who knows?/ Cognitive subject who is at risk What does he know? Why and what?/ Represented object |
| | Knowledge and know-how/ Practices and behaviours | What are the sources, means and types of knowledge related to disaster risk? What are the risk practices and behaviours and how are they justified? How do knowledge and practices at risk define social trajectories? |
| Sociostructural | Threat/vulnerability | How do you explain, according to the population of the different sectors, the relationship between society and nature? What are the natural phenomena associated with disaster risk? What is the perception of frequency, proximity and exposure of natural phenomena associated with disaster risk? How is the natural element represented in the relationship between society and nature? (Threat and appeal) How is this relationship expressed graphically? What is the narrative of vulnerability that relates to disaster risk? From this narrative, how are the physical, social, economic, political and environmental conditions of the municipality represented in the analysis of disaster risk? |
| Socioterritorial | Socio-environmental dynamics: territorialization deterritorialization reterritorialization | How is the spatial experience that takes shape in contexts of dispossession and territorial appropriation related to know-how in the face of risk? How do the materialities and subjectivities that express actions and knowledge in the face of disaster risk manifest themselves territorially? In what way are the relations that characterize the representational phenomenon object of study articulated spatially? How is the socio-spatial agency of urban, rural and peripheral groups characterized? |

*Source* The Author

processual approach: Who knows and from where you know, what and how do you know, why and what do you know? And to what extent is this known in an integrative way?

In this way, the socio-cognitive section describes and analyses the aspects that are directly related to the social study group (cognitive subject), the object represented – which will be expanded on in greater detail in the sociostructural category – the sources, means and types of knowledge, and the practices at risk.

### 3.5.1.1 Piedecuestanos Residents and their Know-How Common to Disaster Risk

The representation of reality rests on and is discovered from the story, that which refers to socially shared knowledge as beliefs, customs and imaginaries, while combining experiences, experience and expectations (personal, family and group). The representation of the object is the representation of a subject, in this case, a social subject. Thus, the richness and multidimensionality of the data makes it possible to obtain information about the cognitive actor, the 'Who knows' – in other words, the inhabitants of Piedecuesta.

Belonging to/being from Piedecuesta is related to feeling part of the place – not just to living in the village, but to actually experiencing the village. Piedecuesta is a socially constructed territory where people can enjoy the public spaces (parks, sports facilities, squares) and participate in the collective activities, at least in relation to the immediate space – be it neighbourhood, sidewalk or sector – and the nearby groups, to make public certain traits of personal life that ensure that they escape from anonymity: the location of the house, family members, occupation, etc.

Generally, a villager of the municipality is legitimized as such from the testimony of its close contacts (neighbours, friends and family), which is the reason why recognizing others is an essential part of the 'being and feeling'. The identity dissertation of 'Who am I?' is constantly set up from 'What is my relationship with that person?' 'Where do I stand?' and 'Where do others situate me?', without obviating the fact that the profession, work they perform, in addition to denoting a type of activity, also indicates the role they serve for society.

'Piedecuesteneidad' is the name that some analysts have given to this feeling of belonging and rooting that rests in that place of the town, now city, where it can be perpetuated despite temporality – in other words, in its own inhabitants. The perception that the cognitive subject has of itself is built within the framework of what it recognizes as the global space that it occupies and in which it develops: subjective territory that contains the object to represent (what and how is it known?) as the means of obtaining knowledge (from where do you know?).

> And when I arrived there, I built my house myself. I still have it under construction, and then that fact of doing things makes people interested and they ask, 'Well, who are you? Where did you come from?' Then they learned that I was a teacher. (Hernando, 56, rural district; Vereda El Volador.)

In this case study, the relevant aspects go beyond the subject-object experience, conjugated in the present. The encounter with that which is represented necessarily has a historical background (personal, family, group), a narrative that connects different biographical social points in a trajectory marked by the identity times: present, past and future. This complexity has been highlighted in the representations of each population group.

## 3.6    The Peri-Urban Population's Representations of Risk: What Do They Know? How do They Know It? And From Where Do They Know?

The periphery sector is mostly formed of people who arrived in the municipality in the last decade of the twentieth century with histories of forced displacement, interdepartmental migration and extreme poverty.[7] These stories are characterized by a remote origin – in time and place – an outcome (mobilization or transfer) and an end (arrival) which begins a new story contiguous to the previous one. Here experience operates as an instrument of knowledge.

> In times of violence about twenty years ago, the situation became very difficult, riddled with many complications in different forms, with many people missing, with dead relatives, stabbed and shot. Then we felt an obligation to emigrate, to leave, to move, though at the time it was not called displacement [...] Back then, we called coming in, here on the side of the bamboos, coming to sector four, the Chinese plan. We left the bus because there was no more road; there was no other entrance. A tree crushed part of my bed, I left it on the way. It was very impressive. I came with my two older children. They helped to carry the suitcases and everything, and that's how we came out here to this small place that my husband had already made. [...] Finally we had roof again, but we had no idea what would happen next. (Eudora, 65, peri-urban district.)

The speech that depicts the risk continues along a similar trajectory, which is why the idea of what constitutes a disaster is distant – very much an event defined in time and space. It is expressed as a set of events that, in parallel, configures a learning process. Experience linked to the generation of knowledge and access to information is a useful tool not only for identifying risk, but also for overcoming it. Risk is hence also perceived as a mobilizing element. Words such as 'organization', 'performance', 'evaluation', 'preparation', 'union' and 'teamwork', predominate in the systematization of associative instruments and group techniques. These concepts indicate the type of practices that are considered essential for confronting problematics that, on the one hand, address a need felt by the collective and, on the other, ought to be prioritized in the community agenda.

It should be noted that mobilization does not imply only moving away from that which materializes the uncertainty, but also generating conditions that modify that feeling, something that relates to the agency of the social subject "that acts, produces and transforms knowledge and practices, as well as their culture and their history" (Serrano 2010: 33). However, the risk reaction also feeds on certain contradictions.

The district of Nueva Colombia has been affected by various hydro-meteorological events, generating concern among the inhabitants of the area and the authorities due to the high degree of exposure. The severity of recent damage

---

[7]Except for families residing in the urban area of the municipality or in other places who have bought land without deeds or with false deeds, taking advantage of the low cost and public services like water and light.

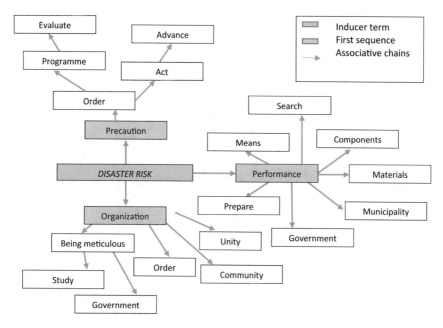

**Fig. 3.4** Free association chart: inhabitants of the district of Nueva Colombia (*Source* Fieldwork data)

arising from these emergencies necessitated the relocation of several families. However, the land that was evacuated has been illegally reoccupied, despite the antecedent, maintaining these dangerous settlement practices (Figs. 3.4 and 3.5).

For this population, exposure to danger is justified by the experiences that mobilized the individual, family or social group to go in search of better living conditions. Consequently, there is a "false perception of security" (Knight 2007:109) that, without denying the knowledge of risk, transfers the threat to neighbouring groups in order to meet other needs which the same inhabitants perceive to be of higher priority: legalization of the land and the delivery of property deeds by the local administration.

> Because the earth slides, because there is a lot of quicksand, it gets loosened and those stones roll down and those are the ones that affect all those small ranches […] No, that doesn't happen here. I am not ungrateful because in this place if it rains, the water drains away everywhere. It rains and after one hour it is dry, again dust. Here I have not been affected by the winter. (Manuel, 78, peri-urban district.)

The risk of permanent exposure to a threat and the need for accommodation to develop a personal, family and community project are disjunctive. Some residents tend to downplay the facts, while others try to get out of a situation which could become dangerous. Thus, the "false perception of security" ends up becoming an argument in favour of staying put, and therefore an essential aspect of social representations of the subject. Such socio-behavioural dilemmas are generated and can more readily be explained socially than psychologically.

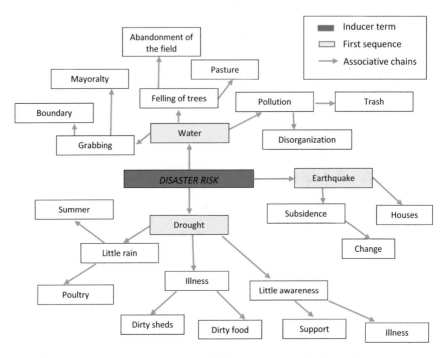

**Fig. 3.5** Free association chart: inhabitants of El Duende trail (rural) (*Source* Fieldwork data)

- Representations of urban and rural population risk: What do you know? How do you know? And from where do you know?

The potentially disastrous fact relates to other places and other times, but may also refer to other times in the same place. The inhabitants of the oldest rural and urban area of the municipality regard the risk as a product of a series of physical, environmental, economic, social and cultural transformations that have not been assimilated as part of daily life, although some of these changes are becoming more and more often manifested. The reflection is born from a similar comparison between the living conditions of yesteryear and those of now, which leads us to talk about the same space in two temporalities: the old Piedecuesta and the Piedecuesta of today. The reflection, of course, leads to long-term expectations that are generated by the evaluation of the current conditions or by the same inhabitants' desire for change: "What we see" and "What we feel". Here we see again the way in which "identity times" operate.

> But I often remember that was very quiet, the neighbourhood. There were no problems of insecurity or anything; you could leave the door open and nothing happened. […] I see it as an immense city, perhaps larger than Bucaramanga; I would like to have good security, so that all the environmental impacts we have caused are then amended. I would like to see it with a subway, and – why not? – with an airport. (Edinson, 30, urban district; San Carlos neighbourhood.)

> Look, it really is very sad for two reasons: there have never before been such uncomfortable summers and there is also the problem of the detriment of water sources and their flows as we're seeing lately. Climate change has been sharply increasing the temperature to such an extent that today it is already practically unbearable. (Gustavo, 73, rural district; Vereda Barro Blanco.)

The risk assessment does not respond exclusively to an objective observation of reality. Thinking of risk scenarios generates a series of emotions, feelings and attitudes that give new meanings to subjects, objects or situations that are evoked in this narrative, forming a judgement or appraisal with regard to these elements. However, in the case of the inhabitants of the municipal and rural areas, for whom the perception of change operates as a source of information, the problems related to citizen security (delinquency) and public health (drug addiction) are directly related to the arrival of a new population that generally experiences highly precarious living conditions. The wrongly placed interns who have settled in the periphery and the Venezuelan migrants, who mostly work hard on poultry farms, belong to population groups that experience premature judgements on the part of other residents.

By contrast, with regard to environmental damage and deterioration of the urban infrastructure, reflection is tilted towards the practices developed by the longer-standing inhabitants of the district and towards a basic awareness of human interference in the cycles of nature.

Knowledge of risk as a result of the transposition of temporal dimensions and a combination of information from different sources[8] is also a feature that has been identified in the accounts of the rural inhabitants. This narrative has, for obvious reasons, a strong relationship with the various dynamics of nature, although it can also be explained by spatial social conditions, as pointed out in some of the inhabitants' testimonies.

The deficit in the supply of water is a topic of vital importance for the rural inhabitants – a matter that is configured as a complex multidimensional problem. For residents of this area, exposure to risk in terms of water shortages does not represent a catalyst for group organization or the search for long-term solutions, a fact that, according to the analysis of field data, is related to the social dynamics of the area and the characteristics of its population.

There are three representative groups: 1) landowners or managers of the estates with neighbourhood disputes – whether about boundaries, water or coexistence; 2) tenants who are in transit or have little time to participate in community activities; and 3) indifferent entrepreneurs in the face of internal dynamics. This panorama has hindered the organization and participation necessary for the confrontation of collective problems.

In the face of long-term droughts, old liquid extraction practices (cisterns or water pumps) have intensified for self-consumption, agriculture and personal hygiene,

---

[8]A selection of different data from experience, perception of change, and institutional information or mass media.

while basic needs are met by the water in family tanks, as the construction of a public or multi-family tank, which would be a viable solution according to the municipality and the residents themselves, has been hampered by the aforementioned conflicts.

> Living without water is not possible. We survive, since some people have cisterns that give a droplet and we get what the mayoralty orders for us when, for example, I cannot use water from the gorge. My brother sends me water from the gorge but it's only for baths and washing. It's not fit to drink or use for cooking because all the water from washing the poultry sheds goes into the gorge. The poultry do have a few holes of clean water that comes out of the sheds but it's mixed with chemicals which filter into the ground and pollute the gorge. We made a small cistern very close to the ravine. The water is rotten and yellow all the time, so it is impossible to consume that water. (Sara, 58, rural zone; trail El Duende.)

How can the social contradictions in the representational study be explained? Since the social representations of risk are neither one-dimensional nor formed by a single type of knowledge, it is logical to assume that, depending to the demands of the context, at least one aspect of that construction of reality will stand out enough to indicate the direction that future practices should take. For example, while continuing to associate the precariousness of the countryside with other social, economic and political factors,[9] climatic alterations have exacerbated the perception of water supply problems in this district. Hopefully, social actions and behaviour in the short and medium term will find a definitive solution to this problem.

However, due to the different nodes of interest that together represent the social construction of risk, decision-making becomes more complex, since several aspects simultaneously require the attention of the cognitive subject. Thus, given the possibility that the donation of land for the installation of the tank might increase the problems experienced by those who have been dealing with long-standing conflicts over land tenure and property boundaries and experiencing difficulties in their relationships with neighbours, it is preferable, from the perspective of the users, to maintain old supply practices, despite recognition of their unsustainability. This is one more example of the way in which socio-behavioural dilemmas are an important factor in understanding attitudes to risk.

## 3.6.1 Anchorage Lines and Objectification: Temporality and Causality of Disaster Risk

The SR of risk relate to the situations and conditions of reality that require a response elaborated from everyday environmental, historical, cultural and contextual knowledge. To maintain a practical objective in the configuration of concrete

---

[9]During the workshops, social cartography also highlighted problems such as the abandonment of the countryside by the national and municipal government, false political promises, the levels of unemployment and the loss of family and community values.

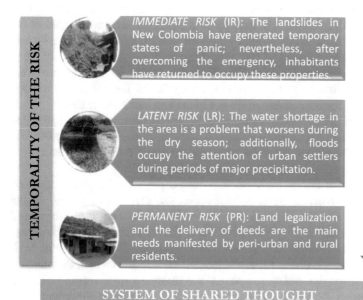

**Fig. 3.6**  Temporality of the risk as an anchorage line (*Source* Author's own preparation based on field data)

realities, the processes of anchorage and objectification are necessary: What effect is known in an integrative way?

On the basis that the object of representation and the anchored SR operate together (Candreva/Paladino 2005; Lara 2005; Serrano 2010), I will focus on the lines of anchorage and social objectification without going into more detail about the configuration of the process. For this purpose, these lines have been studied from the notion of time-cause.

One of the main factors involved in the perception and social construction of disaster risk is the temporality of its manifestations. According to the field findings there are three types:

- Immediate risk (IR), usually related to emergency conditions that temporarily alter or modify daily life;
- Latent risk (LR): intermittent exposure to a problem that appears and disappears in the context of certain contextual triggers; and
- Permanent risk (PR) – the one that assembles the conditions that generate more attention and expectations in the social group (see Figure 3.6).

Knowledge tends to be organized according to the criteria used to evaluate the transitoriness or permanence of the risk. All information that is categorised as permanent risk (PR) is more easily retained, and more likely to become part of shared thought processes. In this arrangement of ideas, the perception of risk plays a more prominent part in both the formation of knowledge and the orientation of the practices; without such measures, daily exposure to the risk would end up normalizing the problem. Conversely, emergencies and accidents are regarded as the result of conflicts in the broad social framework. For example, for the peri-urban inhabitants, the hazards caused by landslides (IR) are due to the saturation of water in the land in the absence of sewers (LR); However, asking for this public service to be provided is difficult in the face of the illegality of the zones they occupy (PR). From this perspective, the solution lies in the regularization of tenure.

The amount of sense and meaning that has been constructed and mobilized around the perception of risk tends towards a narrative of the homogenized collective, articulating the perception of the latent and emergent risks – as a totality – to the consensual universe, without ever making them invisible.

It should be added that in the temporality are expressed, in addition, the psychosocial relations used to structure and concretize the information about the subject – relations of objectification that are manifested, in turn, in the intra- and inter-contextual social territorial processes where the common know-how is organized. Two elements of objectification have been identified for this research: first, the relationship between disaster risk and internal causes; and second, the relationship between disaster risk and causes outside the group.

The members of the three study areas (periphery, rural area and urban area) rationalize the risk according to the origin they attribute to it. Thus, the knowledge gathered about disasters that can be avoided by the individuals themselves and those that cannot naturalizes certain behaviours – of action or omission – by the social actors. A distinction exists, according to this logic, between fully deserved risks (the function of the risk) and non-fully deserved risks, which ends in a reflection on the idea of justice.

According to the testimonies collected, the conditions of "deserved" exposure arise from the lack of belonging to the place and the low environmental awareness of the population, circumstances that generate inadequate citizen practices.

> Disasters occur when we ourselves are to blame for the lack of something we need. For example, I am partly guilty for us losing our transport system. The disaster we are experiencing is that they removed the Metrolínea (Integrated System of Massive Transport of Bucaramanga). There are people who are suffering from the lack of payment for the transport system; they only make losses. Therefore, the people –like me who did not pay are to blame. We made the mistake. (Eudora, 65, peri-urban sector.)

The social introjection of risk and its causes, as can be identified in some work on perception[7] (Schaer 2015; Knight 2007; Bermúdez 1994), also operates in the religious conception that explains the disaster as a divine punishment for the evil behaviour of human beings. In this tone, collective postulates on disaster risk can respond to both the ethical-civic and moral level. In the Piedecuestana narrative the first of these predominates, although the religious element is also apparent in some opinions.

From another angle, the risks that arise from actions (environmental, political, social, economic or cultural) of third parties, oblivious to the daily life of the place, are more difficult to control and predict. Consequently, the behaviour of the groups varies between individual activities that give a temporary solution and organizational strategies that seek to cope with the complexity of the problem.[10] This dual behaviour manifests the dynamic nature of the SR: the agent not only accommodates contextual requirements but can also change that reality by mobilizing the necessary resources.

However, binomial justice-injustice is a notion that has been little explored for issues such as risk or disasters; however, for the individual it represents a cardinal component when it comes to evaluating common know-how.

Within the framework of conditions of exposure, the functionalization of the risk implies the transfer of responsibilities towards the individual or social group. This fact is preceded by the belief of inevitability and justice with regard to the consequences of disaster – for example, environmental pollution or "divine" punishment. In contrast are problems that are identified by the collective as an injustice, as they infringe fundamental rights that must be promptly restored (land legalization). In this scenario, the demand for information on the part of those involved is greater for the risks denominated as "unfair" (PR) than for those considered transient or latent (very surely deserved). This constitutes a fact that has not been incorporated in the municipal management processes.

According to Sen "a calamity would be a matter of injustice only if it could have been avoided, and particularly if those who could have avoided it have failed. In some way, reasoning is nothing more than moving from the observation of a tragedy to the diagnosis of an injustice" (2010: 36). Hence, if an event that could have been avoided happens, it translates into an injustice; therefore, if there is an injustice, there are causalities and responsibilities – as well as people responsible – that cannot be assumed, in their entirety, by the social group (Jerez 2014: 144).

## 3.7  Socio-Structural Dimension

The socio-structural dimension brings together the elements of threat and vulnerability that, in their representational expression, shape the social construction of risk. It takes into consideration the sense of totality that the social group uses to articulate the information of the medium with the knowledge that is born or derived from the experience. Thus, following the narrative of the population, the notions of proximity, exposure and frequency of the phenomenon are interwoven with the representations of the vulnerability (psychosocial, economic, cultural and political), thus giving meaning to common knowledge.

---

[10]For example, the construction of cisterns by rural inhabitants for the extraction and storage of water (temporary action) and the neighbourhood actions in the peri-urban area that seek the legalization of irregular settlements (long-term strategies).

### 3.7.1  The Natural Element: Threat or Resource?

Social cartography has made it possible to identify the most common physical phenomena in the different areas of study: droughts and forest fires in the rural area, landslides in the periphery and overflow of rivers and floods in the urban centre.

Such threats are not isolated from the network of knowledge that has evolved on the environment, which is why attitudes to the natural elements are not anchored only in the negative or the positive, but travel from one area to another, according to the context and the characteristics of the experience. To better understand this point, consider representations of water, an element whose meaning is modified to the extent that it is associated with absence/presence and plenty/scarcity.

According to the inhabitants of the peri-urban area, the excess (presence-abundance) of water in the land on which most of the houses are built constitutes a threat to the stability of the hillsides, a danger that already has a history. According to testimonies, the accumulation of liquid on the surface is mainly due to the discontinuity of the municipal sewer network, a situation that has harmed these communities. The urban population is also highly vulnerable to flooding, which is why there is talk of presence-abundance contexts in the cartographic exercise that analyses the risk. In line with this trend, the natural phenomenon is related to conditions of high frequency and proximity, generating prolonged scenarios of exposure both in the periphery and in the city.

In the case of the rural area, the narrative around the same element (water) is modified in an obvious way, because in prevailing conditions of absence/scarcity, water is conceptualized as a resource of low frequency and increasingly further away, which constitutes a reduction in the risk of flooding but much greater exposure to the threat of drought.

> One of the fundamental requirements of all life is water, and if there is no water, there is no life of any kind. If water is used for human needs – drinking, bathing, food preparation, cleaning – there is not enough for the sustenance of agricultural products and then a serious problem arises because people do not have any produce. So, you have to decide how to use water for survival and maybe resign yourself to starving, or risk your life trying to produce something with minimal resources. (Gustavo, 73, rural sector; trail Barro Blanco.)

The attitude towards natural elements assigned by a social group tends to vary between threat and resource, according to the meaning derived from the experience and without losing sight of the object of social representation (disaster risk). Hence, the perception of physical phenomena in risk scenarios produces two tendencies: the greater the frequency, proximity and exposure, the more the natural element corresponds to the notion of threat; conversely, in the face of low frequency and proximity, and high exposure, the most common associations correspond to the resource dimension (Fig. 3.7).

The society-nature link refers to a discourse on a temporary space – "transactional" totality, according to Alba (2004: 117) – which corresponds to the

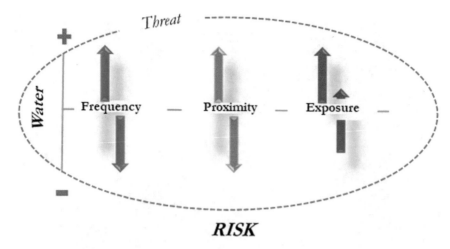

**Fig. 3.7** Notions of frequency, proximity and exposure of the natural element (threat/ resource) (*Source* Author's own preparation based on field data)

binomial territory-identity from which the natural basis is not independent. The proximity, the level of exposure and the frequency of the event are not fully objective knowledge, since these notions are modified according to the construction that is made of the inhabited territory, a construction concerning identity.

In the mapping – part of the social cartography – carried out by the rural inhabitants (Figure 3.8), two important aspects are highlighted in reference to the territorial frontiers: first, the boundary of the sidewalk in family plots and poultry farms that operate in the area, demarcating private and business-owned areas; and second, the exclusion of the rural path of the territory called Piedecuesta – which is the name of the municipality of which the rural path is also part – at the point that actually refers to the urban centre.

The emphasis on property boundaries and the trajectories of the rivers and ravines makes sense given the problems of tenure, neighbourhood conflicts and water-hoarding that have occurred in the zone. The cartography indicates that rural inhabitants try to reach agreements about the boundaries of familiar land, the extension of the adjacent poultry farms and the trajectories of the rivers and the ravines, agreements that are implicitly evidenced in the indications drawn on the map: cartography clearly points out the boundaries between land, each plot bears the name of the owning family or the name of the farm, and different colours are used to differentiate the courses of rivers and roads.

Following this analysis, the lack of attention on the part of the local administration is symbolized in the geographical exclusion that the residents make of the rural path in relation to the territory called Piedecuesta (municipal urban centre), a symbolic separation manifesting the inequalities experienced by the rural people in comparison with the city (see the work of Vélez/Rátiva/Varela 2012).

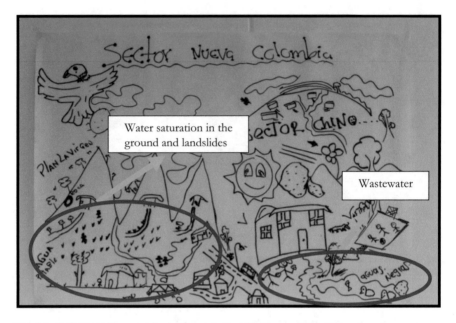

**Fig. 3.8** Social mapping: periphery (Nueva Colombia) (*Source* Field data)

On the other hand, in the case of the peri-urban map (Fig. 3.9), the lateral perspective of the drawing is highlighted, a characteristic that operates as a contrasting perspective:

- Between the flat terrain and the slopes, an aspect that allows the scale of threats related to the topography of the place (landslides by water saturation, rock falls, landslides) to be calculated; and
- Between the natural landscape and the socially constructed landscape, an aspect that refers to the notion of appropriate space or territory

The landscape that results from the relationship nature-society is a landscape different from the one that precedes the process of territorializing, an appropriation that exceeds the instrumental objectives because it also generates affective loads, valuations and an identity built around the territory (Giménez 2005). This factor is identified in the map elaborated by the residents of Nueva Colombia (periphery), where elements of the natural environment and elements of the urban landscape are highlighted, as well as the positive and negative results of this interaction: human intervention in nature generates problems such as pollution of the environment and deforestation of the mountains, but also promotes development through the opening of roads and the construction of sports facilities. The appropriation also concerns the sense of belonging, something which has not been legally recognized, so that the 'security' of tenure is based on the daily activities which generate the 'possession' of land and influence assessments of the threat.

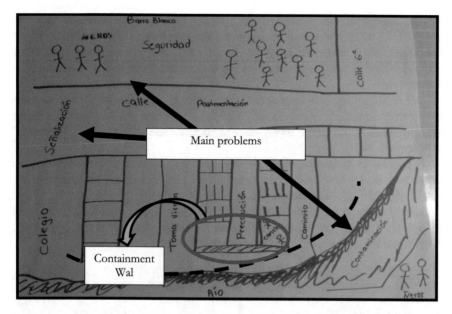

**Fig. 3.9**  Social mapping: urban sector (Barrio Blanco neighbourhood) (*Source* Field data)

Finally, the urban cartographic product emphasizes the location of those areas considered insecure, contaminated or dangerous, conditions related to disaster risk (pollution, insecurity, road problems). The map depicts the wall built by the city hall to reduce flood exposure arising from the proximity of the river, which is also interpreted by residents as an indication to demarcate the borders of the colony. The objective is to distinguish between the "other" risk areas located outside the sector and those dangers that are very close to public areas of interest (school, sports facilities, parks).

### 3.7.1.1  Vulnerability Represented

The conditions of vulnerability are realities that are reinterpreted and discussed daily. They have a material and objective dimension of importance, but they rarely manifest themselves as an absolute value in collective thinking. The forms of vulnerability are interwoven in unsafe environments, feeding each other; thus, uncertainty is mobilized between territories (geographical blur), and shared and propagated from diverse perspectives that make up the continuous – not fragmented – narrative of risk.

I will begin by addressing the various manifestations of social vulnerability. It is striking that the most recurrent problem at the time of discussing the topic of disaster risk is the generalized context of insecurity and delinquency themes that are not directly related to the dynamics of nature.

Theft, homicide and other crimes are associated with the sale and consumption of narcotic drugs. The areas with the greatest conflict arising from the consumption and sale of drugs are the periphery settlements; however, the populations of other sectors are worried about these issues spreading to the long-established neighbourhoods and sidewalks, since they cannot ensure that micro-trafficking activities are carried out exclusively by residents of the irregular settlements.

In line with this, it is important to note that, in the narrative of risk, sources and effects do not always share the same territory, an aspect that is linked to the consolidation of symbolic limits configured from the discourse of "otherness" (Colombani 2008).

In relation to the displaced people and migrants, there is a unilateral concept of otherness which tells the story from the urban centrality of the processes of distinction/identification. However, within the belts of poverty, the notion of "us" and "others" and the delineation of "the other" and "the others" are also present, both to heterogenize the population and to mark the times – identity – of the dialogue: what was and what is.

> Every day there are new strangers sitting around smoking *basuco* or watching to see what they can steal at night. One cannot leave anything outside because after dawn things mysteriously disappear. Here it used to be very good because there were no more invasions, but now there are invasions here and there and we believe all these vicious people come from there. That is all. (Manuel, 78, periurban district.)

In this argument, otherness relates not only to the characterization of subjects, but also to the elaboration of temporal images that paradoxically converge in a shared reality. According to the information collected, the scenarios that lead this "other" to engage in the consumption and sale of drugs are not very different from the conditions that most of the population must face. For example, the lack of opportunities for young people, dropouts, and the crisis of family authority in the face of the long hours worked by parents are causes that are inserted into the context of labour and economic precariousness in the perceptions held by a large number of the residents. There is, then, a discursive paradox that lies in the heterogenization of the individual in homogenization schemes of the context.

According to the testimonies, the widespread transgression of rights linked to citizen security, decent housing, health, education and employment (social vulnerability) is explained by the malfunction of the local management apparatus, a highly damaging political and institutional vulnerability which is detrimental to the effective experience of democracy.

The trust/political mistrust is related, according to the experts, to citizens' perceptions of the fulfilment/non-fulfilment of institutional duties, as well as the costs of the democratic system (Palazuelos 2012). In the case of risk management, the losses and costs are not only economic, but also human. The failure to fulfil the functions of planning, development and evaluation materializes in the increase in the population exposed to risk in the municipality, mainly through the physical vulnerability of their properties, either because of the location or because of the inferior quality of the materials.

Another fundamental component is the socio-environmental crisis manifested by the people. With regard to this, two types of problems are differentiated: 1) global processes that are manifested locally (climatic variability); and 2) effects generated by the particular synergy of the context (pollution, misuse of natural resources, water grabbing, etc.).

The climatic variability experienced for some decades is related, according to the testimonies, to the street child phenomenon and climatic change, macro processes whose effects are potent in a local context of environmental susceptibility.

> The climate has been increasing temperatures sharply to such an extent that today it is almost indescribable. Most flora and fauna species have disappeared as a consequence of climate change; all these impacts have been generated as a result of our neglect, of our bad attitudes towards the environment. (Gustavo, 73, rural sector; trail Barro Blanco.)

The economic and material losses increase during the atypical periods of rain or drought, especially among the peri-urban and rural population, because of both the damage caused during emergencies and the effects on the means of family subsistence. In this context, the economic vulnerability before the disaster exacerbates the environmental crisis in the most precarious sectors of the municipality, a condition that reinforces the experiences of susceptibility in other areas of social life (health, education, housing, work).

All those material and symbolic manifestations of the failures in the social structure are configured as daily disasters that are rationalized, internalized and communicated as such (disasters). Thus, the vulnerability represented refers to this realistic and meaningful narrative that combines the various experiences of shared and differential susceptibility, a fundamental factor in understanding the process of social construction of risk. The vulnerability represented is a dynamic, re-created and critical dimension that, in the case of the study area, has been synthesized in the following associative network (Fig. 3.10):

## 3.8 Conclusions

This study of representations of disaster risk highlights three distinct key elements: 1) the content of the social construction of risk; 2) the processes from which the risk is generated; and 3) the broad social base that contains the risk. The epistemological background of Moscovician thought speaks of the subjectivity in origin, nature and function as a social aspect that, without obviating the cognitive component, focuses on the exchanges that occur in daily life, beyond the psyche.

In this understanding, the generation of institutional strategies for risk management should not ignore the common sense that composes and organizes social know-how, nor the contextual (material and symbolic) conditions that precede the development of an incident.

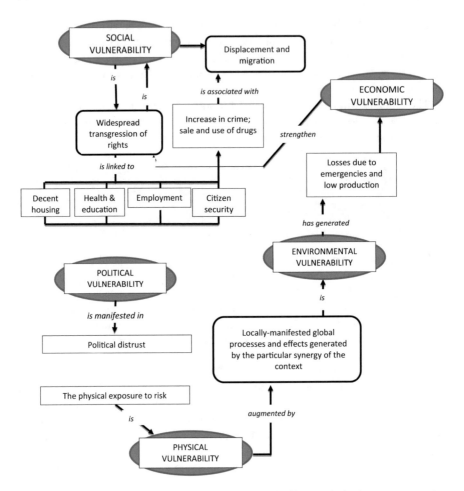

**Fig. 3.10** Vulnerability represented: associative network (*Source* Author's own preparation based on field data)

It is logical to think that the materialization of risk can influence the modification of the shared thinking system. However, if the information generated during emergencies – which is only part of the process of forming and integrating knowledge into the common-sense system – is not assessed in conjunction with more complex and defined links of everyday knowledge behavioural changes are unlikely to take place in time. This would explain the contradictions that are usually present in social reactions to problems that cannot easily be solved by traditional interventions.

Disaster risk, as a social construction, refers to the conditions of vulnerability and threat represented, to the sociocognitive processes of collective thought, and to the temporal space intentionality that recreate the Territory-Identity relationship, aspects that can contribute to new ways of approaching research on the topic.

# References

Alba, Martha de, 2004: "Mapas mentales de la Ciudad de México: una aproximación psicosocial al estudio de las representaciones espaciales", in: *Estudios Demográficos y Urbanos*, 55 (january–april): 115–143.

Anderson, Mary and Woodrow, Peter, 1989: *Rising from the ashes: development strategies in times of disasters* (Boulder: Westview Press).

Banchs, María Auxiliadora, 2000: "Aproximaciones procesuales y estructurales al estudio de las representaciones sociales", in.: *Papers on Social Representations*, 9: 3.1–3.15.

Bermúdez, Marlen, 1994: "El manejo institucional y percepción de la población en el terremoto de Limón", in: *Revista Geolóica de América Central*, special issue: 221–224.

Bronfenbrenner, Urie, 1987: *La ecología del desarrollo humano* (Barcelona: Paidos).

Caballero, José Humberto, 2007: "La percepción de los desastres: Algunos elementos desde la cultura", in: *Gestión y Ambiente*, 10,2 (August): 109–116.

Calixto, Raúl, 2008: "Representaciones sociales del medio ambiente", in: *Perfiles Educativos*, 30,120: 33–62.

Candreva, Ana; Paladino, Celia, 2005: "Cuidado de la salud: el anclaje social de su construcción. Estudio cualitativo", in: *Universitas Psychologica*, 4,1 (January–June): 55–62.

Colás, Pilar, 2006: "Género y contextos sociales multiculturales: educación para el desarrollo comunitario", in: Soriano, Encarnación (Ed.): *La mujer en la perspectiva intercultural* (Madrid: La Muralla): 20–43.

Colombani, María Cecilia, 2008: "A propósito de Dioniso y Apolo: mismidad y otredad: el juego de las tensiones", in: *Nuntius Antiquus*, 2 (December): 25–40.

Departamento Administrativo Nacional de Estadística (DANE), 2018: *Estimación y proyección de población nacional, departamental y municipal total por área 1985–2020* (Bogota: DANE).

Douglas, Mary; Wildavsky, Aaron, 1982: *Risk and culture: an essay on the selection of technological and environmental dangers* (Berkeley: University of California Press).

García, Virginia, 2005: "El riesgo como construcción social y la construcción social de riesgos", in: *Desacatos-Revista de Antropología Social*, 19 (September–December): 11–24.

Giménez, Gilberto, 2005: "Territorio e identidad: breve introducción a la geografía cultural", in: *Trayectorias*, 7,17 (January–April): 8–24.

González, Edgar; Valdez, Rosa, 2012: "Enfoques y sujetos en los estudios sobre representaciones sociales de medio ambiente en tres países de Iberoamérica", in: *CPU-e, Revista de Investigación Educativa*, 14 (January–June): 1–17.

Hollisch, Gisele, 2014: "Las representaciones sociales y las ideas previas de los alumnos", in: *Memorias del Congreso Iberoamericano de Ciencia, Tecnología, Innovación y Educación, 12–14 de noviembre de 2014*, Buenos Aires, Argentina.

Jerez, Deysi, 2014: "Prevención y mitigación de desastres en Colombia: racionalidad comunicativa en políticas públicas" (MSc dissertation, National Autonomous University of Mexico [UNAM], Faculty of Political and Social Sciences).

Jerez, Deysi, 2018: "Construcción social del riesgo de desastres en el Municipio de Piedecuesta, Santander (Colombia): dimensiones socio-representacionales" (PhD dissertation, UNAM, Faculty of Political and Social Sciences).

Jodelet. Denise, 1986: "La representación social: fenómeno, concepto y teoría", in: Moscovici, Serge: *Psicología social II* (Barcelona: Paidós): 469–494.

Lara, Eric, 2005: "Sonaron siete balazos: narcocorrido: objetivación y anclaje", in: *Trayectorias*, 7 (January–April): 82–95.

Lavell, Allan, 1996: "Degradación ambiental, riesgo y desastre urbano: problemas y conceptos: hacia la definición de una agenda de investigación", in: Augusta Fernández, Maria (Ed.): *Ciudades en riesgo: degradación ambiental riesgos urbanos y desastres* (Panama City: La Red/USAID, Colombia).

Lavell, Allan, 2004: "Vulnerabilidad social: una contribución a la especificación de la noción y sobre las necesidades de investigación en pro de la reducción del riesgo", in: *Memoria del Seminario Internacional Nuevas Perspectivas en la Investigación Científica y Técnica para la Prevención de Desastres* (Lima: Sistema Nacional de Defensa Civil/Perú; Save the Children/ Suecia): 48–55.

Molfi, Eneida M., 2000: "Deconstrucción de las representaciones sobre el medio ambiente y la educación ambiental", in: *Tópicos en Educación Ambiental*, 2,4: 33–40.

Nuño, Bertha, 2004: "Modelos de toma de decisiones con los que intentan resolver el consumo de drogas ilegales adolescentes consumidores y sus padres que acuden a tratamiento a CIJ en Guadalajara" (PhD dissertation, UNAM, Faculty of Psychology).

Palazuelos, Israel, 2012: "La desconfianza en los partidos políticos y la percepción ciudadana de desempeño gubernamental: México ante América Latina", in: *Revista Mexicana de Análisis Político y Administración Pública*, 1,1 (January–June): 79–107.

Reigota, Marcos, 1990: "Les représentations sociales de l'environnement et les pratiques péd-agogiques quotidiennes des professeurs de Sciences a São Paulo-Brésil" (PhD dissertation, Louvain-la Neuve: Catholic University of Louvain, UCL).

Rizo, Marta, 2006: "Conceptos para pensar lo urbano: el abordaje de la ciudad desde la identi-dad, el habitus y las representaciones sociales", in: *Bifurcaciones: revista de estudios cultur-ales urbanos*, 6: 1–13.

Serrano Oswald, Serena Eréndira, 2010: "La construcción social y cultural de la maternidad en San Martín Tilcajete, Oaxaca" (PhD dissertation, UNAM, Institute of Anthropology, Faculty of Philosophy and Letters).

Schaer, Andrea, 2015: "Modelo teórico de comportamientos sociales frente al riesgo y el desas-tre", in: *Párrafos Geográficos*, 14,2: 97–125.

Starr, Chauncey, 1969: "Social benefit versus techonological risk", *Science*, 165, 3,899 (September): 1,232–1,238.

Vélez, Irene; Rátiva, Sandra; Varela, Daniel, 2012: "Cartografía social como metodología par-ticipativa y colaborativa de investigación en el territorio afrodescendiente de la cuenca alta del río Cauca. Cuadernos de geografía", in: *Revista Colombiana de Geografía*, 21,2 (July–December): 59–73.

Villarroel, Gladys, 2007: "Las representaciones sociales: una nueva relación entre el individuo y la sociedad", in: Fermentum. *Revista Venezolana de Sociología y Antropología*, 17,49 (May–August): 434–454.

# Chapter 4
# Confluences between Social Representations Theory and the Psychology of Active Minorities

Aline Reis Calvo Hernandez

## Introductory Comment

**Pedrinho  Guareschi**

*It is not always easy[1] to select and nominate someone who stands out from others, especially when we are dealing with teamwork and partnership. Therefore I beg the pardon of the other participants when indicating Aline as someone who has led work on social representations (SR) and continues to advance research in this field. As you can see when reading the chapter, and as the author herself makes very clear in her writing, all the work developed by the research and reflection group 'Ideology, Communication and Social Representations', in addition to the purpose of always seeking to be plural, was always within a spirit of absolute partnership, seriously based on a premise of the educator Paulo Freire (1987), who inspired the ethical dimension of learning for us: "there is no one who knows more or less: there is different knowledge".*

Aline Reis Calvo Hernandez is an associate professor in the Department of Basic Studies at the Education College of Universidade Federal do Rio Grande do Sul. She has a doctorate in Social Psychology and Methodology from Universidad Autónoma de Madrid and a Masters in Education from Pontifícia Universidade Católica do Rio Grande do Sul. She is also leader of the research group Political Psychology, Education and Stories of the Present (CNPq). Email: aline-hernandez@hotmail.com.

[1]Pedrinho Guareschi is a professor and researcher at the Federal University of Rio Grande do Sul (UFRGS), and an international lecturer with many decades of experience in social representations theory. He was previously a professor at the Pontifical Catholic University of Rio Grande do Sul (PUC-RGS). He is also a lecturer in *stricto sensu* at the Pontifical Catholic University of Paraná (PUC-PR) and an associate researcher at the Carlos Chagas Foundation (CCF). Email: <pedrinho.guareschi@ufrgs.br>.

© Springer Nature Switzerland AG 2021
C. Prado de Sousa and S. E. Serrano Oswald (eds.),
*Social Representations for the Anthropocene: Latin American Perspectives*,
The Anthropocene: Politik—Economics—Society—Science 32,
https://doi.org/10.1007/978-3-030-67778-7_4

*If I asked Aline Hernandez for the job of introducing her chapter, which, by the way, is very well written and extremely serious and honest, it is because she represents, both in her writings and in her investigations, a happy synthesis of the many concerns that gave life, enthusiasm and pleasure during the thirty-four years of our happy group journey of discussion and investigation.*

*I allow myself to highlight just a few points that summarize the group's activities, which Aline admirably represents:*

*The importance of emphasizing that SR are practices. I have no more detailed knowledge of the material produced in light of SR in other parts of the world, but at least here, in the context of Latin America, the use of SR as a practice did not previously stand out in studies and research. It is interesting to draw attention to that here, as in academia the preference has always been – and still is – given to the theoretical dimension, even in situations like ours, in which concrete action/ practice was historically fundamental. It is even surprising that no attention was given, from the beginning, to Moscovici's statement (2012: 39) in the first paragraph of his seminal book: "It is the specificity of SR – symbolic substance – and their creativity – the practice that produces such a substance – that differentiates them from the sociological and psychological notions with which they are compared and from the phenomena that correspond to them". And it is your creativity, as Moscovici himself emphasizes (Marková 2006; Moscovici 2003), that makes SR a Theory of Innovation too.*

*In both her doctoral work and subsequent research activities Aline has always pleaded the importance and indispensability of placing emphasis on the dimension of practice, and this was even more evident in the effort expended in making the group willing to translate into Portuguese Moscovici's book* Psychologie des Minorités Actives [Psychology of Active Minorities], *in which activity she was the central actor.*

*Another point in which Aline collaborated to ensure that SRs were taken in their broadest scope, and also in their complexity, was to call attention to the dimension that is always present, but little remembered and discussed, in SR: the dimension of cognitive polyphasia. In her opinion, when discussing André Guerra's doctoral thesis (Graduate Programme in Social and Institutional Psychology, UFRGS, 2020), "A Social Psychology of Law", with its suggestive subtitle "Investigations on a dissident magistracy: spectres of resistance", which was theoretically illuminated by Serge Moscovici's theory of Active Minorities, Aline draws attention to the fact that certain actors can live with cognitive dissonances and manage to survive in highly contradictory situations, being able to envision dissent in favour of social change.*

*Finally, I would like to emphasize a methodological innovation in the author's work, with the use and analysis of images and their semiologies as an empirical corpus for her studies in SR. In her doctoral thesis, and later in her postdoctoral research, she drew attention to the importance of narrative and reading images published by the media, especially in times when it becomes hegemonic in the media.*

*For this reason, and much more, as you will see from the writing that follows, I am sure that the employment of SR will remain central in the field of social psychology. Personally, I share Moscovici's suggestion (Moscovici 2003; Veronese/ Guareschi 2007) that the study and theory of SR can be considered as the unifying space for the area of social psychology.*

# References

Freire, Paulo, 1987: *Pedagogia do oprimido* (Rio de Janeiro: Paz e Terra).
Marková, Ivana, 2006: *Dialogicidade e representações sociais: as dinâmicas da mente* (Petrópolis, RJ: Vozes).
Moscovici, Serge, 2012: *A psicanálise, sua imagem e seu público* (Petrópolis, RJ: Vozes).
Moscovici, Serge, 2003: *Representações sociais: investigações em psicologia social* (Rio de Janeiro: Vozes).
Silva, André Luiz Guerra da, 2020: "Uma psicologia social do direito: investigações sobre uma magistratura dissidente: espectros da resistência" (PhD thesis, Universidade Federal do Rio Grande do Sul, Programa de Pós-Graduação em Psicologia Social e Institucional da Orientador).
Veronese, Marília; Guareschi, Pedrinho Arcides, 2007: *Psicologia do cotidiano: representações sociais em ação* (Petrópolis, RJ: Vozes).

**Abstract** This chapter presents the state of studies, research and publications on social representations in Brazil, and especially in the state of Rio Grande do Sul. In order to do this, three objectives are pursued: presenting the research groups and the production areas in social representations in the country; discussing the work that has been taking place in the south of Brazil, especially in social representations, active minorities and politics; and delineating the perspectives and challenges of research into social representations in the country. Qualitative methodology was used, through an exploratory bibliographical, descriptive and interpretative study. First the output of research groups registered in the Directory of Research Groups and Lattes Platform of the National Council for Scientific and Technological Development (CNPq) is presented in order to highlight some descriptive statistics and map the scenario of the groups and production areas in the country. Subsequently, an analysis is made of the relationship between Social Representations Theory and the Psychology of Active Minorities, two fundamental theoretical proposals in Moscovici's work (a relationship which is still little explored in Brazil) in order to deepen the epistemological discussion about social representations and their importance in sociological social psychology. Finally, the commitments that are made, the possible horizons, and the challenges that must be faced by social representations researchers in Brazil are discussed.

**Keywords** epistemology · social representations · active minorities · politics

## 4.1 Introduction

It was a pleasure and joy to receive the invitation from Pedrinho Arcides Guareschi, a renowned Brazilian researcher in the field of sociology and in studies of social representations, to write this text. The proposition of the book, as you already know, is that a researcher in the area nominates the person among their collaborators who stands out in terms of innovation and advancement in studies and research on social representations. Therefore, I reflected for a few days on the challenges and guidelines of this text at a time when Social Representations Theory completes approximately forty years of exposition on the Latin American subcontinent.

According to Sá and Arruda (2000), the coming of social representations to Brazil is due to the return of Maria Auxiliadora Banchs to Venezuela after completing her doctorate under the guidance of Serge Moscovici in 1979. In 1982, Banchs invited Denise Jodelet to Caracas, and that same year she came to north-east Brazil to teach a methodology course in social representations and collaborate in the elaboration of research into mental health at the Federal University of Paraíba (UFPB).

In Brazil this historical-political moment was in commotion because of this social saturation and the beginning of a struggle for the recovery of democratic freedoms overwhelmed by the military civilian dictatorial regime begun in 1964. After the 'Years of Lead', the pressure of the 'Economic Miracle',[2] the 'disappearance' of political activists, the Institutional Acts and the intensification of censorship and political exile, the regime was weakened. In 1982, the city and state elections demonstrated widespread rejection of the regime by the people and a strong tendency toward democratic openness in motion for "*Diretas Já*" (direct elections now movement). In this respect, there was a theoretical-epistemological lack in the Brazilian universities that would support the struggle and strengthen the resistance.

> (...) in Paris, in the 70s... the resistance to repression was debated, a fact that was present in many of our countries; they would analyse the masses in movement, the active minorities, the environment, the body, mental illness, and also the relation between these phenomena and the diffusion of ideas, always within a psychosocial perspective. During the effervescence of those years (in which hatched right in front of us the feminist and environmental movements), the theoretical-methodological debates and life in Paris had ties of affinity and affection among the seminar attendees and between them and their professors. The Theory of Social Representations emerged as a possible answer to problems that used to distress us and were present in the lives of our compatriots and in our professional practice in our countries of origin. (Sá/Arruda 2000: 14).

Undoubtedly, the return of these Latin American and Brazilian researchers from France was instrumental in the propagation of Social Representations Theory

---

[2]The Years of Lead correspond to the period between 1969 and 1973, when the well-known economic miracle took place in Brazil, involving a paradoxical increase in GDP, but also an increase in inflation, concentration of income and social inequality.

across the subcontinent and Brazil, but I emphasize the ontological and ethical dimension of the theory and its contribution to the understanding of social phenomena in the perspective of social change. It is a theory originating from Genetic Psychology, and is consequently proposed with the intention of revisiting the genesis of processes and social phenomena in order to understand (or transform) them in depth.

Moscovici's seminal work *La Psychanalyse, son Image et son Public* (1961) was widely circulated in academic circles and gained strength in Brazil as an important Social and Community Psychology reference. After eighteen years, the author published *Psychologie des Minorités Actives* (1979), which, curiously, is not as well-known in Brazil, despite adding valuable elements to sociological analyses of power, social influence and political activism. In this work Moscovici presents fundamental theoretical presuppositions in order to formulate a paradigm of the change in counterpoint to the science of conformity, social homeostasis and the governance of the conducts promulgated by the North American Functionalist paradigm.

In spite of not drawing an explicit parallel between the two works, Moscovici mentions that the power of active minorities lies in the formulation of consistent behaviour, of enabling representation in the public sphere in favour of assertive social and historical changes. Later I will explain more deeply the importance of this work and analyse its confluences with Social Representations Theory.

Following this brief introduction, I present the lines of analysis that will be addressed in the text. Firstly, I will present an overview of the research and studies on Social Representations in Brazil nowadays, analysing the certified research groups, their regional distribution and the project areas. Then I will go into specifics about the importance and output of two research groups from the state of Rio Grande do Sul, namely the group 'Ideology, Communication and Social Representations' led by Pedrinho Guareschi, and the group 'Political Psychology, Education and Stories of the Present', led by me. This is when I will deepen a theoretical-epistemological debate on social representations, active minorities, power and social influence. Finally, I will present some possible challenges and horizons in the field of social representations in Brazil.

## 4.2 Scenarios of Studies and Research in Social Representations in Brazil

In 2014 the group of researchers Martins, Carvalho and Antunes-Rocha published some detailed research about the scientific output on social representations in Brazil. To this end, they undertook a bibliographic search of the Directory of Research Groups and Lattes Platform of the National Council of Scientific and Technological Development (CNPq). Created in 1992, this database works as a repository of information about the country's research groups in different areas of knowledge. Leading research faculty members fill in a set of data in order to register their groups in the directory and the Higher Education Institution (IES) must certify them. The database lists the name and main research topic of the group, institutional and regional identifiers, research lines, researchers, partner institutions and equipment. After being filed the data needs to be updated periodically.

In order to carry out the parameterized search on the website, the researchers decided that the expressions 'social representation' and 'social representations' should be contained in the titles of the research groups in a temporal period of searches from July to September 2012. After analysis of the results, the duplicated listings were excluded, along with those which did not mention Social Representations Theory on their individual pages. As a final result they found 172 research groups in the country.

While searching for the social representations research groups, the authors recorded that the first research group in Brazil, entitled 'Physics Teaching', was founded in 1967 by Dr Marco Antonio Moreira of the Federal University of Rio Grande do Sul. They stress that this historical data is notable because it provides a point of comparison with the diffusion of Social Representations Theory in the country, which dates from the mid-1980s. In 1987, only twenty years after the founding of the Physics Teaching group, the 'Ideology, Communication and Social Representations' group was founded by Dr Pedrinho Arcides Guareschi, initially at the Pontifical Catholic University of Rio Grande do Sul (PUCRS) and currently attached to the postgraduate programme in psychology at the Federal University of Rio Grande do Sul (UFRGS). 1993, the year in which the Directory of Research Groups was created, marks a positive inflection point in the registration and popularization of research groups in social representations in Brazil, in addition to pointing to exponential growth in the registration of groups and scientific projects in the area.

The results show a predominance of research groups in public institutions (88%), the rest being distributed among private and non-profit institutions. Regarding the distribution of groups by Brazilian region, the survey indicates that 34% of the groups are in the South East, followed by 30% of groups in the North East and 23% in the South of the country. The Mid West represents 8% of the research groups, most of which are located in institutions of the Federal District, and the North region accounts for only 5% of research output. Researchers point out that the Higher Education Institutions (IES) with the most research groups in social representations are the Federal University of Rio Grande do Sul (UFRGS) with nine groups, the Federal University of Pernambuco (UFPE) with eight groups and the Federal Rural University of Rio de Janeiro (UFRRJ) with seven groups.

Regarding the concentration areas of the research groups, data indicate that more than 60% belong to the field of human sciences, followed by health sciences with 17%, applied sciences with 12% and arts and literature with 7%. The areas of exact sciences and agrarian sciences are not very significant, adding only 2% of groups between them (Martins/Carvalho/Antunes-Rocha 2014). This statistic denotes the significance of Social Representations Theory in the field of human sciences, especially in the areas of education and psychology, social sciences, sociology, history and anthropology, applied sciences, communication, social service and law, being a theory concerned with the analysis of sociological phenomena and the changes produced in societies. Looking at the lines of research, the authors point out that only two groups highlight studies and research related to the epistemological and methodological aspects of Social Representations Theory, which is an emerging topic that lacks research in the country.

When reviewing the database in 2018, it was found that there are only two groups registered as international, although we know that the internationalization of Brazilian research in social representations is an urgent agenda for the development and advancement of the country's scientific production, in addition to the exchange of knowledge with other scientific societies. We also found only two groups registered as interdisciplinary, evidencing the interface of Social Representations Theory with multiple areas of science in addition to social psychology.

## 4.3  Confluences between Social Representations Theory and the Psychology of Active Minorities

In historical terms, the second research group to be founded in the country is entitled 'Ideology, Communication and Social Representations' and is led by Dr Pedrinho Arcides Guareschi. Currently the group has an institutional affiliation with the Federal University of Rio Grande do Sul (UFRGS), linked to the postgraduate programme in psychology. The group began its activities in 1987 and has made a major contribution to scientific output on social representations. The vast scientific production of Guareschi (more than ninety articles published in national and international periodicals, more than thirty books edited and compiled, more than ninety chapters in edited books, more than seventy articles in newspapers and more than 500 articles published in conference proceedings was always developed, discussed and endorsed by him in partnership with his research group. Guareschi has always shown himself to be an organic intellectual committed to the democratization of science and through sociological social psychology engaged with the problems of social groups, believing that science can and should provide society with answers and complement innovations and technologies.

Throughout 2006 and 2007 I had the opportunity to develop my postdoctoral research, "Memories of Psychology in Rio Grande do Sul in the decades of Political Repression: Active Minorities, Social Representations and Activism", under the guidance of Dr Pedrinho Guareschi and Dr Helena Scarparo, who were then professors of the postgraduate programme in psychology at the Pontifical Catholic University of Rio Grande do Sul. In addition to the research, every Tuesday we had study afternoons, when doctoral, postdoctoral and college scholarship recipients met to read, discuss or translate texts in social representations, social communication and political sociology.

During this period, the in-depth study of some of the works was remarkable, but here I will highlight Moscovici's *Psychologie des Minorités Actives* (1979), which we translated from Spanish to Portuguese, compared with the French and English edition (Hernandez/Accorssi/Guareschi 2013). The book was published by *Editora Vozes* in the Social Psychology Collection. Although it was not very widely read in Brazil, perhaps because it was not translated into Portuguese, it is of fundamental relevance to social psychology. When the book was launched at the University of Évora, Portugal during the XI International Conference of Social Representations in 2012, Moscovici himself emphasized that this was a fundamental work in his intellectual production, a transgressive work where he refutes the Functionalist and Positivist paradigms that dominated social psychology until around 1970 and proposes instead a new paradigm of innovation and change from social minorities. As the back cover announces:

> Serge Moscovici, an important figure in the field of Social Psychology committed to the study of minorities, analyses in this work three themes of first magnitude in current Sociology: the problem of change, the role of social influence in relation to transformation and the integral meaning of minorities as factors of change. Among the merits of this study are highlighted: the success of its challenge to the traditional theorizing of social influence and the author's courage in becoming a solitary critic against already consolidated positions, discovering a new role for active minorities that no longer needs to resign itself. Against a Social Psychology which defended that conflict, change and dissent compromised the balance of the system and should be eliminated, Moscovici dares to say what no one has said. With such boldness, Social Psychology offers the possibility of seeing in the conflict an open door to transformation. In this book, Moscovici plays the role of an active minority: he challenges the hegemonies of the emerging norm and innovates in Psychology when he questions a series of premises that had hitherto been consolidated. (Hernandez/Accorssi/Guareschi, 2013).

It is a work on dissent from the perspective of political sociology, of dissent as an important political action in the style of behaviour of active anti-hegemonic minorities from 'non-conformist' social groups. The formulation of a 'dissident psychology' is a factor of innovation and social change, as cogently argued by Moscovici. When it comes to majorities and minorities, leaving is always more costly than staying, and disagreeing is always more courageous than conforming.

Nevertheless, in *Psychologie des Minorités Actives*, Moscovici does not make explicit the relationship between the activism of active minorities and social representations. However, establishing this relationship is fundamental to the study of social representations of phenomena and social groups. Thus, I will dwell on the analysis of a set of indicators of the relationship between the two theories.

### 4.3.1 The Process of Representing: The Polyphasia of Social Knowledge

Social representations work as a collective strategy of transforming the unfamiliar into the familiar through the elaboration of a consensual universe of senses and meanings that allow us to make phenomena, objects, persons and events known in the social context in which we participate (Moscovici 1981, 1984). The emergence of new phenomena, behaviours and concepts requires social groups to engage with them, so that they are understood, named, meant and even transformed.

For Jodelet (1986) a social representation arises when we formulate words or expressions and associate them with the images that allow us to understand a previously unknown phenomenon or social behaviour. Hence, daily and historically we create new schemes of thought that are coupled with pre-existing ones through a process of interpretation and reconstruction of social 'objects'. Social representation is a form of knowledge socially elaborated and shared, with practical orientation focused on the construction of a common reality in the social set.

Social representations are, in themselves, a collective and communicational phenomenon, because they cover the social phenomena of signification, anchoring them in words and images inaugurating concepts. These collective formulations over realities and social phenomena form the historical and cultural baggage of human groups. It is the interaction and processes of social mediation that provide subjects with the capacity and the need to formulate concepts and, consequently, these representations guide the matrices of thoughts, attitudes and actions. The work of representing reality translates into a sociocognitive effort to understand it, to attribute meanings to the permanent relation between the externality of the (social) subject and its interior (I). According to Jodelet (1986), it is about the I-Other-Object relationship, which is elaborating modes of knowledge whose function is to facilitate social communication and guide behaviours.

In order to operate, social representation uses two processes: objectification and anchoring. Objectification is the translation of an abstract phenomenon into an image, into a concept. Anchoring is when this 'new' conceptual scheme is linked, anchored to a collective referential and the concept formulated becomes meaningful for a group, propagating itself in the social imagery.

One of the source ideas of Social Representations Theory was strongly influenced by the socio-interactionist paradigm in conceiving that the interiorization of psychological functions (elementary or complex) occurs through social interactions, with thinking and language being activities that modify the internal and external realities of the subjects. In 1925, Vygotsky clarified the concept of 'activity' as those changes of context in the subject and vice versa, the processes of mediation between the human being and their contexts, activity being the function that enables the development of cognitive abilities from socio-historical changes that we experience in society.

It is the relationships between subject/context that formulate the learning networks and processes of significations of knowledge, allowing the communication of individual and collective experience. In this sense, the processes of intelligibility and the elaboration of explanatory systems on reality are based on cultural polysemies.

In Vygotsky's work (1998), knowledge is presented as an element derived from different cognitions or 'inputs'. In presenting the Genetic Development Plan, influenced by his readings of Piaget, Vygotsky proposes four ways to think about the production of knowledge, namely: phylogenetic, from biological and ances-tral inheritances of the human species; ontogenetic, centred on the passages, stages that human development traverses throughout the life cycle; sociogenetic, anchored in the productions and historical-cultural interactions and of the situated knowledge; microgenetic, according to the psychological and existential experi-ences and livingness of each subject (intra-subjective).

However, the variations in the forms of knowledge raise the question as to which explanatory references should be used. Since the advent of modern sci-ence there has been a binary and hierarchical scale of organization of knowledge, whose referent describes development as superior/civilized/complex or inferi-or/primitive/simple, the first triad being the most highly valued (Jovchelovitch 2008).

The different types of learning, cognitive development and psychological functions (affection, language, memory, etc.) are integrated and interdependent systems in the internalization of social experiences. Different modes of knowl-edge depend on the context of their production and aim to respond to different objectives. Moreover, the multiple forms of knowledge do not appear in different groups or contexts, but coexist side by side in the same context, social group or person. People will make use of one form of knowledge, or another, depending on the particular circumstances in which they find themselves and the particular interests they possess at a given time and place. Cognitive polyphasia refers to a state in which different types of knowledge and rationalities live side by side in the same person or collective (Jovchelovitch 2008).

The theoretical proposition underlying the concept of cognitive polyphasia is the idea of knowledge that is formulated in the dynamics of intersubjective social interactions and cultural contexts. On that account, the act of knowing can only be understood if it is linked to the social relations from which its logic and rational-ity derive. Knowledge comes to be seen from its plasticity-elasticity, being able to present as many rationalities as required by the infinite variety of sociocultural situations that characterize the human experience.

Thus, people and groups will seek one way or another of knowing and learning according to the problems and demands of the social context and their configura-tions. These different forms of knowledge coexist and do not exclude each other. This coexistence contests the centric formulation in which complex forms replace simple forms. Instead of abandoning previously known primitive forms of knowl-edge, human communities continually seek the resources that different knowledge offers them and invent new cognitions.

The recognition of the coexistence of different rationalities in the same group of people not only problematizes the effects of the classical definitions of cogni-tion, but contributes to the extension of human understanding and thus leads to the production of a rationcination that, instead of denying, is able to communicate with its own differences.

At this point in the discussion, I will delineate a first approximation between Social Representations Theory and Psychology of Active Minorities: the conflict between science/rationality and common-sense knowledge/social knowledge, since beliefs, practical knowledge of quotidian, popular lore are associated with the absence of rationality or even irrationality.

The study of local knowledge and the deepening of situated epistemologies have been treated as a marginal theme. Marková (2006) was very interested in the study of 'common sense' knowledge and demonstrated that such knowledge often generates scientific distrust, is of low academic interest and is conceptually inferiorized. Nevertheless, this type of knowledge guides the human being in everyday situations and through life, being social knowledge and community experiences integrated with scientific knowledge.

When one thinks of the epistemologies of situated matter, one agrees that the different types of knowledge must be recognized for what they are, for their multiple meanings and functions, understood as legitimate, and for the differences they express. It is a vision in opposition to that of modern psychology: cerebral, of the mental individual, of the governance of conduct. In order to become aware of the situated epistemologies of social groups, it is necessary for them to speak about their beliefs, thoughts and actions, to go through ethnographic paths and to understand how these groups and people elaborate their cognitive and affective schemes regarding social reality.

Social knowledge is accompanied by a variety of cognitive goals, from the search for knowledge to the exercise of power. I believe that this knowledge of the social sense, besides being fundamental to scientific knowledge, should not be considered inferior or superior to academic knowledge, but different (Freire/Faundez 2002).

Moscovici demonstrated the importance of everyday knowledge in order to give visibility to the existence of the diversity of knowledge among the various social groups that compose societies within society. In addition to the interaction between different types of knowledge, the knowledge is transformed and permeates the imagery and cognitions of social groups beyond those who originally elaborated it.

## 4.3.2 The Process of Representing Oneself: Active Minorities, Influence and Sociopolitical Change

Social representations are possibilities of communication, processing and orientation of everyday relationships. In this sense, they weave the threads that stitch cognition, emotion and conduct together, with the intention of forming universes of opinion. Here I will focus on the second link between Social Representations Theory and the Psychology of Active Minorities: the process of representation is also the process of representing oneself or the group, and thus a process of social influence.

Being the social qualifier of representations – the producer and organizer of conduct and social communication – the material and symbolic context in which individuals and groups stand is interpreted – elaborated – through codes and values related to specific social positions. Therefore, the different representations of social groups are being shaped in the same sociopolitical context, in an arena of antinomies which come into alignment (and opposition).

Most of the materials from which social representations are elaborated are stored in the common cultural background of a society, through values that are considered fundamental, beliefs and widely socialized norms – in other words, cultural references that together form the collective memory. This sociocultural framework, whose ideology is partly primordial, guides the mentality of an era and provides the basic categories which generate the representations. At a more general level, the sources of determination of social representations meet at the set of economic, social and historical conditions that characterize society and the system of values and beliefs that circulate in it.

These beliefs spread through networks of interaction, through which cultural elements are combined and recombined once again, giving rise to specific representations. We can consider these shared sets of representations to be the ones which give a group its particular social identity, since living in a context of interpretation and a set of ideas establishes a link with the objects that affect us and the recognition of a collective identity.

Social representations elaborated around a particular reality emerge in a given context through the production of social meanings. It is a social, collective and interactive construction mediated by social, historical and cultural relations whose pillars are interpretation and social behaviours. The production of social meanings is a dialogical, intersubjective practice, a communicational phenomenon where language bases social and discursive practices right there where they are born and become effective: in everyday life (Spink 1999).

Discursive practices are languages in action so as to guide everyday practices. With this orientation, the discourse prescribes rules, linguistic regularities (aesthetic utterances, forms and content) and also confrontations between the numerous voices that meet in a social context or in a sociopolitical arena. It is precisely this tension between prescription and rupture that produces 'strangeness' in the speech and the moments of resignification of new meanings about the same reality.

Guareschi uses as an example of active minorities the case of the Landless Workers' Movement (MST)[3], the most expressive social movement in Latin America linked to conflicts over land and territorial disputes. Since its inception in 1985 during the period of Brazil's redemocratization, the Movement has pursued three main goals: to fight for land, to fight for agrarian reform, and to fight for social change in the country. According to the MST, its members want to be producers of food, culture and knowledge. And beyond that they want to be builders of a socially just, democratic, equitable and harmonious country (Movimento dos Trabalhadores Rurais Sem Terra 2018). Organized in twenty-four Brazilian states, MST has been making its way towards the development of sustainable agricultural production, from cultivation on unproductive lands.

---

[3]Movimento dos Trabalhadores Rurais Sem Terra. *Nossa História*, at: <http://www.mst.org.br/nossa-historia/84-86>

MST's activist performance – or, in terms of Moscovici (2011), the consistency of its style of action – deserves to be highlighted as an example of an active minority in the Brazilian political scenario. Since the movement emerged, it has adopted an internal logic of formulating discourses and actions. These actions are formulated and operated from a basic, collective and cooperative structure. All actions are discussed and planned in the community. Some behaviours are repeated, giving consistency to the activist praxis: the actions of formation and study of the references of struggle, with a Marxist perspective and Liberation Theology; the mystics,[4] which are actions with a strong symbolic and affective content that prepare the groups for collective actions; the marches occupying the main streets of Brazilian state capitals, followed by encampments in front of public power headquarters; the occupation of unproductive lands, followed by a period of encampment, needed for the elaboration of land use projects; and, finally, a period of negotiation with the National Institute for Colonization and Agrarian Reform (INCRA), which registers the families in the list of beneficiaries of the National Programme of Land Reform (PNRA), where the settlements of MST are located.

Guareschi always stresses that the behavioural styles of MST are consistent and, consequently, effective, since nothing has been done in governmental terms regarding agrarian reform in Brazil, except the policy of democratization of land ownership and food production for the internal market that has been developed and operated by the MST.

> Behaviour style is a new and rather familiar concept. It refers to the organization of behaviours and opinions, to the development and intensity of its expression; in a word, a 'rhetoric' of behaviour and opinion. Behaviours in themselves, like the sounds of a language taken individually, have no meaning of their own. Only combined according to the intentions of the individual or the issuing group, or according to the interpretation of those to whom they are addressed, might they have a meaning and arouse some reaction. The repetition of the same gesture or the same word may, in some cases, reflect an intransigence and rigidity; in others it might express the certainty. [...] What does all this imply? Simply that a whole series of behaviours offers two aspects: the first – instrumental – defines its object and provides information relative to that object; the other – symbolic – gives information about the state of the agent and source of behaviours – defines it (Moscovici 2011: 117).

The behavioural styles of MST were consolidated in the public sphere through a discursive and activist rhetoric in favour of agrarian reform and food production. The same set of public actions was repeated and consolidated over the years, combining a specific model of production, an economic model of income distribution among workers, and respect for the environment.

There is an argumentative substrate in the social representation that emerges as a way of understanding and representing an era. After twenty years of military dictatorship and periods of strong political repression, the MST destabilizes the

---

[4]Mystics are collective celebrations that precede the actions of formation and struggle. It is common for mystics to mix religious elements and symbols of struggle, such as a song or prayer and the MST flag, T-shirt or cap. It is an action of collective affective strengthening.

logic of representative political spheres in relation to agrarian reform. The MST questions the accumulated knowledge about the agrarian identities related to the latifundia of Brazilian colonial history and as forms of life and work in society.

According to Arendt (1994: 59), politics is the public space for primacy, the place of speaking and doing in benefit of plurality. The human being is political in its capacity for action, relationships, entrepreneurship and projection, for venturing into something new, and for its willingness to create and transform. The political arena is also the place of heteronomy production, a place of tension and dissent, and hence of political action where this ability to communicate and take part in political action puts antinomies and dissenters in confrontation.

The philosopher insisted on the meaning of public life as this dimension of different senses, that locus of social significance where it is important to be seen and heard from different angles. For Arendt, power emerges when people and groups unite and act together. This is the third point of confluence between Social Representations Theory and the Psychology of Active Minorities: the political dimension of representation, the ability to act as a result of of dissent and innovation.

I consider that politics has an aesthetic strength, based on the sentient world and the relations between people, institutions and systems. The aesthetic is a way of organizing politics in accordance with sensitivities and cognitions: to imply, to show, to construct the intelligibility of events (Rancière 2005). According to Moscovici (2011) it is the "rhetoric" of the political event and its intensity.

However, in the dominant social imagery of the most diverse Latin American societies politics seem to oscillate between legal representation and aesthetic representation, the first being the nuclear idea. Averse to this hegemonic thinking of assigning political activity to a dominant elite is the dimension of participatory politics, of anti-hegemonic political groups that struggle for social change.

The political community is more than the major political elite. This idea of power in conflict between minorities and majorities is clearly expressed by Moscovici (2011). For him minorities do not translate into quantitative terms, because in countries like Brazil, for example, minorities are the popular majorities. The term 'majority' indicates a place of political status, a class, an elite occupying places of legal-representative power.

For the author, minorities are subjects or groups that occupy the place of the border, the frontier, and minority representation introduces the new discourse, a new space of possibilities, power and action. Active minorities are not interested in taking the place of the centre, and generally operate at a certain distance from the channels of institutional politics. Examples are social movements, collegiate bodies, communtary, participative forums, etc. The social representations elaborated by active minorities produce strangeness and contestation.

Minorities resist and struggle when they recognize that something is not working well, that it is necessary to react, to formulate new patterns of confrontation. Therefore, minorities present in the public sphere a series of interpretations, arguments and behaviours that serve as a political instrument of manifestation of antagonisms and end up exerting social influence through new collaborators.

Here, once again I will echo Hannah Arendt when she says that the human condition is action, and it is impossible not to act. In this sense, to think about a Psychology of Minorities is to think about a Psychology of Dissent. The action of dissident groups is to inaugurate, disagree, make something appear in the public sphere for the first time, add to the world. Political dissent has this ability to produce new and multiple meanings. In the terms of Rancière (2005), politics will only be democratic if we encourage the multiplicity of manifestations within the community.

In the opposite sense of social conformity, deviant groups suggest new representations and social meanings of the phenomena. We have seen the example of the MST in the matter of agrarian reform, in the production of healthy foods, in the proposal of a sustainable agriculture in dialogue with the environment. Moscovici (2011) demonstrated the power of dissent from social groups that do not conform to norms. He explained that social influence will come about through the formulation of rhetoric and consistent behaviours that demonstrate political effectiveness. Dissent acts as a demand, an active minority litigation in motion. Thus, the dissident subject or group operates from a new representation, because it must say the word, take a stance in the political field.

The political field is a field of publicity, of antagonism and conflict, of opposition from groups in tension, be they institutional or social. It is a territorial arena, where there is a struggle of forces of emancipation versus forces of subordination. The public sphere accommodates these manifestations and relationships of tension, antagonism, plurality and diversity.

For Arendt, the public sphere is the space to reveal relationships of oppression and hostility, a place that always excludes people or themes, a place where irreconcilable and divergent values are debated. This locus of conflicting powers regulates social action – life in society – because in it circulates the ideological, symbolic, economic power and the strategies of influence of some groups over others. For Moscovici (2011), therefore, consensus requires levels of exclusion.

This idea of social influence and change profoundly questions the logic employed by Liberalism that inaugurates modernity and the false sensation of reasonable and rational consensus called democracy that seems to exclude no one. In social sciences this epistemological contribution of the liberal, individualistic and cognitivist logic was dominant in social and political psychology until the mid-1970s. In the North American theoretical contribution (Functionalism) politics were relegated to the rational and instrumental dimension and what was at stake was a dispute for resources and opportunities in the political arena. In this perspective, power was reduced to the mental capacity and action of isolated individuals and groups.

The epistemology of active minorities proposes a counterpoint to such a view. In deeply criticizing the Functionalist theory, Moscovici (2011) proposed to think of power from its relational dimension, considering the different forces and systems in dialogue. For him, minorities face the weight of dissent, of social nonconformity, of questioning the emerging norm, and from there they formulate new interpretations and representations of the world, committed to the extension of the participatory spaces. Dissident groups will propose an alternative interpretation to the one in place and will have the task of confronting leaders and their 'herds'.

The insurgency of active minorities is part of the history of humanity. Minorities account for the movements of insubordination to the historically determined, the effects of colonization and the institutionalization of knowledge and power. But democracy presupposes the equitable and balanced distribution of power, where representative and participatory forces complement each other. The distribution of the spaces of political power envisages the effective participation of society in its multiple and plural dimension in the elaboration of social policies.

It is interesting to analyse the idea put forward by Niederle and Grisa (2013) about the phenomenon of public action – the participation of the individuals, groups and sectors involved in public policy – which is the perspective of a democratic constitutional state.

> In opposition to the classic model of public policy conceived by a centralized state, acting on well-defined and delimited sectors, there is a growing analysis that seeks to approach the mechanisms of "public action", a definition that accentuates the set of interactions between several actors that participate in the construction, implementation, monitoring and evaluation of public policies at their most varied levels. (Niederle/Grisa 2013: 98).

It is not a question of reducing the role of the State, but of considering it as one of the instances of political interface, since the ontological triad proposed by Claus Offe (1999) – state, market and society – is less and less defined. Participating in the implementation, monitoring and evaluation of public policies is, in fact, participating in politics without encountering rough bureaucracies, with the possibility of reformulating, adjusting or expanding initiatives.

Nowadays in Brazil there is an expansion of social consultation mechanisms, such as councils, collegiate groups, and participatory forums, where resource disputes involve more and more struggles for legitimation and recognition (Honneth 2003). These struggles highlight new ideas, representations and values that challenge established institutions and require the formation of new commitments to guide action by the state and civil society. In addition, in light of this context, new challenges arise to the analysis of politics, provoking the construction of innovative approaches that integrate the different components of public action.

This enlargement of the instances of participation and social representation is due to a renewed political context, a left-centre political management that, in the last three governmental administrations, increased access to education, social assistance, democratization of admission to public universities, expanded the Brazilian middle class, and, consequently, strengthened the affirmation of collective identities. The political process was strengthened through possibilities of articulation, action and reflection. In this respect, Moscovici (2011) states:

> We can begin, for example, by distinguishing between 'top' innovation and 'bottom' innovation. The first comprises the changes introduced by the leaders, that is, by the people who have the necessary authority to impose new behaviours on their supporters or to persuade them to accept deviant behaviour […] the bottom process of change and innovation results from action of a minority that does not have a privileged status (Moscovici 2011: 182).

Nevertheless, the delimitation of the boundary between State and society emerged with brute force when the country watched in astonishment the impeachment of President Dilma Rousseff in 2016 by the National Congress and, on the eve of the new electoral period, the conviction[5] of Luís Inácio Lula da Silva in January 2018. In both cases, many jurists contested the delegations and accusations against the President and the candidate, alerting them to the articulation of a well-executed political coup between the Chamber of Deputies, the National Congress and the Judicial System.

A detailed analysis of the conflict involves, first of all, building a map, a kind of cartography of the conflict: revisiting emerging political events (beyond the media facts), which marked the lives of the social groups involved, analysing the scenarios, the spaces where the events unfolded, the social actors (people, groups, institutions involved), the relations of forces. Second, analysing the conjuncture (events, scenarios, actors) in relation to the macro-structural plan (the national historical context and its economic relations), raising the major political issues of the moment, the institutional and social forces that are directly involved in the big questions, the actors (people, leaders, social groups) that represent these forces.

In this scenario we had, on the one hand, the Workers' Party (PT), the trade union bases and social movements fighting for the expansion of democratic rights and spaces, and, on the other, the National Congress, the ruralist party, the evangelical benches and the extreme right parties overthrowing social and popular policies in favour of a minimal state whose economy is based on agribusiness, state privatization and the resumption of morals and good manners touted by the logic of the traditional family. The criminalization of social movements has been found to be strong in the states and municipalities, which weakens public mobilization by propagating the social imaginary that the Brazilians do not like, do not understand and do not get involved with politics.

The conflict always situates its forces, its antagonistic fields in opposition, for it is a tenacious rivalry between divergent fields. In the sphere of ethics, orientated to reflect on conflict resolution, the first stage would be the exposition of the antinomies, where the groups expose their ideas in opposition, and the dialogicity of the conflict, where they use discursive strategies. The second would be the instance of negotiation, where proposals are presented, a stage in which there is hardly any consensus, but tacit agreements on both sides. When the negotiation is not effective, the legal process of the conflict occurs or the closure of the same by the usurpation of rights, the use of violence or criminalization.

---

[5]The work *Comments of an announced sentence: the Lula process* recently launched in Brazil (Proner/ Cittadino/Ricobom/Dornelles 2017) brings together over 120 analysts from different areas of knowledge who analyse the intricacies of Lula da Silva's conviction process, based on their controversies and inconsistencies, including the invention of accusatory evidence designed to convict Lula for his 'populist' politics and prevent him from running for election in 2018, for being the top candidate for the presidency of the country.

For Arendt (2007), power ends where violence begins, because negotiating power, in political terms, means avoiding violence. It is a clash between opponents, typical of a plural society, where the human condition is the condition of action, the public sphere, and contributing to society.

## 4.4 Possible Horizons in an Arena of Challenges

Throughout his work, Moscovici pointed out that social psychology had moved away from the field of mass psychology, political activism, and the study of social crisis on account of considering them to be 'popular' rather than scientific subjects. He also pointed out that this disregard was dangerous, since it separated psychology from its social function, from thinking of the strange and the processes of rupture as producers of social change, and ignored their transformative political potential. There is an argument surounding social psychology from Vygotsky to Martín-Baró: the need to look at what happens around individuals in order to understand what they do and, not infrequently, what they think and feel.

In this chapter I presented the state of studies and research on social representations in Brazil and discuss some theoretical and conceptual operators in relation to Social Representations Theory, the psychology of active minorities, and, in particular, the dimensions of power, conflict and social influence in an approach to sociological social psychology.

Initially I made a point of analysing the modern impacts of the Functionalist theory on the social sciences, which reduced the analysis of conflict to a sociological variable that should be controlled or eliminated in favour of social conformity. For years, this tendency obstructed the analysis of the conflict, demands and protests of the sociological field as objective dimensions of social change.

However, renewed theoretical impulses made this conception turn into an approach that began to analyse conflict as a salutary aspect in democratic societies, and started to understand the cultural elements: the psychological and symbolic processes of daily encounters and exchanges of knowledge about the realities. Conflict began to be evidenced as a network of powers that intersect in social relations, involving opposing groups, collectivities and human institutions. In this case, politics – as an activity – comes to be seen as an active dimension of social groups, beyond the State and its constitutional arrangements of law, policing, and administration.

I therefore made a point of drawing attention to the work of Moscovici (2011) as an important sociological contribution that categorizes the episteme of social minorities as a form of knowledge capable of producing strangeness, problematizing, innovating, and changing current flows and orders through the recognition of differences and placement of conflict and new styles of behaviour. According to Bhaba (1998: 23–24):

> [...] the broadest significance of the postmodern condition lies in the awareness that the epistemological limits of those ethnocentric ideas are also the enunciative frontiers of

other dissonant voices and stories, even dissidents, minority groups. It is in this sense that the frontier becomes the place from which something begins to be present in a movement not dissimilar to that of the ambulant, ambivalent articulation, from beyond.

Throughout the text I have highlighted relevant aspects of active minorities and their powers of representation and social influence. I emphasized power-knowledge as a human quality scattered capillarily in society: power as a capacity of action, to act in the public sphere, and power as a political field, as a stage of disputes between correlations of forces. In the latter case, I mentioned the power of civil society and active minorities as a dissident and articulating power, capable of instituting structural and symbolic changes, based on interpretative analytical work that operates from the margins, unlike the governmental or majority elites in terms of political status that conceive the State solely as a direct dominion or command, i.e. regard the State as legal government.

In this regard, I would like to reiterate that the State must be considered the public sphere itself, as a locus where civil society operates through persuasion, through social influence devices, triggering conflicts that have the prospect of change. It is in the public sphere that channels of speech, persuasion and action are created. There, in the field of tensions and conflicts, power will be questioned and transformed. The State is then an arena of disputes and negotiation, where political reforms must go through the exercise of participation, through protest, negotiation and dissent. When these rights are denied, a state of exception is established, often backed up by the legal authorities.

I have pointed out that the power of active minorities is the power to represent and influence and it is effective when they succeed in triggering, questioning, and challenging the power orchestrated from top to bottom by representative-institutional elites without effective popular participation. Thus, a very important dimension of conflict analysis is the power of negotiation, when opposing sides bring into the political arena their ideas, strategies and dissent, bargaining for possible, if temporary, agreements. It is not a question of entering into consensus, but of deliberating and negotiating agreements. When negotiation does not take place and the legitimacy of litigation is overcome, the political dimension is diminished and violated.

The permeability points between the two classic theories of the French current of social representations constitute analytical indicators for those who intend to set foot further in the field of social conflicts and disputes in the political arena. These indicators give an account of the psycho-sociological aspects of the objective and subjective conditions of political phenomena in macro and micro-social terms, of the historical dimension of the institutional political organization, and of the political behaviour of a society.

Nevertheless, it is still a challenge in the Brazilian scene to continue studying the relationship between the two theories formulated by Moscovici. Regarding the psychology of active minorities, the country lacks research that places political elites, parties and personalities on show and also analyses the relationships that elites establish with social minorities in action. The time has come to investigate the correlations of forces established in the political arena in a country in the midst of a political conflict that is undergoing an administrative dictatorship disguised as democracy.

There is a huge controversy between the structural and epistemic approaches to active minorities in Brazil. While the idea that the minority viewpoint is considered a sign of sanity and social innovation is already familiar in the field of political psychology, this awareness is not shared by governmental elites. We are currently experiencing a period of strong social repression, where minorities, social movements and union foundations are considered a factor of social insanity and are strongly criminalized. Thus, studying the elites and their political behaviour in relation to social minorities is an urgent task for social and political psychology researchers in Brazil. It is also vital to study in depth the interactions between the various actors and sectors that participate in the implementation, monitoring and evaluation of public policies, as these dimensions of the governance of public action are so important to the political life of a country and the guarantee of democracy.

If Latin American social psychology is genuinely concerned about the extension of democratic spaces and values and with the deepening of political practices, it must abandon its attempt at neutrality and commit itself to understanding, denouncing and confronting the social problems that affect societies. According to Martín-Baró (1986: 19), "Latin American psychology must decentralize its attention to itself, take care of its scientific and social status and propose an effective service to the needs of the popular majorities." The perspective has to be horizontal and participatory, and include the groups that make up the popular majorities.

Both intellectual work and the production of knowledge and social technologies must operate as instruments of denunciation and intervention. The relevance of the complaint lies in the possibility of retrieving and making visible historical facts, social events and cultural antinomies from different perspectives. The production of knowledge becomes intervention insofar as it favours the engendering of social practices which, through the capacity of reflection, criticism and action, can revisit and understand the past, re-signifying it in the present, in order not to passively repeat it.

# References

Arendt, Hannah, 2007: *A condição humana* (Rio de Janeiro: Forense Universitária).

Arruda, Angela; Sá, Celso Pereira, 2000: "O estudo das representações sociais no Brasil", in: *Revista de Ciências Humanas*, special issue: 11–31, at: <> (19 May 2020).

Bahba, Homi, 1998: *O local da cultura* (Belo Horizonte: Editora UFMG).

Freire, Paulo; Faundez, Antônio, 2002: *Por uma pedagogia da pergunta* (Rio de Janeiro: Paz e Terra).

Hernandez, Aline Reis Calvo; Accorssi, Aline; Guareschi, Pedrinho, 2013: "Psicologia das minorias ativas: por uma psicologia política dissidente", in: *Revista Psicologia Política*, 13,27: 383–387, at: https://dialnet.unirioja.es/servlet/articulo?codigo=7431659 (19 May 2020).

Honneth, Axel, 2003: *Luta por reconhecimento: a gramática moral dos conceitos sociais* (São Paulo: Ed. 34).

Jodelet, Denise, 1986: "La representación social: fenomenos, concepto y teoría", in: Moscovici, Serge (Ed.): *Psicologia Social II* (Barcelona: Paidos): 478–494.

Jovchelovitch, Sandra, 2008: *Os contextos do saber: representações, comunidade e cultura* (Petrópolis: Vozes).

Martín-Baró, Ignacio, 1986: *Hacia una psicología de la liberación* (El Salvador: UCA Editores).

Martins, Alberto Mesaque; Carvalho, Cristiene Adriana da Silva; Antunes-Rocha, Maria Isabel, 2014: "Pesquisa em representações sociais no Brasil: cartografia dos grupos registrados no CNPq", in: *Psicologia: teoria e prática*, 16,1: 104–114, at: https://www.redalyc.org/pdf/1938/193830151009.pdf (27 April 2021).

Moscovici, Serge, 1979: *El psicoanalisis, su imagen y su publico* (Bueno Aires: Huemul).

Moscovici, Serge, 1981: *Psicología de las minorías activas* (Madrid: Morata).

Moscovici, Serge, 1984: "The phenomenon of social representations", in: Farr, Rob; Moscovici, Serge (Eds.): *Social representations* (London: Cambridge University Press).

Moscovici, Serge, 2003: *Representações sociais: investigações em psicologia social* (Rio de Janeiro: Vozes).

Moscovici, Serge, 2011: *Psicologia das minorias ativas* (Petrópolis, RJ: Vozes).

Niederle, Paulo André; Grisa, Cátia, 2013: "Ideias e valores: a análise da ação pública a partir das interfaces entre a abordagem cognitiva e a economia das convenções", in: *Revista Política e Sociedade*, 12,23 (January/April): 97–136.

Offe, Claus, 1999: "A atual transição histórica e algumas opções básicas para as instituições da sociedade", in: Bresser Pereira, Luis Carlos; Wilheim, Jorge; Sola, Lourdes (Eds.): *Sociedade e Estado em transformação* (São Paulo/Brasília: Editora Unesp/Enap): 119–145.

Proner, Carol; Cittadino, Gisele; Ricobom, Gisele; Dornelles, Joao Ricardo, 2017: *Comentários a uma sentença anunciada: o processo Lula* (Bauru: Canal 6).

Rancière, Jacques, 2005: *A partilha do sensível: estética e política* (Rio de Janeiro: *Editora* 34).

Spink, Marie Jane, 1999: *Práticas discursivas e a produção de sentidos no cotidiano: aproximações teóricas e metodológicas* (São Paulo: Cortez).

Vygotsky, Lev, 1998: *A formação social da mente* (*São Paulo:* Ed. Martins Fonte).

## Other Literature

Movimento dos Trabalhadores Rurais Sem Terra: *Nossa História*, at: http://www.mst.org.br/nossa-historia/84-86.

# Chapter 5
# Relationships between Beliefs and Social Representations: A Brief Theoretical Reflection

Renata Lira dos Santos Aléssio

## Introductory Comment

Maria de Fátima de Souza Santos

*The Laboratory of Human Social Interaction (LabInt) is[1] a research group registered in CNPq that has included the field of social representations and practices among its lines of research since 1990. According to Carvalho/Antunes-Rocha (2014), the two oldest research groups that include Social Representations Theory as one of their lines of research are in the psychology research field: the "Ideology, Communications and Social Representation" group (UFGRS), created in 1987 and led by Pedrinho Guareschi, and the "Laboratory of Human Social Interaction" (UFPE), created in 1990, led by Maria Isabel Pedrosa and Maria de Fátima S. Santos, with the social representations and practices line of research.*

*Since the beginning, the LabInt has also been a formation centre for researchers. Undergraduate, masters and doctoral students have a place for continuous collaborative discussion and research development. There are weekly meetings to discuss theoretical and methodological aspects of Social Representations Theory*

---

Renata Lira dos Santos Aléssio is an Associate Professor in the Department of psychology at the Federal University of Pernambuco (UFPE). Email: renatalir@gmail.com.

---

[1]Professor Maria de Fátima de Souza Santos is a Senior Researcher in the Department of Psychology at the Federal University of Pernambuco. She studied her undergraduate degree in psychology at the Federal University of Pernambuco and her PhD in psychology at the Université Toulouse le Mirail. She conducts research in the areas of social psychology and developmental psychology, with the emphasis on social representation, working mainly on the following topics: Social Representations Theory, violence, adolescence, old age, health and social practices. She currently coordinates the Center for Studies in Social Representations Serge Moscovici. Email: <mfsantos@ufpe.br>.

*and its relationship with other social psychology theories or to discuss ongoing research, whether at the undergraduate, masters or doctoral levels or even the advisors' research. Therefore, this environment stimulates the researcher's forma-tion as well as being a rich space of social interaction regulated by affection and respect.*

*It was in this collaborative environment that Renata Lira dos Santos Aléssio emerged as a researcher. Initially as an undergraduate psychology student, Renata Aléssio stood out in the social psychology class by proposing, alongside a few of her classmates, a study with the Landless Workers' Movement (MST, Movimento dos Sem Terra) using Social Representations Theory. In 2001, with a scholarship funded by CNPq for Scientific Initiation Students, she developed, in the LabInt, a subproject entitled "Childhood and violence in the field: social representa-tions and educational practices in rural settlements". This project resulted from a CAPES/PROACAD programme on human development and violence that we had developed in association with the University of Brasilia (coordinated nationwide by Angela Almeida) and the Federal University of Espírito Santo (local coordina-tion by Zeidi Trindade). It resulted in a working partnership and friendship that lasts to this day.*

*During the entire undergraduate course, Renata Aléssio received a scholarship from the scientific initiation programme and worked on different research pro-jects with social representations as a theoretical basis. At the end of her under-graduate course, still under my guidance, she presented a monograph on the social representation of violence in Cordel's literature on Cangaço, obtaining, in 2004, second place in the Dante Moreira Leite Award – Psychology and Social Communication: Production of Subjects, Subjectivity and Cultural Identities, pro-moted by the Federal Council of Psychology.*

*At that moment, the LabInt was participating in a CAPES-COFECUB project, which involved the University of Brasilia, the Federal University of Pernambuco, the Catholic University of Goiás and the University of Aix-en-Provence (the Laboratory of Social Psychology directed at that point by Jean-Claude Abric). By the end her undergraduate course, resulting from achievements during her forma-tion, Renata Aléssio went to the University of Aix-en-Provence where she obtained her Master's degree (receiving the Alban scholarship) and her doctorate (funded by CAPES), under the supervision of Dr Thémis Apostolidis. She continued her work in the area of Social Representations Theory, turning her attention to sci-entific dissemination, health and bioethics matters. Her subject of study began to integrate scientific dissemination, particularly the common-sense knowledge that is produced on embryo research. In her thesis, she developed five studies that not only showed the public clash over embryo research, but also "clarify the social and symbolic dynamics that sustained the research subjects' decision making", as noted by Jodelet (2011: 25). Her thesis opened the way for the discussion of a sensitive subject – embryo research – and the theoretical discussion of the rela-tionship between content and process, beliefs and social representations, besides being a great example of methodological triangulation that allows us to deepen and validate the results of each piece of research.*

*On her return to Brazil, the UFPE psychology postgraduate studies programme (PPGPSI) requested a Regional Scientific Development (DCR)-FACEPE/CNPq scholarship for her. By receiving the scholarship, she was inserted in the PPGPSI, under my supervision, and developed research, taught classes and advised master's students. However, her institutional commitment also led her to engage in the undergraduate psychology course. In the same area that she had been working on her doctorate degree, she developed the project 'Biotechnologies and the Human Embryo' using a psychosocial approach with Social Representations Theory. After a period as a DCR scholarship student, she enrolled into tender at UFPE and was approved, which allowed the continuity of her research, teaching and community intervention work, which uses Social Representations Theory for technical support.*

*Currently, Renata Aléssio is focusing on the relationship between Social Representations Theory and other theories and/or social psychology concepts. Her work continues that done in the LabInt, bringing, however, a new perspective on the theory, on the research methods and the subjects to be investigated. Her chapter 'Relationships between beliefs and social representations: a brief theoretical reflection', presented in this book is an example of this theoretical search to clarify concepts and relationships between concepts in psychology and in other areas, demonstrating her openness to knowledge, the necessary curiosity and courage not to settle for a theory that is already familiar, which are fundamental characteristics in a researcher. I consider Renata Lira dos Santos Aléssio one of the best examples of successful educational policy, as the country invested in her via scholarships from her undergraduate studies to final training as a postdoctoral student, in public universities and resulting in a serious researcher, committed to the institution and to the production of knowledge in her area.*

*Her work allows us to reflect, from a theoretical perspective, on the relationship between beliefs and knowledge about new controversial themes in today's society to prompt us to think about the conceptions of the epistemic subject underlying the different theoretical conceptions. From discussion on topics which currently include beliefs in the Flat Earth and in the 'weakness' of maternal milk, and disbelief in evolution theory, she leads us to reflect on the epistemological perspective that makes it possible to rethink the notion of object in the relationship between representations and practices, and raises questions that may lead to Latin America's own research agenda in the field of Social Representations Theory.*

# References

Jodelet, Denise, 2011, "Ponto de vista: sobre o movimento das representações sociais na comunidade científica brasileira", in: *Temas em Psicologia*, 19,1: 19–26.

Martins, Alberto Mesaque; Carvalho, Cristiene Adriana da Silva; Antunes-Rocha, Maria Isabel, 2014: "Pesquisa em representações sociais no Brasil: cartografia dos grupos registrados no CNPq", in: *Psicologia: Teoria e Prática*, 16,1: 104-114, at: https://www.redalyc.org/pdf/1938/193830151009.pdf (19 May 2020).

**Abstract** It is common in studies regarding social representations to think of beliefs as a class of elements which exist in representational contents of the most diverse objects. This chapter will provide, in an abridged form, Moscovici's conceptual ideas regarding the relationship between beliefs and representations. The next part presents two different theoretical perspectives that encompass that relationship. The conclusion suggests avenues to be explored in the context of the development of Social Representations Theory in Latin America.

**Keywords** Beliefs · Common sense · Science · Doubt

## 5.1   Introductory Elements for Reflection

Moscovici (2003) stated that social psychology is a science dedicated to study of the conflict between individuals and society, but that nevertheless it stands out more for the way it analyses phenomena than for a specific domain or area of study. The psychosocial approach, according to Moscovici, is characterized by a three-sided relationship, alter – object – ego, subverting the division between objective/subjective or rather individual/collective which was a hallmark of social psychology of the 1960s and 70s. Representational theory had at that time the virtue of transforming social psychology into a sort of anthropology of contemporary life, particularly since the construction of representational phenomena is one of the specificities in today's societies, typically marked by a plurality of ways of life, by the speed of communications and by the development of science (Jodelet 2001). The concept of social representations would then function as an important tool for analysing the relationship between psychology and anthropology because of its capacity to overcome existing contradictions surrounding the notion of culture in social sciences, and more specifically the supposed rationality attached to cultures and the irrationality of the epistemic subject (Jodelet 2002a).

In the context of contemporary societies, acting in a rational way means basing ideas, actions, and daily practices on scientific knowledge (Moscovici 1995). Science should bring light to dark alleys, eradicate diseases, defeat ignorance, and triumph full of reason. However, we realize quite perplexed how human beings nurture 'absurd', 'magical' and 'superstitious' beliefs, scientific and technological developments notwithstanding. We see in the twenty-first century movements like the Flat Earth Society, composed of people who don't believe the Earth is spherical, but flat. Also, there are people who don't believe that man has walked on the moon or still believe in creationism against evolution. Following Moscovici (2012: 22), we could ask: "How can human beings, endowed with rationality […], be misled by ideas which experience and reason have already falsified?"

This resistance to science is not confined to ordinary people. Researchers and those who possess high levels of scientific knowledge are sometimes sceptical about global warming and its anthropic causes, despite the efforts on the part of the scientific community to prove that climatic changes caused by human beings have harmful effects and could be mitigated by urgent political measures. Kahan (2015) showed, in a sample of Americans, that belief in the idea of global warming linked to fossil fuels is stronger in those with a higher level of general training in science. If the sample is cross-referenced against political views, an interesting phenomenon can be observed. For the group of Liberals and Democrats, the same previous effect is observed: the higher the scientific training someone has, the more they agree with the anthropic hypothesis for global warming. However, among Conservatives and Republicans, the contrary is observed: the higher the scientific training someone has, the less they agree with this hypothesis. Can't science make them change their views?

The expected conclusion that people would develop favourable attitudes towards particular "objects" (beliefs, situations or things) because they have 'correct' information had already been challenged by Moscovici (1976) when he studied representations in psychoanalysis. Kahan (2015) states that regarding certain things which are important to the social groups to which people belong, accepting science-based evidence calls into question their own familiar, religious and political identities. In other words, there's a high risk of 'exclusion', and in view of that, the group's beliefs prevail. This is a real example of how social anchoring has an influence on the process of constructing and maintaining social representations: it makes the object a continuum of pre-existing ideas. Regarding objects whose origin can be explained by science, that anchoring activates a network of values, beliefs and norms about what is socially accepted (Joffe 2002; Wagner/Kronberger 2002).

Another example of an idea which is accepted as true despite scientific evidence to the contrary is the belief in 'weak' breast milk. Shared conceptions about how the baby sleeps and cries support that belief. The 'proof' that breast milk is of poor quality comes from the perception that the baby doesn't sleep enough or cries too much. It is a persistent belief in many communities, despite campaigns designed to show that breast milk can nurture a baby perfectly well. That belief is among the most prevalent reasons for the replacement of breast milk with cow's milk, the latter often considered stronger and more nurturing than the former, even though some human babies are allergic to it (Marques/Cotta/Priore 2011). Are women who believe that human breast milk is weak incapable of grasping scientific knowledge? Can't they understand the information provided in campaigns?

The examples presented show the existing duality in western culture between knowledge and beliefs. According to Moscovici both categories are not only different one from another, but are also exclusive and opposed; they represent two very distinct ways of thinking about human nature. So this is one of the contrasts in the debate concerning beliefs in social psychology: rationality of knowledge versus irrationality of beliefs. Regarding Social Representations Theory, Moscovici (1992) rejects a hierarchy of knowledge, showing that knowledge and belief are manifestations of different arrangements of the same mental tools, each in accordance with the context and functionally orientated specificities.

According to Jovchelovitch (2002), the idea of the irreducibility of social representations as a way to discover their functions in social life was inspired by Lévy-Bruhl's thought about the discontinuity of logical thinking and the coexistence of different rationalities. It is in that context that Moscovici advocates the idea of "mental forms": religion, science, magic, and ideology. Each mental form generates valid knowledge from available information. They differ with regard to (1) how different combinations of the same cognitive operations are employed (classification, comparison, implication, for instance) and (2) the value scale linked to the social hierarchy of each these forms.

According to Moscovici (2003: 348), "knowledge and beliefs are opposed but paired concepts, such as reason and faith. They may have the same content, but different qualities." Progress in science, considered central to modernity, is used to illustrate such a relationship. Progress in knowledge is subject to constant doubt and falsification in search of validating arguments. Progress as a belief is related to the trust or scepticism we have as human beings. Regarding knowledge, it is reasoning that matters – unlike beliefs, in which opposed images are juxtaposed. Moscovici (2003: 349) states:

> I wonder if we can really understand individuals' and groups' mental attitudes, if we disregard the hybrid mix between faith and knowledge; that is, what we consider true for believing in and what we believe in for considering it true.

Where the author approaches "post-scientific" common sense, marked by a coexistence of rationality and beliefs, we observe the growth of "a new magical thought", for "the times when science flourishes are also the times when flourishes magic" (Moscovici 1992: 312). We should stress that all through his work, Moscovici states that scientific thought never replaces magical thought. Magic is defined as a power to mould the world according to one's will. In this sense, among the best studied forms of "magical thought" stand either the causal attribution or information processing. From this viewpoint, human beings are seen as cognitively limited and incapable of reasoning properly about the information around them or about the observation they make about the world around them. An instance of this is "fundamental attribution error", a perspective in which human beings tend to explain an event from a personal point of view or from dispositional causes while disregarding environmental elements that could explain it.

As an example of that, someone might explain John's difficulties in getting a job by saying he is not lucky with interviews, or doesn't make the best of himself, or is incapable of searching for a job, ignoring the fact that economic crises may happen from time to time. The tendency to personalization is characteristic of liberal societies in which individuals stand as the foundation and individualism is the dominant ideology, expressing a "cultural arrangement" (Moscovici 1986). This way of processing information and assigning causes, these 'mistakes' or 'biases' aren't a failure of reasoning, but part and parcel of a coherent social life, connected to "previous attachments towards a conceptual system, an ideology, an ontology or even a point of view" (Moscovici/Hewstone 2005: 556).

For Kalampalikis (2006), there is a paradox in modernity, which is to consider humans to be rational beings, though they engage in different forms of irrationality in a society where science prevails. The author echoes Kakfa when he says logic is left aside by a man willing to live. He also reminds us that the verb 'believe' has several meanings, depending on the idiom employed. In French or Portuguese, for instance, we can believe in (trust in; affirm something exists) or believe that (represent something). There is a huge conceptual diversity in the field of social psychology, as well as in the field of social representations. Flament and Rouquette (2003: 19) affirm that beliefs

> constitute one of the opaquest notions when applied to the research of social thinking. They seem to be a limit for behaviour evaluation, as they are circumscribed by its content and by the cognitive process that leads to that content: obvious to those who share them, they are preposterous to others, and no argumentation can be construed between both groups.

This classical opposition – belief/ knowledge – is based on two different conceptions of the relationship between beliefs and representation. The first approach, inspired by Moscovici's conception of the three-sided psychosocial perspective (ego-object-alter), points to a division between knowledge-based representations and belief-based representations, while admitting, however, that beliefs and knowledge can coexist as representational contents. That is the perspective proposed by Marková (2006) in her book *Dialogicality and Social Representations*. The second approach considers beliefs to be a form of knowledge. I mention here, among others, the viewpoints of Denise Jodelet (2002b), Thémis Apostolidis (2002), Nikos Kalampalikis (2006), and Sandra Jovchelovitch (2008). Both approaches are be summarized below.

## 5.2 Belief-based Representations, or rather Knowledge-based Representations?

Marková (2006) proposes two different relationships between beliefs and representations, according to Moscovici's three-sided approach. These relationships emphasize either the Ego/Alter relationship or the Ego/Object relationship. According to the author (2006: 230):

> beliefs can be defined as mental states in a considerable length of time. Beliefs are generally rooted on culture, tradition, and language. They are marked by a firm and rigid conviction and are very often soaked in passion. Beliefs are socially rooted, so the object's fixation originates from Alter instead of the Object itself. That means that the believer does not seek proofs or evidence regarding the object.

According to this viewpoint, the object rests in the 'background' of the three-sided relationship and serves as a way for the beliefs' actualization 'without an object'. By way of illustration, Marková writes about beliefs brought about by the emergence of AIDS, such as the risk of contagion from handshaking or AIDS as a punishment. These beliefs are not aimed at a specific object; rather, they are available in our cultural structure. It should be stressed that illnesses that are believed to be a punishment for sin (moral responsibility) are those included in identity exchanges in a radical alterity – the other seen as strange and inferior regarding their own group (Jodelet 2005). In this sense, the object can't be just a background. The object is a unique entity construed in a relationship with Alter, that can be menacing, for instance.

While belief-based representations stand between Alter and Ego, the knowledge-based representations stand between Ego and the Object. The common-sense knowledge is based on experience, talks (first-hand knowledge, according to Moscovici and Hewstone [2005]) or else has its source in the transformation of scientific knowledge into common sense (second-hand knowledge). Knowledge, according to Markovà (2006: 232), "is the examination of the phenomenon's nature in the most complete way possible, and independently of other people". One who knows, as opposed to the believer, is ready to argue, prove, and criticize. The content of belief-based representations and knowledge-based representations may be the same; the difference lies in the way 'truth' is searched for. Belief-based representations arise from consensus. Knowledge-based representations are shared by groups that are in search of evidence. For Markovà, beliefs are like habits, while knowledge is an ability.

In her opinion, representations are almost never based exclusively either on beliefs or knowledge, but it is possible to see how one prevails over the other. Regarding social practices, belief-based representations would be more resilient to changes and could produce even more social categorization and exclusion, while knowledge-based representations would be paramount for the reduction or eradication of discrimination and exclusion. I would especially like to discuss the idea that beliefs produce discrimination, the meaning of which is very similar to the meaning of prejudice, as well as the notion of discrimination as a manifestation of cognitive biases. According to this perspective, prejudice is the expression of mistaken, hurried, biased, and incomplete judgements about someone, resulting from 'mistaken' or 'false' beliefs. Regarding the AIDS example mentioned before, beliefs about punishment would not be based on discrimination or prejudice. Rather, they would be 'false beliefs', as a product of categorization and social differentiation, implicit in a view of homosexuals as a radical alterity. These beliefs revisit a hierarchical power relationship in a normative context in which homosexuality was considered a pathological condition.

In general, regarding this perspective which separates belief-based representations and social-representations based on knowledge, beliefs are seen as 'flawed' or biased, close to cognitive entities evaluated according to criteria required by scientific knowledge, by a probabilistic reasoning model, as Apostolidis (2002) states. The other perspective, on the contrary, sees beliefs as a form of knowledge.

## 5.3  Beliefs are a Form of Knowledge

According to Jodelet (2002b), both beliefs and representations belong to social thinking. These two forms possess their own logic, are collectively elaborated, and reveal psychological and social functions. One function common to both beliefs and social representations is their expressive function: people who share a belief reveal something about themselves or about the group they belong to. Beliefs are not monolithic formations; they can be divided into religious, moral, magic, and be based on scientific authority. For Jodelet (1986), the very notion of representation presupposes an accession and participation process which makes it resemble beliefs.

For her, the notion of belief is widely shared among different cultures, in which roots lies the idea of economy (2002b). Not by accident, the financial system employs expressions such as 'having credit', 'credibility', etc. Jodelet (2002b) notes that beliefs imply relationships of negotiations, accession, trust, and power. In this conception, believing in what we know becomes necessary to act in the world. She cites as an example the 9/11 terrorist attacks. After the attacks, the CIA (Central Intelligence Agency) commander was summoned to explain why American authorities were not able to prevent those attacks in the face of so much information. He answered: "We knew it was possible. But we didn't believe in what we knew." Regarding truth, the belief's dynamics are connected to faith, and are marked by the phenomenon of accession or consent. In this perspective, god was created by the believer, and the believer ascribes power to what they believe. For that reason, beliefs are the manifestation of the dynamics of the power behind social thinking: if beliefs do not rule the knowledge of reality, the belief only can generate reality into knowledge.

Jodelet (2002b: 162) defines belief, affirming that "the specificity of the notion is linked to cultural, symbolic, and economic aspects in social exchanges, and to the dimensions of trust, love, power, influence and above all the notion of faith, which is an important accession factor". Beliefs are alive, colourful and based on participation and social commitment, being important for the form of knowledge they result in (Moscovici 2001). This conception is far from the Anglo-Saxon idea of attribution biases, information processing, and is strongly limited to an intraindividual reach, a private world-view that does not bear any relation to the Social Representations Theory approach. The research on social representation "concerns a thought modality of its constitutive aspect – the processes – and its constituted aspect – the products or contents" (Jodelet 2001: 20). The processes seem to be linked to more universal phenomena, while the contents are linked to social contexts, and, for that reason, they are specific to a given culture.

In research concerning contents, the analysis should be concentrated on how they evolved socially, how they are organized "in terms of meaning's mediation and work about references" (Moscovici/Vignaux 1994: 29). From the point of view of representational processes, we should study the role of "social factors in the formation and functioning of common-sense knowledge, and identify the systems of interpretation and collective thinking", leaving behind models of the "lonely thinker" kind, especially because thinking is a social matter, a human activity linked to communication, to traditions, and to socially validated procedures to solve ad hoc problems. Individuals do not choose beliefs according to criteria of truth and falsehood, but because beliefs furnish them with criteria for the construction of the social reality (Moscovici 1992).

According to Kalampalikis (2006: 233) "beliefs have an economic role on ordinary life, a barter economy that leaves room for creation in a credit environment which produces different practices that are sometimes contradictory". Different belief systems can coexist and be organized around a unifier principle which generates internal compatibilities and hierarchy – for instance, the coexistence of both a system of beliefs which is supported by values and a system of pragmatic beliefs for medical doctors who are dedicated to palliative care in France (Jodelet 2002b). The idea of adhering to palliative care may clash with the profession's image of healing's success. In this context, doctors employ technically palliative care to terminal patients. When death approaches, medication administered to lower pain starts being represented as lethal cocktails. Curative function is then menaced.

In this context, Jovchelovitch (2008: 213) challenges the difference between beliefs and knowledge. She employs the concept of cognitive polyphasia to show that "knowledge is a plural and malleable phenomenon which contains different epistemic forms and even different forms of rationality". Social Representations Theory thus accommodates the dilemma between knowledge and culture, integrating experiences that have been accumulated by a society's culture in a world of plural life (Jodelet 2002).

The concept of cognitive polyphasia can be illustrated by the modern problem of research involving human embryos. The human embryo is a socially constructed object, and as such confronts us with a representational functioning based on different kinds of thinking. While elaborating and expressing social representations about human embryos, we see demonstration-based knowledge being employed, and at the same time a representational state which is a mixture of the universe of science and that of common sense. This state of cognitive polyphasia is manifested in situations of social change (Moscovici 1976; Wagner/Hayes 2005), such as controversial debates about the regulation of research involving embryos (Aléssio 2012). When I analysed legislative debates in Brazil and France regarding the adoption of laws around embryo research, I observed a dynamic association between common-sense, religious, and scientific knowledge, all mixed together in the representational content about human embryos. Especially in Brazil, this polyphasic state supported attitudes against and pro embryo research. How could such a phenomenon possibly happen? For its interpretation, we need to understand the role the polyphasic state assumes in the space of social exchange.

Those opposed to authorizing embryo research would try to play a role in protecting shared fundamental representations about what 'human' and 'family' mean, as it can be observed in a Brazilian Federal deputy's speech:

> Scientists are aware that from the very moment when fecundation of an ovule occurs, a new life emerges. Just a few minutes after fecundation, forty-six chromosomes appear and then there is the miracle of life. This is a scientific truth and man can't and is not allowed to destroy life's grace, divine gift, deeper than whatever human hands could ever conceive. Embryos are live human beings. They are evolving, transcendent beings, possessing body and soul. Embryos are different subjects from their parents and, since their conception, they also possess their own genetic and permanent identity, the genome.

From this excerpt, we can perceive how scientific knowledge about the human embryo's chromosomal composition is used to explain the "miracle of life". Religious knowledge is used to state that this "scientific truth" is a gift that can't be modified by human hands. The idea that the embryo's nature corresponds to a divine "will" is rooted in the Catholic Church's teachings, according to which the nature's order is the divine order, and for that reason it cannot be profaned by human hands (Luna 2002). It should also be stressed that the notion of a human embryo as a person – a live human being – is based on the premise of it having a "genetic identity" and soul. This conception of life as a "code" or "message" helps to transform the human embryo into an autonomous personality, an abstract entity, an individual "defined exclusively by their genetic blueprint (distinct individual biogenetic potential)" (Porqueres I Gené 2004: 140).

To the research's supporters, their expression of science-based knowledge and at the same time of religious values had less to do with conservation than with a project to change the law:

> For that purpose, God provided men with intelligence, to develop medicine, to reach progress and not to step back to a fundamentalism without any basis, for more important than life is healing people, taking them off their wheelchair so they can find beauty in walking. And so many people present at this historical evening will be able to witness how National Congress took a big step forward to change the conception of the history of true life.

This excerpt shows how a mixture of science and religion is used in a pragmatic argument in favour of the importance of healing pathologies, such as paralysis. "Free will" is a religious principle that appears in an implicit form through the idea of "human intelligence given by God", man being responsible for employing his wit for well-doing. The "raise up and walk" expression, borrowed from Jesus' parable of healing a handicapped man in the Bible, is employed as an argument for conducting research involving embryos, and to legitimize the desire to heal as a motive for embryo research.

Religious values play a significant role in the public sphere in Brazil, setting up a normative framework deeply rooted in the functioning of collective life. Therefore those deputies in favour of authorizing embryo research mixed together science and values – a fusion of universes that functioned to protect their own identities and religious attachments, as they sought to advocate a new practice contested by religious people.

An important task to be pursued regarding social representations is to stress the functional aspect of cognitive polyphasia, for what really matters is how their relationship stands in the context of action and communication (Moscovici 2001). The scientific knowledge proliferation about human embryos was not capable of eliminating common-sense knowledge, stirring the deployment of multiple and ill-assorted images. Growing embryos *in vitro* is socially permeated with an ambiguous cultural status, with the frozen and stored human embryo becoming an "artifact" which lies between "nature" and "culture" (Strathern 1992). The importance assigned by different social actors (jurists, patients' associations, religious institutions, scientists) is still a sign of the social impacts linked to recent developments in the field of medical science, especially since the research with human embryos involves a whole set of emotions, generating at the same time hope, fascination, distrust, fear, and anxiety.

In the context of practices not yet totally accepted by a society, such as palliative care or the research with human embryos, beliefs are based on future-anticipating representations in an ever-changing social world. For Kalampalikis (2006), the driving force of social representations is based on beliefs. Beliefs are assigned the role of projecting and connecting with the past, regarding what can be kept and what can be changed. What we have then is a less 'pessimistic' perspective regarding beliefs.

## 5.4   Final Considerations

Both perspectives here summarized may point to different conceptions of the epistemic subject, while highlighting the object in the relationship between representations and practices. I agree with Jodelet that every social representation is the representation of something or someone. It is not a mere copy of real or ideal, or a subjective part of the object, or an objective part of the subject. It is the process by which relationships are established (Jodelet 2005). This means priority should be assigned to intersubjective and social approaches in the research of representations. In other words, the connection with the object is an intrinsic part of the social place and must be interpreted according to this framework; it is not an internalized object, but rather a relationship – representations express the way groups think of themselves in their relationship with objects which have an impact on them (Moscovici 1986).

Epistemological problems about the nature of beliefs in modern societies are either social or political problems (Moscovici 1995). Moscovici shows that societies are undergoing a changing process, from "conceived" societies to "lived" ones. In lived societies, there is a continuity of knowledge, rituals and beliefs with symbolic thinking that unites these forms of knowing (Moscovici 2002). Conceived societies produce a stigmatized thinking: knowledge detached from belief, progress detached from tradition, science disconnected from beliefs. Acting in a rational way is acting based on knowledge that is opposed to beliefs – "prove it and we'll believe it" – citing Condorcet. He states that "the tendency to rationalize notwithstanding, modern society is like any other society, a machine of making gods" (Moscovici 2003: 343).

The times we are living in reverse Condorcet's maxim. We live in an era of convictions, of the 'post-truth' and the 'believe and we'll prove it'. As Moscovici declares in an interview to Markovà, "proofs pro or against are secondary to beliefs". We have founded a society not because we are rational beings. Society is built on contract and negotiation, and reason is a by-product of its formation (Moscovici 1995). The paradox of modern societies is that they increasingly require from us faith and belief, despite rationalization. According to him, "democracy and rationality would be less vulnerable today if they were endowed with the indisputable power of a collective faith" (Moscovici 1995: 24). The question brought about is how central individuals are in the creation of our representations. Individuals are autonomous and self-sufficient entities which guarantee the relationship between knowledge and action (Moscovici 1992). It is individualism itself that leads to the "new magical thinking", for supposing an irrationality of the individual immersed in the culture's rationality.

For Kalampalikis, beliefs must be conceived as a mode and not as thought's content, translating a "trusted knowledge followed by rules of action" (2006: 229). The author suggests that social psychology should be centred on the pair belief/disbelief and not belief/reason. This is a productive suggestion for theoretical work that could be developed in Latin America, considering critiques directed to that continent's scientific production, which is characterized by several empirical qualitative studies, with a modest theoretical contribution to the field of social representations, albeit its potentialities (Doise 2002). Following the clue of 'doubt' could lead us to reconsider the link between knowledge and belief (Duveen 2002): how doubt, confrontation and conflict play a role in the production of social representations.

Moscovici (2012) states that the whole of our intellectual and collective life happens in the permanent tension between resistible and irresistible ideas. Resistible ideas are adapted according to situations; they are a project of rationality in contemporary societies and conceive culture as a kind of 'shared mistake'. Irresistible ideas are those that prevent us from thinking in another way, ideas which are imposed and impossible to be discarded, for we depend on them to live and act socially. What resistible and irresistible ideas occur in the context of Latin America? These are several unanswered questions which deserve our attention in future research projects.

# References

Aléssio, Renata Lira dos Santos, 2012: *Représentations sociales et embryon humain: une approche psychosociale comparative Brésil/France* (Aix-en-Provence: Université D'Aix-Marseille).
Apostolidis, Themis, 2002: "Représentations d'autrui dans le contexte d'une relation intime: remarques topologiques sur les croyances", in: *Psychologie et Société*, 5: 13–41.
Doise, Willem, 2002: "Da psicologia social à psicologia societal", in: *Psicologia: Teoria e Pesquisa*, 18,1: 027–035.
Duveen, Gerard, 2002: "Construction, belief, doubt", in: *Psychologie & Société*, 5: 139–155.
Flament, Claude; Rouquette, Michel-Louis, 2003: *Anatomie des idées ordinaires, comment étudier les représentations sociales* (Paris: Armand Colin).
Jodelet, Denise, 1986: "Fou et folie dans un milieu rural français: une approche monographique", in: Doise, Willem; Palmonari, Augusto (Eds.): *L'étude des représentations sociales* (Neuchatel: Delachaux & Niestlé): 171–192.
Jodelet, Denise, 2001: "Um domínio em expansão", in Jodelet, Denise (Ed.): *As representações sociais* (Rio de Janeiro: EdUERJ): 17–44.

Jodelet, Denise, 2002a: "Les représentations sociales dans le champ de la culture", in: *Social Science Information*, 41,1: 111–133.

Jodelet, Denise, 2002b: "Perspectives d'étude sur le rapport croyances/représentations sociales", in: *Psychologie & Société*, 5,1: 157–178.

Jodelet, Denise, 2005: "Formes et figures de l'altérité", in: Sanchez-Mazas, Margarita; Licata, Laurent (Eds.): *L'autre, regards psychosociaux* (Grenoble: Presses universitaires de Grenoble): 23–47.

Joffe, Helen, 2002: "Social representations and health psychology", in: *Social Science Information*, 41,4: 559–580.

Jovchelovitch, Sandra, 2002: "Re-thinking the diversity of knowledge: Cognitive polyphasia, belief and representation", in: *Psychologie et Société*, 5,1: 121–138.

Jovchelovitch, Sandra, 2008: *Os contextos do saber representações, comunidade e cultura* (Petrópolis, RJ: Vozes).

Kahan, Dan, 2015: "Climate-Science Communication and the Measurement Problem", in: *Political Psychology*, 36: 1–43, at: https://doi.org/10.1111/pops.12244.

Kalampalikis, Nikos, 2006: "Affronter la complexité: représentations et croyances", in: Haas, Valérie (Ed.): *Les savoirs du quotidien: transmissions, appropriations, représentations* (Rennes: Presses Universitaires de Rennes): 225–234.

Luna, Naara, 2002: "As novas tecnologias reprodutivas e o estatuto do embrião: um discurso do magistério da Igreja Católica sobre a natureza", in: *Gênero*, 3,1: 83–100.

Marková, Ivana, 2006: *Dialogicidade e representações sociais: as dinâmicas da mente* (Petrópolis: Vozes).

Marques, Emanuele Souza; Cotta, Rosângela Minardi Mitre; Priore, Silvia Eloiza, 2011: "Mitos e crenças sobre o aleitamento materno", in: *Ciência & Saúde Coletiva*, 16,5: 2,461–2,468.

Moscovici, Serge, 1976: *La psychanalyse, son image et son public* (Paris: Presses Universitaires de France).

Moscovici, Serge, 1986: "L'ère des représentations sociales", in: Doise, Willem; Palmonari, Augusto (Eds.): *L'étude des représentations sociales* (Neuchatel: Delachaux & Niestlé): 34–80.

Moscovici, Serge, 1992: "La nouvelle pensée magique", in: *Bulletin de Psychologie*, XLV, 405: 3,301–3,324.

Moscovici, Serge; Vignaux, Georges, 1994: "Le concept de thêmata", in: Guimelli, Christian (Ed.): *Structures et transformations des représentations sociales* (Lausanne: Delachaux & Niestlé): 25–72.

Moscovici, Serge, 1995: "Modernité, sociétés vécues et sociétés conçues", in: Dubet, François; Wierviorka, Michel (Eds.): *Penser le sujet autour d'Alain Touraine* (Paris: Librairie Arthème Fayard): 57–72.

Moscovici, Serge, 2001: "Why a theory of social representations", in: Deaux, Kay; Philogène, Gina (Eds.): *Representations of the social : bridging theoretical traditions* (Oxford: Blackwell): 8–36.

Moscovici, Serge, 2002: "Pensée stigmatique et pensée symbolique: deux formes élémentaires de la pensée sociale", in: Garnier, Catherine (Ed.): *Les formes de la pensée sociale* (Paris: Presses Universitaires de France): 21–54.

Moscovici, Serge, 2003: Representações sociais: investigações em psicologia social (Petropólis: Vozes).

Moscovici, Serge; Hewstone, Miles, 2005: "De la science au sens commun", in: Moscovici, Serge (Ed.), *La psycologie sociale* (Paris: Presses Universitaires de France): 545–572.

Moscovici, Serge, 2012: *Raison et cultures* (Paris: Éditions de l'École des hautes études en sciences sociales).

Porqueres I Gené, Enric, 2004: "Individu et parenté, individuation de l'embryon", in: Héritier, Francoise; Xanthakou, Margarita (Eds.): *Corps et affects* (Paris: Odile Jacob): 139–150.

Strathern, Marilyn, 1992: *After nature: English kinship in the late twentieth century* (Cambridge: Cambridge University Press).

Wagner, Wolfgang; Hayes, Nicky, 2005: *Everyday discourse and common sense: the Theory of Social Representations* (Hampshire: Palgrave MacMillan).

Wagner, Wolfgang; Kronberger, Nicole, 2002: "Discours et appropriation symbolique de la bio-technologie", in: Garnier, Catherine (Ed.), *Les formes de la pensée sociale* (Paris: Presses Universitaires de France): 119–150.

# Chapter 6
# Social Representations of Justice as Developing Structures: Sociogenesis and Ontogenesis

Alicia Barreiro

## Introductory Comment

José Antonio Castorina

*It is a pleasure to introduce Dr Alicia Barreiro. I met her in 2001. She was an*[1] *undergraduate student when she joined a research project I was directing at Universidad de Buenos Aires, Argentina. Since then, I have been guiding her academic endeavours. First, I supervised her research thesis for the Educational Psychology Master Programme at the Faculty of Psychology – Universidad de Buenos Aires (2008). Then, I was her thesis advisor for the Doctorate Degree in Educational Science at the Faculty of Philosophy and Letters (2010) – Universidad de Buenos Aires. There, I also directed her post doc in 2011. She was under my care as a junior researcher at Consejo Nacional de Investigaciones Científicas y Tecnológicas (CONICET – Argentina) from 2012 to 2014 when she became Assistant Researcher. Since 2014, we have been co-directing a research project at Universidad de Buenos Aires.*

*Her research agenda has resulted in outstanding book chapters and articles published in national and international journals. The following chapter is an*

Alicia Barreiro is professor of psychology at the University of Buenos Aires and at the Latin American Faculty of Social Sciences (FLACSO-Argentina), and an endowed researcher at the National Scientific and Technological Research Council (CONICET-Argentina). Email: avbarreiro@gmail.com.

[1]Prof. José Antonio Castorina is a senior researcher and Director of the Institute of Research in Educational Sciencies (IICE) at the Faculty of Philosophy, University of Buenos Aires, and has links with the National Pedagogical University (UNIPE) and CONICET. He gained his PhD in Education at the Universidad Federal do Río Grande do Sul (Brazil). He is a Doctor *Honoris Causa* at the National University of Rosario (Argentina) and an Honorary Professor at the University of San Marcos (Peru). Email: ctono@fibertel.com.ar.

© Springer Nature Switzerland AG 2021
C. Prado de Sousa and S. E. Serrano Oswald (eds.),
*Social Representations for the Anthropocene: Latin American Perspectives*,
The Anthropocene: Politik—Economics—Society—Science 32,
https://doi.org/10.1007/978-3-030-67778-7_6

*excellent example of her work. She has been able to draw a sequence of empirical works articulating Social Representations Theory with developmental psychology, in particular in the moral domain of knowledge. There is an argumentative thread that binds her work, i.e. individuals' unresignable constructive activity to appropriate social representations in their social practices. The methodological rigour and clarity in conceptual elaboration are the most salient features of these studies. I was also able to witness her independent thinking and ample reflexivity when we shared several studies attempting to link ideas from developmental psychology and Social Representations Theory.*

*As a researcher, she started out as a developmental psychologist interested in Social Representations Theory (SRT). Duveen exceptionally acknowledged Piaget's constructivism in SRT (Castorina 2010) and studied how social representations influence the individual's cognitive appropriations. Moreover, Duveen even included ontogenesis in social representations of gender (Duveen/Lloyd 1990), the constitution of semiotic function claimed by Piaget, giving rise to studying the cognitive process implied. This would serve in the analysis of Barreiro's work.*

*Barreiro works on complementing both disciplines to study the development of the notion of justice. Her critical position on Piaget's viewpoint on moral thinking is closely related to her own data analysis. Dr Barreiro upholds some of Piaget's core tenets, but she modifies the universality and immanent tendency towards equilibrium in moral development to a 'more advanced' point of arrival. Analysts of SRT have agreed on the difficulties this theory faces, such as the fuzziness of some of its definition and the demand to reconcile notions with other disciplines and schools of thought (Sammut/Andreoli/Gaskell/Valsiner 2015). Regarding the articulation of developmental psychology, empirical work, theory construction, and epistemological thought should be combined. Barreiro has produced and analysed data with great creativity while looking for methodological consistency and adopting a clear stance on the conceptual issues of Social Representations Theory. In this way, she has traced an interesting and fruitful path aimed at systematically binding SRT with developmental psychology.*

*She has analysed the controversies over social knowledge in different disciplines as well as the concepts and definitions used in an attempt to elucidate the ontological and epistemological assumptions and the relational epistemological frame that enables interdisciplinary dialogue. Barreiro achieved this in her empirical study on the development of the representation of justice, using typical procedures from each discipline. Thus, she could convincingly show that developmental psychology can become a tool for understanding the ontogenesis of social representations, and she also showed that social representations can be seen, from a developmental psychology standpoint, as constraining individual elaborations of social notions.*

*In this way, she understood the meanings of the ontogenesis of social representations of justice in terms of dialectic inference studied by Piaget (1980): subjects are able to independently articulate the meaning field before, such as retributive and utilitarian justice giving rise to new meanings that subsume them. It was key in her study to address the personal process of argument elaboration as a conceptual construction that, at the same time, is the appropriation of pre-existing social representations. In sum, we can see an active assimilation that relativizes and integrates previous notions in a genesis that overcomes them. Yet, they do not disappear in the development, as this researcher expertly shows.*

*It is not a minor matter; this researcher admits that discourse analysis is not sufficient to grasp the subjects' elaboration of meanings when interacting with others. This finding opens a new research path that challenges the rigid discipline borders while keeping their relative autonomy.*

*Studying the representation of justice in newspapers of Buenos Aires and university students unions showed the existence of social representations of justice with a clear retributive tone, a hegemonic social representation for the city's inhabitants. Prominent in the newspaper articles and the university students unions, distributive justice is not present. So why is there such unilaterality? Barreiro got inspiration from some tenets of cultural psychology (Valsiner 2014) that claim that societies actively produce not only meanings in the world but also lack of meanings. Thus, through the meaningful construction of social representations associated with a process of news dissemination, the meanings of retributive justice stand out and other modalities of justice are invisibilized because they threaten the status quo of Argentinian society. Regarding the sociogenesis of social representations of justice, when selecting their own objectivization mechanism, some aspects of the object will be represented and others will not. In this way, the tension stemming from the power relations in groups and between groups rises.*

*To sum up, how has Barreiro contributed to this issue? She enhances the constructivist side of SRT when including nothingness as an aspect of the objectification of a social representation. This paves the way for reconsidering the relationship between social representations and ideology in the field of SRT. It is not only a 'horizon' in the attempt to elaborate social representations, but also intervenes in its constitution since hegemonic social representations – or most of them – impose the dominant group's viewpoint. This view is ideological as it legitimizes relationships of domination. Finally, the theoretical elaboration of the 'construction of nothingness' is adequately sustained in the empirical research because it facilitates the discussion of social repression which is crucial in the relationship between social representations and social sciences.*

# References

Castorina, José Antonio, 2010: "El análisis meta-teórico en la investigación psicológico-educativa", in: Alzamora, Sonia Gladis; Campagnaro, Liliana (Eds.): *La educación en los nuevos escenarios socioculturales* (Santa Rosa: UNLPAM).

Duveen, Gerard; Lloyd, Barbara (Eds.), 1990: *Social representations and the development of knowledge* (Cambridge: Cambridge University Press).

Piaget, Jean, 1980: *Les formes élémentaires de la dialectique* (Paris: Gallimard).

Sammut, Gordon; Andreoli, Eleni; Gaskell, George; Valsiner, Jan, 2015: *The Cambridge handbook of social representations* (Cambridge: Cambridge University Press).

Valsiner, Han, 2014: *An invitation to cultural psychology* (California: Sage).

**Abstract** Social injustice has been present in Argentina, as in other Latin American countries. Retributive justice has become a daily matter of debate since the fear of crime and the degradation of political and social participation have led to the weakening of social bonds and fractures in the sense of community. Reflections and discussions on diverse – even opposite – ways to understand and promote justice have a long tradition in social sciences that can be traced to Ancient Greece. The justice-injustice opposition could be considered a theme that permeates the entire history of Western culture (Marková 2000), leading, within the same society, to the coexistence of different ways of understanding justice based on different ideological groundings (Campbell 2001). Social representations (henceforth SR) are the product of everyday exchanges and are constructed as a way to understand social objects that challenge the cultural available meanings and require a symbolic coping process. The ontogenesis of SR is the process through which individuals reconstruct SR while appropriating them and, in so doing, develop diverse social identities (Duveen/Lloyd 1990). This chapter presents a set of studies aimed at understanding the sociogenesis and ontogenesis of SR of justice, as well as the relationship between the two processes.

**Keywords** Justice · Sociogenesis · Ontogenesis · Social representations

## 6.1 A Developmental Approach to Justice as Social Representation

Social injustice has been present in Argentina like in other Latin American countries. Wealth distribution is highly unequal and one in three Argentinians lives in poverty (Salvia/Donza 2017). This extreme social inequality has resulted in an increase in violent events and disruptive behaviour, and has triggered in most Argentinians the fear of being a victim of crime (Muratori 2017). Retributive justice has become a daily matter of debate since the fear of crime and the degradation of political and social participation have led to the weakening of social bonds and fractures in the sense of community. This loop increases fear, violence and distrust of others (Bergman/Kessler 2008).

Reflections and discussions on diverse – even opposite – ways to understand and promote justice have a long tradition in social sciences that can be traced to Ancient Greece. The justice-injustice opposition could be considered a theme that permeates the entire history of Western culture (Marková 2000), leading, within the same society, to the coexistence of different ways of understanding justice based on different ideological groundings (Campbell 2001).

The notion of justice is polysemic. However, in everyday life, people appeal to justice to legitimize their interests and criticize the existing power relations in the confrontation of others (Starklé 2015). Participation in social debates demands individuals' knowledge of what is at stake. This is possible only if the participants share some common frames of symbolic reference. Social representations (henceforth SR) are the product of everyday exchanges and, as a form of collective knowledge, are meaningful structures that provide members of a social group with a shared code to communicate about the daily life phenomena and the challenges they face (Marková 2012; Moscovici 2001). SR are constructed to understand social objects that challenge the cultural available meanings and require a symbolic coping process (Wagner 1998). During this collective meaning-making process the media uphold the discourse that accompanies this process by diffusing SR (Moscovici 1961).

Since the dynamic process of social representing implies a temporal dimension, SR should be studied as developing structures. Although they can be described at any specific moment in time, analysis of them requires a developmental perspective inasmuch as they constitute both the process and the product of social knowledge construction. There are two analytical levels in the study of SR dynamics: sociogenesis and ontogenesis (Duveen/Lloyd 1990). SR sociogenesis refers to the communicational process that leads to the construction of SR. This constitutive process involves tensions and meaningful negotiations in the social groups' present as well as in their past because SR are the outcome of historical, political and cultural circumstances (Kalampalikis/Apostolidis 2016). The ontogenesis of SR is the process through which individuals reconstruct SR while appropriating them and, in so doing, develop diverse social identities (Duveen/Lloyd 1990).

Hence, the research presented in this chapter complements the methodological and theoretical approaches of social psychology and developmental psychology (Castorina 2010). As Duveen (1994) pointed out, social psychologists are challenged to adopt a developmental perspective on the sociogenesis of social representation challenges, whereas developmental psychologists need to explain how individuals become social actors via the appropriation of the social representation that configures their cultural environment and, in turn, contributes to the development their social identity. Both disciplines study the same phenomena, but developmental psychology is focused on individuals whereas social psychology follows a collective approach. Although research would be enhanced by complementing both perspectives, this has been the exception rather than the rule (e.g. Barreiro 2013a; Barreiro/Castorina 2017; Leman/Duveen 1996; Lloyd/Duveen 1990; Psaltis/Duveen 2006; Psaltis/Duveen/Perret-Clermont 2009; Psaltis/Zapiti 2014).

In sum, I present here a set of studies aimed at understanding the sociogenesis and ontogenesis of SR of justice, as well as the relationship between the two processes. According to previous studies, people tend to deny inequality and to justify unjust social systems (Jost/Hunyady 2005). Hence, it is relevant to study how justice is represented in Argentinian society, because attitudes towards social order – acceptance or contestation – are usually organized in accordance with SR of justice (Starklé 2015). Specifically, my colleagues and I analysed the sociogenesis of SR by considering the media's diffusion, and the particularities of the objectification and anchorage process according to power relations in an unequal social context and the individual's ideological positionings. In the same vein, we analysed the appropriation process of such SR by individuals during childhood and adolescence, focusing on both the cognitive mechanisms that may be involved, and the possibilities set by the development of thought.

## 6.2   Social Representation of Justice Ontogenesis: Diffusion and Ideological Positionings

According to Social Representations Theory (SRT), social groups understand their cultural environment through a meaning-making process. The main function of SR is to establish a code of social exchanges in the dialogical interaction between the members of a social group. The continuous adjustments and negotiations that occur in social relations encompass the development of shared meanings that function as action patterns (Wagner 2015). Thus, SR refer to the practical knowledge that links individuals with objects in a triple sense (Jodelet 1985). They emerge from dialogical interactions framed by institutions; they are constructed in everyday social practices; and they are used by individuals to act upon others or to adjust their behaviour to social expectations. Hence, SR as emergent meanings create the social object that only exists as the outcome of people's interactions over time in a specific context (Wagner 2015).

Through this representational activity, a single object can take on different meanings depending on the context and social position in the social group to which it belongs, thus yielding multiple realities (Jovchelovitch/Priego-Hernández 2015; Marková 2012). Although SR are shared knowledge, this does not imply that all members of a group think in similar ways. The organization of individual knowledge is influenced by common principles and shared common grounds in a social group. SR enable people to communicate with each other even when they disagree (Andreouli/Chryssochoou 2015). Consensus and conflict are both essential parts of any social order and are constitutive of any social representation.

Hence, SR allow individuals to create and know their social reality via a meaning-making process carried out by two dialectically related mechanisms: objectification and anchoring (Moscovici 1961, 2001). Through the objectification process, the unfamiliar becomes a collectively constructed meaningful structure which is accepted as the 'real' object in the eyes of a social group. This mechanism selects some aspects of the represented phenomenon and constitutes a figurative core that transforms abstract realities into concrete images. Following the central core theory (Abric 2001; Moliner/Abric 2015), more salient and consensual meanings constitute the nucleus that organizes the representation and has a twofold purpose: to give meaning to the other elements that make up the social representation by creating or transforming them, and to organize the social representation by determining the links between its elements.

The anchoring process refers to the conventionalization of the unknown by categorizing and comparing it with seemingly suitable cultural available categories. By doing this, groups classify things and label them with familiar names to obtain an accurate picture of them (Moscovici 2001). The resulting meaningful structure needs to be integrated into the changing social context. Therefore, anchoring modifies the peripheral elements of SR which express differential attitudes to the object, i.e. the different meanings that the same object can have depending on the particular features of a given group of individuals (Doise/Clemence/Lorenzi-Cioldi 1992)

Social positioning refers to the diverse positions people can adopt in a network of significations. In other words, the organizing principles that constitute the core of a social representation may result in different or even opposed positions taken by individuals in relation to common reference points. Positionings could be drivers since individual's viewpoints are different, but they are not idiomorphic since there is a meaningful frame of reference that legitimizes the actor's point of view (Sammut 2015). They can be understood as a way of processing information to align individuals' thinking with society's. So ideological assumptions play an important role because they are the background from which SR are "cut out" (Jodelet 1991). Any ideology is a combination of belief systems, discourse and social power that supports a specific vision of the social world (Eagleton 1997). Hence, the differences in ideological assumptions may be expressed in different positionings about the same social object in the same social group.

During this representing activity, social groups select some elements or characteristics of the represented object to be included in the social representation while others are not included. As Duveen pointed out: "A representation, then, is not only a way of understanding something, it is also always a way of not understanding something" (1998: 461). Power relations between and within social groups are crucial factors in determining if something is represented or not. The meanings that prevail in this struggle among representational fields within the social arena constitute a positive representation, a specific symbolic structure that occupies the place of the real object in individuals' everyday life. From a developmental approach based on cultural psychology assumptions, it could be claimed that the other possible representations become nothingness and remain as the dark side of the positive social representation or the non-present parts of that structure that makes it conceivable (Valsiner 2014). Since they challenge the dominant vision of the social world and, in that sense, they are perceived as threats by social groups, some meanings of the representational field can be repressed or excluded If there are no SR of some emotionally decisive object, it may signal their overwhelming affective presence in the social group's daily life. What cannot be anchored in the collectively available system of meanings and values would be repressed, transforming its ontological quality into nothingness (Barreiro/Castorina 2016). This constructive activity is neither conscious nor voluntary and determines what subjects will and will not recognize as real. Hence, it is possible to describe a scale in the collective construction of nothingness during the social representing process that goes from eclipsing some parts of the object to obscuring their existence.

As previously mentioned, SR are constructed to face social phenomena relevant to a social group and demand a symbolic coping process (Wagner 1998). The media uphold the discourse that accompanies this process by diffusing social representations (Bauer 1994; Jovchelovich 1994; Moscovici 1961; Wagner/Hayes 2005). Diffusion is defined by Moscovici (1961) as a communication system in which the issuer attempts to establish a relationship of equality and equivalency with the public by adapting to it. The purpose of the written press is to guide the community by being a mediator between the information and the audience. Therefore, newspapers become bridges between the representational object and their readers. A spiral movement leads from the individual's representations constituted by the act of diffusing information to the reconstruction of the SR (Duveen 2008).

In the diffusion process, the press tries to match the assumed taste and vocabulary of its readership as much as possible because the news must be appealing (Moscovici 1961). People prefer speaking with others and reading newspapers that share their own opinions and tend to confirm their own beliefs instead of challenging them (Wagner 1998; Wagner/Duveen/Farr/ Jovchelovitch/Lorenzi-Cioldi/ Markovà/Rose 1999). At this point, the anchorage process adds the new unknown elements to the previously familiar meaningful categories, and simplifies complex notions to make them accessible to the community. This process is carried out by inducing analogies – transferring properties and evaluations from one already known phenomena to the new one (Marchand 2016). In the same vein, the familiar images used by the media to illustrate what they are trying to explain contribute to the objectification process.

Therefore, different ways of presenting an object become customary communication forms which turn the object into a cultural, cognitive and linguistic sign in a given place and time. In Moscovici's opinion (1961), the crafting of messages and their adaptation to specific cultural norms imply recognition of the mediating role played by diffusion among social groups and their value systems. Likewise, diffusion as a form of communication is targeted not at a defined group of people but directed at the masses, defined as a set of different groups of socially conjoined people. There is a close relationship between the attitudes and values of a group and the characteristics or patterns of communication sustaining it, giving rise to heterogeneous psychosocial forms of organization (Duveen 2008). Moreover, diffusion is characterized by fellow-feeling as the bond that keeps the social group united, while in propagation and propaganda, the other communication systems that Moscovici (1961) describes, the collective bonds are communion and solidarity, respectively (Duveen 2008).

In sum, diffusion is not only a channel to transmit statements and values but also constructs contents and realities, wielding influence over people's behaviours and opinions (Wagner 1998). This form of communication tends to create the social reality of the object of representation, helping to shape a kind of message and to confer social value upon it (Moscovici 1961).

Framed in this theoretical background, my colleagues and I carried out a study to understand the diffusion process of SR of justice in the most-read newspaper in Argentina and its relations with its readers' SR (see Barreiro/Gaudio/Mayor/ Santellán/Sarti/Sarti 2014 for more details on study). It should be said that the analysed newspaper belongs to a media group that also owns television stations and other printed media. Consequently, the SR diffused in this newspaper are also spread via other media. We analysed the content of the excerpts of news ($n = 318$) where the term *justices* appeared in twelve Sunday editions. After the analysis, two different ways to represent justice were identified:

*Retributive:* This refers to getting what one deserves for a previous action or deed ($n = 293$). For example: "Anyway, I'm not too optimistic about what Pino [political figure] says. I think it's very unjust to us. We may have made mistakes, but the party is a set of

ideas, principles and values. I would like him to say which UCR [political party] principles make it impossible for him to meet us". (*Clarín* newspaper 2011)

*Distributive:* This refers to the distribution of goods or resources among a group of people
($n = 25$). For example: "We should aim to sustain over time some core policies to promote economic development and just distribution of wealth". (*Clarín* newspaper 2011)

The analysis shows the pre-eminence of retribution over other possible meanings of justice in the newspaper articles.

Concurrently with the newspaper analysis, we studied its readers' SR of justice
($n = 246$; aged 18 to 53). To this end, we used the word association technique (Rateau/
Lo Monaco 2016; Wagner/Hayes 2005) in a self-report questionnaire. Participants were
asked: "Please write the first five words that come to your mind when you think about
the word *justice*". This technique is widely used in research into SR because it enables
researchers to discern its semantic field and hierarchical structure. We then examined
the frequency and average range of the word association using the Evoc[1] programme
(Verges 1999). In this way, we distinguished between high-frequency, low-frequency
and idiosyncratic terms by analysing the corpus distribution (see Barreiro/Gaudio/
Mayor/Santellán/Sarti/Sarti 2014 for more details on the analytical procedure).

The highest frequency and the most quickly associated words showed the most
prominent, agreed-upon meanings of the term *justice* and therefore form the core
of the SR (Verges 1999). These words were: *lawyer, right, equity, balance, equality, injustice, judge, trial, law, freedom, power, late* and *truth*. With the exception
of the term *equity*, we may interpret that all the associations have a clearly retributive meaning. In fact, in the strict sense, even *equity* refers to retribution through
distribution; that is, each person should get what they deserve (Campbell 2001).
Likewise, the terms *lawyer, judge, trial, law* and *freedom* refer to the social institution charged with imparting justice.

The first periphery around the core is made up of the high-frequency but not so
quickly associated elements (Verges 1999). They include *punishment, order, security, society* and *corruption*. These words were highly frequent but not as prominent as those in the SR core because of their association range.

Similarly to the meaning field of the SR identified in the newspaper articles, the
retributive sense was salient to its readers' SR. Therefore, we can glimpse a mutual
constitutive circular movement between the SR of justice diffused by the newspaper
and its readers' representation. This is the social construction of the representational
object by both the most widely read newspaper, influencing people's opinions, taking shape in a kind of message regarding justice and conferring social value over
it, and the individuals who symbolically cope with this object in their everyday
interactions, using the information spread by the press among other elements. We
should stress that in both the excerpts from the newspaper and its readers' SR of
justice, there was no link to social issues like the distribution of goods or resources.
In this sense, as was mentioned before, we consider that societies not only actively
produce meanings about the world but also produce the lack of meanings. This
absence signals the presence of an absence because it invisiblizes the Argentinian
social distribution problems and it is also the outcome of a constructive process. In

this sense, we might think that the process of symbolic coping that this newspaper propagates through the diffusion process brings to the fore meanings linked to retribution and denies other ways of understanding justice which might be threatening to the status quo, given the sharp inequalities in Argentinian society.

According to this study, retributive meanings prevailed during the meaning-making process by which this social representation was diffused by the newspaper and reconstructed by the inhabitants of Argentina. For them, only one possible meaning of this object becomes evident: retribution. This meaning denies the distributive meaning of justice since it could challenge the current social order. In this case, the object helps to crystallize a specific meaning functional for dominant social groups and becomes a *hegemonic* social representation (Moscovici 1988).

In the face of the relevance of the retributive meaning in the social relationship, we decided to move forward and analyse the relationships between this semantic content and its anchorage in a specific ideological belief strongly related to retributive conceptions: the belief in a just world (Lerner 1988). According to such a belief, everybody gets what they deserve, and everybody deserves what they get. This belief is the foundation stone of a meritocratic social system and denies social injustices and justifies the current social order (Jost/Hunyady 2005). The results of our study did reveal the existence of different social positioning in the peripheral elements of the social representation according to the individual's levels of such an ideological belief (see Barreiro/Castorina 2015 for more details on this study). Participants with higher levels of belief in a just world represented justice as something beyond society and everyday life, as a matter for the judges who work to maintain social order and peace. Conversely, individuals with lower levels of belief in a just world think of justice as a social institution that regulates people's freedom, with some negative effects on their daily functioning. The different social positioning identified in the peripheral elements showed the outcome of the anchorage in the ideological belief. Peripheral elements forge ties between individuals' everyday practices and the core of a social representation, allowing it to adapt to the specific conditions of each context. Those participants with high levels of belief in a just world may use it to justify the status quo by denying that justice is a matter of daily social relations, representing it as something that works independently of the social interactions framed in a specific social and historical context. We concluded that these results illustrate how ideology – as a broad frame – intervenes as the background of the constitution of the social representation, enabling and at the same time denying other possible meanings linked to justice.

## 6.3 The Ontogenesis of the Social Representation of Justice

The previous analyses have shown that the social representation of justice in Argentinian adults and in the most-read newspaper highlights a retributive meaning and occludes the problems of social wealth distribution. In accordance with the developmental approach which my colleagues and I adopted to study SR, it is necessary to understand the ontogenesis of this social representation. In other words, we tried to elucidate how this social representation was appropriated by children and adolescents during their socialization process depending on their cognitive development and what cognitive mechanisms may explain such reconstructive processes. In doing so, we complemented methodological and theoretical assumptions of SRT with those proposed by developmental psychology.

SRT is a constructivist theory based on the relational interaction between individuals and the representational object, in other words, both are co-constructed in a developmental process (Duveen 2002). Individuals become social actors during socialization, but in this construction and reconstruction process novelty emerges, i.e. the development of new understandings about the social objects as well as the transformation in individuals' psychological structures (Duveen/Lloyd 1990). However, there is novelty only for the actor, as a construction of a new understanding of his or her world, and not for their social group (Duveen 2007).

Developmental psychology has a long tradition of studying the representation of justice, starting with Piaget's (1932) research into children's moral development. He considered distributive justice as the most rational moral notion since it is based on reciprocity, one of the main properties of the operational stage. In Piaget's theory moral development is conceived as a non-stop trajectory from less to more rational thought stages that involve a more objective knowledge or are closer to the 'real' object (Marková 2012; Psaltis/Duveen/Perret-Clermont 2009). Conversely, according to recent developmental psychology findings, there is no such incremental tendency because the understanding of justice presents a u-shaped pattern (Smetana/Villalobos 2009). Retributive-based ideas of justice shift their focus to strict equality at the beginning of adolescence and later to equity in middle adolescence. At the same time, adolescents are more capable of coordinating and integrating divergent aspects of social domain situations and can better manage abstract concepts of law and rights.

Against this background, we carried out a study with children and adolescents ($n = 216$; aged 6 to 17) (see Barreiro 2013a for more details on the empirical study). For data collection, individual interviews were performed following the Piagetian *méthode clinique* (Duveen/Gilligan 2013; Piaget 1926). The previously established interventions of the interviewer began with the question: "Please tell me something that has to do with justice that happened to you, or that you have seen, heard or whatever." Once the participants had constructed a narrative answering this question, the interviewer formulated all the necessary questions to understand the interviewees' representations of justice and the accounts given to support their attitudes. This methodological approach – classic in developmental psychology – was selected for two main reasons. On the one hand, we attempted to study the social representation of justice in children's and adolescents' everyday life. Their personal experience constitutes a base from which to analyse the developing relationship between the individual and the social environment (Hedegaard 2012; Hviid 2012). The children and adolescents may have narrated a situation that did not involve them directly, but the recalled situations bore affective and cognitive connotations for them because the expressed the participants' cultural, moral order (Miller 2006). On the other hand, this methodological strategy ruled out spontaneous associations about justice, like the word association technique used to assess the SR among adults. In this study, this non-reflexive strategy is complemented with the argumentative process enabled by the dialogical structure of the interview that provides information about the interviewees' position on justice.

In this way, via the analysis of similarities, differences, and recurrences in participants' narratives, seven analytical categories were constructed to describe the different representations of justice identified:

a. *Utilitarian representation:* In this category, the answers included justice seen as "something that enables everyone to be happy" or "what does good to people", 'good' being synonym for happiness.

Victoria (6;05)[2]: *Just is going to my aunt's house on Saturdays.*[3] [And why is it just?] *Because I like it.*

Lorena (16;01): *Something that an individual wants or something that we want society to become.* [Could you give an example?] *For example, what is being done about justice influences all problems that we have; justice must be done to solve the problems.* [Which problems are you referring to?] *Poor people... Everything that is being done to make their situation better. It is unjust that nobody takes care of making their situation better so that all people can be well.* [Who has to take care of them?] *Ah... the Government and the organizations, to correct things."*

This utilitarian representation of justice is present in a high percentage across ages six to seventeen as a basic representation or integrated with the other representations, as explained later. However, there is an age-related reconstructive process; for example, Victoria (6;05) defines justice as something that brings happiness to her, while Lorena (16;01) considers that justice would solve social problems and ensures that everybody is fine. The interviewees think of justice as something that is good for people and enables them to live happily. Yet the youngest children focused their answers on themselves or on things they like to do. Aged nine and onward narratives tended to include a social perspective by considering collective or other people's happiness and including different social settings or institutions (e.g. school, sports club). It seems that children from nine to ten years old are capable of thinking of justice as a social relation system that includes them by coordinating different domains of social experience. Probably this change is facilitated by cognitive development that enables individuals to extend their understanding of the social

---

[2]Indicates participants' age in years and months.

[3]The following notations were used in the transcripts: "[ ]" indicate interviewer's words, *italics* indicate interviewees' words.

world complexity. Their more frequent participation in different social situations or institutions with more complex organization allows them to include and coordinate the personal, moral and conventional dimensions of the same social situation (Smetana/Villalobos 2009).

b. *Retributive representation:* This category includes the answers which show an understanding of justice as a relationship between actions performed or individual merits and punishments or rewards received. Justice is understood as a balance between what is deserved and what comes out:
   Daniela (8;06): "*For example, it is unjust if he was fired from his job for nothing. [And how would it be just?] It would be just if he was fired because he did something wrong*".
   Clearly Daniela (8:06) thinks of justice as situations in which the persons involved get what they deserve. In contrast, outcomes that are not deserved are unjust. This representation tends to appear in narratives constructed by children from nine years old onwards, due to the development of the capacity to consider the perspective of others and include it to judge the appropriateness of a punishment or reward. The presence of this representation increases with age.

c. *Distributive Representation*: This category refers to justice as a distribution based on equal norms applied to all people involved without any favouritism or preference.
   Agustin (16;07): "*For example, when a teacher evaluates the exam results, it is just because she takes the same basis to evaluate us; almost always, she applies the same criteria. For me Justice is a synonym of equality, that everybody has the same conditions, possibilities and could get to the same point. It is unjust, for example, that some people are born with everything and others with nothing. Some people are very wealthy while others are very poor. I think that is unjust*".
   Agustin's (16;07) narrative shows that he thinks of justice in terms of the egalitarian treatment that some teachers give to their students. The opposite, "injustice", refers to inequality because we are all equal and deserve to have the same opportunities. However, there are very few individuals in our study who shared this meaning of justice.
   In the course of cognitive development, these basic representations, which at first emerge as independent argumentation systems, merge to form a dialectical movement from independence to articulation, and also relativization (Piaget 1980). This gives way to four different representations: (d) utilitarian representation in a retributive situation (e) utilitarian representation in a distributive situation; (f) distributive representation in a retributive situation; (g) utilitarian representation in a situation of retributive distribution. In this way, on being integrated with utilitarian representation, retributive representation attains the highest frequency in the age group from ten to seventeen. That means that for most of the adolescents participating in this study, justice is what allows people to live happily and the way to achieve this is through punishment. Due to the relevance of this representation, it will be the only integrated representation that we are going to explain and exemplify here (see Barreiro 2013a for more details on the rest of the categories).

d. *Utilitarian representation within a retribution situation:* The narratives described situations of punishment or reward, but both were conceived merely as methods to achieve happiness for everyone, which is the meaning of justice for most of these adolescents:

Juan (11;03): *When you fight with a guy and he says, "No, no, no, he hit me", the people in charge of justice, in this case the teacher, must seek the truth ... so your name will be clean while the name of the other becomes spoiled (...)* [What do you think justice means?] *Justice is a way to bring order, (....) is a way to make things right and people who are committed to justice try to promote good over evil, to make people live better.*

Despite the fact that the three basic representations can be integrated, none is abandoned across this developmental process. They remain recognizable in individuals' narratives. In this respect, contrary to many studies on the cognitive development of moral judgements (Damon 1990; Kohlberg 1981), the results suggested that the meaning-making process does not occur due to a conflict between contradictory representations, but as a result of the coexistence of different meanings that are not opposite (see Barreiro/Castorina 2017). In Piaget's (1980) theory, the term 'dialectics' acquires a specific meaning as it is used to refer strictly to the constructive process of the emergence of novelty in conceptual systems. The establishment of new concepts assumes that previously existing ones are included in the new one constructed by this movement. Thus, a spiralling process of meaning construction takes place. Piaget (1980) thereby defined dialectics regarding a non-deductive inference leading either from one conceptual system to another more advanced system, irreducible to the former, or to a conclusion from premises that do not include it. Thus, in order to explain how individuals derive knowledge from other knowledge while interacting with the objects of knowledge, Piaget claimed that deductive inferences, characteristic of thought in its structural facet, alternate with dialectical inferences that facilitate interpretation of the dynamics of cognitive development. Through this process, which allows movement from one system of meanings to another that surpasses and includes it, it is possible to conceive the emergence of cognitive novelty. From this point of view, the inferential process leading to the construction of novelties can be studied without referring to the contradictions that took place in a previous logical step.

Thus, the SR of justice (utilitarian, retributive and distributive) formerly independent of each other are integrated – from the age of around nine to ten years onwards – and their properties become dependent on one another. At the same time, they constitute an example of conceptual relativization, in that the properties that characterize them are defined by their relationships with the remaining elements in the system that integrates them. In the case of SR of justice, both retributive and distributive justice become a method for utilitarian justice. They are, in other words, turned into a strategy for achieving happiness for the greatest number of people. This is how the construction of new meaning, in the field of justice as an object of representation, follows the path from the initial independence of its characteristics and properties to integration into a more complex representation. This dynamic of meanings provides an explanation for the development process that facilitates a broader and more abstract understanding of the SR of justice pertinent to their social group.

According to this data, the ontogenesis of SR of justice which involve a meaning-construction process mainly manifests itself in the process of integration and dialectical relativization of the representations (Barreiro/Castorina 2017). It could be argued that the integrated representations express the development of novelties, as the construction of a new form of representing justice includes and transcends the three basic representations (utilitarian, retributive and distributive). As previously stated, more complex representations not only define justice (for example, justice is making people live happily), but also a method of achieving it (for example, administering punishment or rewards according to personal merit).

In another vein, these integrated representations refer to a broader field of the phenomena. Subjects think of the way social or institutional systems involving different individual and social roles work beyond their direct personal experience. The latter is what smaller children base their representations on. The fact that none of the representations of justice is abandoned during development indicates the strong continuity of the collective meaning in the individual conceptualization processes. Moreover, the process of construction of meaning, in this case of integrated representations, can be considered the result of a genuine inference. For instance, it goes from isolated representations of retributive and utilitarian justice to a new representational unit that includes and transcends them. Children are trying to comprehend the SR available in their environment in an attempt to understand their social world. As there is neither a distance nor a causal relation between actions and beliefs, children infer the SR by abstracting them from their early interactions. This process involves not only a compulsory imposition of collective meanings but also an active meaning-making process that will enable children "to get a feeling for the world's constitution and the repetitive collaboration patterns" (Wagner 2015: 21). SR are overarching structures across different patterns of social interaction, and they must be comprised of one's own particular behaviour as well as that of others. Also, SR are dynamic fuzzy structures, and they are inferred by children from stable patterns of correlation across the elements of the unity that cannot be defined out of the social context.

Furthermore, this inferential process is not about replacing previous representations that are less epistemologically valid than others closer to 'reality', since all of them are present within children of the age range considered (six to seventeen). Besides, there are no scientific criteria to establish such a judgment about the epistemic validity of justice representations, because, as explained in the introduction of this paper, there is no consensus on the most adequate sense of justice among philosophers. Justice is a social object, which has been constructed and transformed in the course of a social group's history, legitimizing different social practices (Campbell 2001). Moreover, considering SR as cognitive tools might explain both the mechanisms for their transformation when individuals appropriate them and the way they are abstracted by children in their everyday interactions with adults and peers.

Also, the most frequent representation of justice in adolescents, in which retribution is a method to achieve social welfare, is coherent with the social representation of it identified in adults and in the most-read newspaper. Those have revealed that the hegemonic social representation of justice has a retributive and institutional meaning. However, the aforementioned integration-differentiation process shows that individuals are active agents who reconstruct the hegemonic social representation through their cognitive abilities. More specifically, the retributive social representation of justice is not appropriated simply through compliance with a framework of constrained relations, because the representations become more abstract and complex in relation to the participant's age and social experiences.

Thus, it could be concluded that children and adolescents appropriate such a retributive social representation of justice in the socialization process. It should be noted that the retributive representation of justice does not correlate to any specific general stage of thought development since it is present in all age groups. Rather, retributive representation is enabled by its inference and enactment in specific interactions and discussions, making its particular meaning more salient than other possible meanings of justice. Another interesting issue emerging from the results obtained is the very low frequency of distributive justice in all age groups. This may be explained by the sociogenesis of the social representation. As already mentioned, the distributive meaning of justice might be denied because it would threaten the current social order. However, the appearance of this social representation in some children and adolescents may indicate the presence of an *emancipated* social representation (Moscovici 1988), resulting from the circulation of knowledge in a minority group. However, new research is needed to better understand the characteristic of such a social group and its position in the social field.

## 6.4 The Need for an Interdisciplinary Approach to SR: Social and Developmental Psychology Complementarity

This chapter presented a set of studies dealing with the development of SR and illustrated how developmental psychology could provide insights that offer SRT a more complex understanding of such a process. The collaboration between both theoretical and methodological perspectives when studying the development of social knowledge is possible because their ontological and epistemological philosophical assumptions are compatible (Castorina 2010). That is, they share a relational conception of the world that aligns the conceptualization of their units of analysis as constitutively related elements.

Specifically, utilising both disciplines allows us to study the relationship between collective meaning-making processes that construct SR and the individual's conceptualization activity in appropriating them without considering the former as platonic abstractions and the latter as the result of a cognitive construction disembodied from the personal experience with the social environment. SR are meaning structures constructed in a sociogenetic process that take place over time. However, the contributions of developmental cultural psychology made it possible to interpret the absence of a distributive social representation of justice in the newspaper and adults as the active social construction of nothingness. This developmental dynamic is expressed in how the tension between what is evident and what is neglected allows the construction of a social representation that denies threatening meanings of a social object. My colleagues and I proposed that the process of collective symbolic coping, together with the most-read newspaper, diffuse a retributive social representation of justice, denying other possible ways of understanding a polysemic social object because some meanings can threaten the status quo. If the individuals started to consider the current differences in the distribution of goods and social benefits among different social groups, it would jeopardize the current social order. The social construction of nothingness to repress the problems involved in wealth distribution may lead to the prioritization of the meanings that legitimize the punishment of actions that could threaten the current functioning of the social system.

Nevertheless, while individuals reconstruct the hegemonic retributive social representation of justice, during their socialization novelty emerges. Focused on the ontogenesis of SR, this chapter has tried to show the insights provided by the dialectical inferential processes to account for the individual's reconstructive activity during the process of assimilation of SR. This cognitive mechanism allows us to capture the transformations of collective meanings throughout the process by which individuals actively appropriate them. In other words, individuals reconstruct the SR of justice to understand it, depending on their cognitive abilities and their personal experiences with this social object. As Duveen pointed out: "The knowledge of the social world is limited by the structures we have availed for apprehending it" (Duveen 2002: 141).

Summing up, the meaning-making processes involved in the development of SR of justice imply the construction of novelty within two different types of constraint. One refers to the social guidance of individuals' inferential activity that enables them to think of justice in retributive terms, disabling other possible meanings such as distribution. The other is cognitive because individuals reconstruct or understand SR of justice according to their age-related thought capacity.

# References

Abric, Jean Claude, 2001: "A structural approach to social representations", in: Deaux, Kay; Philogène, Gina (Eds.): *Representations of the social: Bridging theoretical traditions* (Malden: Blackwell Publishing): 42–47.
Andreouli, Eleni; Chryssochoou, Xenia, 2015: "Social representations of national identity in culturally diverse societies", in: Sammut, Gordon; Andreouli, Eleni; Gaskell, George; Valsiner, Jaan (Eds.): *The Cambridge handbook of social representations* (Cambridge: Cambridge University Press): 309–322.
Barreiro, Alicia, 2013a: "The ontogenesis of social representation of justice: personal conceptualization and social constraints", in: *Papers on Social Representations*, 22: 13.1–13.26.

Barreiro, Alicia, 2013b: "The appropriation process of the belief in a just world", in: *Integrative Psychological and Behaviorial Sciences,* 47: 431–449.

Barreiro, Alicia; Castorina, José Antonio, 2015: "*La creencia en un mundo justo como trasfondo ideológico de la representación social de la justicia*", in: *Revista Colombiana de Psicología,* 24,2: 331–345.

Barreiro, Alicia; Castorina, José Antonio, 2016: "Nothingness as the dark side of social representations", in: Bang, Jytte; Winther-Lindqvist, Ditte (Eds.): *Nothingness: philosophical insights into psychology* (New Jersey: Transaction Publishers): 69–88.

Barreiro, Alicia; Castorina, José Antonio, 2017: "Dialectical inferences in the ontogenesis of social representations", in: *Theory & Psychology,* 27,1: 34–49.

Barreiro, Alicia; Gaudio, Gabriela; Mayor, Julieta; Santellán Fernandez, Romina; Sarti, Daniela; Sarti, María, 2014: "*Justice as social representation: diffusion and differential positioning*", in: *Revista de Psicología Social,* 29,2: 319–341.

Bauer, Martin, 1994: "A popularização da ciência como imunização cultural: a função de resistência das representações sociais", in: Guareschi, Pedrinho; Jovchelovitch, Sandra (Eds.): *Textos em representações sociais* (Petrópolis: Vozes): 229–256.

Bergman, Marcelo; Kessler, Gabriel, 2008: "Vulnerabilidad al delito y sentimiento de inseguridad en Buenos Aires: determinantes y consecuencias", in: *Desarrollo Económico,* 48,190/191: 209–234.

Campbell, Tom, 2001: *Justice* (New York: Palgrave Macmillan).

Castorina, José Antonio, 2010: "The ontogenesis of social representations: A dialectic perspective", in: *Papers on Social Representations,* 19: 1–19.

Damon, William, 1990: *The moral child: Nurturing children's natural moral growth* (New York: Macmillan).

Doise, Willem; Clemence Alain; Lorenzi-Cioldi, Fabio, 1992: *Représentations sociales et analyses de donnes* (Grenoble: Presses Universitaires de Genoble).

Duveen, Gerard, 1994: "Crianças enquanto atores sociais: as Representaçôes Sociais em desenvolvimento", in: Guareschi, Pedrinho; Jovchelovitch, Sandra (Eds.): *Textos em representações sociais* (Petrópolis: Vozes): 261–296.

Duveen, Gerard, 1998: "The psychosocial production of ideas: Social representations and psychologic", in: *Culture & Psychology,* 4,4: 455–472.

Duveen, Gerard, 2002: "Construction, belief, doubt", in: *Psychologie & Societé,* 5: 139–155.

Duveen, Gerard, 2007: "Culture and social representations", in: Valsiner, Jaan; Rosa, Alberto (Eds.): *The Cambridge handbook of sociocultural psychology* (Cambridge: Cambridge University Press): 543–559.

Duveen, Gerard; Gilligan, Carol, 2013: "On interviews: a conversation with Carol Gilligian", in: Moscovici, Serge; Jovchelovitch, Sandra; Wagoner, Brady (Eds.): *Development as social process* (London: Routledge).

Duveen, Gerard; Lloyd, Barbara, 1990: "Introduction", in: Duveen, Gerard; Lloyd, Barbara (Eds.): *Social representations and the development of knowledge* (New York: Cambridge University Press): 1–10.

Duveen, Gerard, 2008: "Social actors and social groups: A return to heterogeneity in social psychology", in: *Journal for the Theory of Social Behaviour,* 38,4: 369–374.

Eagleton, Terry, 1997: *Ideología: una introducción* (Paidós: Barcelona).

Hedegaard, Mariane, 2012: "Children's creative modeling conflict resolutions in everyday life as central in their learning and development in families", in: Hedegaard, Mariane; Aronsson, Karin; Højholt, Charlotte; Ulvik, Oddbjorg (Eds.): *Children, childhood and everyday life* (Charlotte: Information Age Publishing Inc.): 55–74.

Hviid, Pernille, 2012: "'Remaining the same' and children's experience of development", in: Hedegaard, Mariane; Aronsson, Karin; Højholt, Charlotte; Ulvik, Oddbjorg (Eds.): *Children, childhood and everyday life* (Charlotte: Information Age Publishing Inc.): 37–52.

Jodelet, Denise, 1985: "La representación social: fenómenos, conceptos y teoría", in: Moscovici, Serge (Ed.): *Psicología social II* (Barcelona: Paidós): 17–40.

Jodelet, Denise, 1991: "L'idiologie dans l'étude des représentations sociales", in: Aebischer, Verena; Deconchy, Jean-Pierre; Lipiansky, Marc (Eds.): *Idéologies et représentations sociales* (DelVal: Cousset Suisse): 15–29.

Jost, John; Hunyady, Orsolya, 2005: "Antecedents and consequences of system-justifying ideologies", in: *Current Directions in Psychological Science*, 14,5: 260–265.

Jovchelovich, Sandra, 1994: "Vivendo a vida com os outros: intersubjetividades, espaço público e representações sociais", in: Guareschi, Pedrinho; Jovchelovitch, Sandra (Eds.): *Textos em representações sociais* (Petrópolis: Vozes): 229–256.

Jovchelovitch, Sandra; Priego-Hernández, Jacqueline, 2015: "Cognitive polyphasia, knowledge encounters and public spheres", in: Sammut, Gordon; Andreouli, Eleni; Gaskell, George; Valsiner, Jaan (Eds.): *The Cambridge handbook of social representations* (Cambridge: Cambridge University Press): 163–178.

Kalampalikis, Nikos; Apostolidis, Thémis, 2016: "Le perspective sociogénétique des representations sociales", in: Lo Monaco, Gregory; Delouvée, Sylvain; Rateau, Patrick (Eds): *Les représentations sociales: théories, méthodes et applications* (Louvain-la-Neuve: De Boeck Supérieur): 79–84.

Kohlberg, Lawrence, 1981: *Essays on moral development: The philosophy of moral development* (San Francisco: Haper & Row).

Leman, Patrick; Duveen, Gerard, 1996: "Developmental differences in children's understanding of epistemic authority", in: *European Journal of Social Psychology*, 26,5: 683–702.

Lerner, Melvin J., 1998: "The two forms of belief in a just world: some thoughts on why and how people care about justice", in: Montada, Leo; Lerner, Melvin J. (Eds.): *Responses to victimizations and belief in a just world* (New York: Plenum): 247–270.

Lloyd, Barbara; Duveen, Gerard, 1990: "A semiotic analysis of the development of social representations of gender", in: Duveen, Gerard; Lloyd, Barbara (Eds.): *Social representations and the development of knowledge* (New York: Cambridge University Press): 27–46.

Marchand, Pascal, 2016: "Les représentations sociales dans le champ des medias", in: *Les représentations sociales: théories, méthodes et applications* (Louvain-la-Neuve: De Boeck Supérieur): 281–392.

Marková, Ivana, 2000: "Amédée or how to get rid of it: Social representations from a dialogical perspective", in: *Culture Psychology*, 6,4: 419–460.

Marková, Ivana, 2012: "Social representations as an anthropology of culture", in: Valsiner, Jaan (Ed.): *The Oxford handbook of culture and psychology* (New York: Oxford University Press): 487–509.

Miller, Joan, 2006: "Insights into moral development from cultural psychology", in: Killen, Meanie; Smetana, Judith (Eds.): *Handbook of moral development* (Nueva Jersey: Erlbaum): 375–398.

Moliner, Pascal; Abric, Jean Claude, 2015: "Central core theory", in: Sammut, Gordon; Andreouli, Eleni; Gaskell, George; Valsiner, Jaan (Eds.): *The Cambridge handbook of social representations* (Cambridge: Cambridge University Press): 83–96.

Moscovici, Serge, 1961: *La psychanalyse: son image et son public* (París: PUF).

Moscovici, Serge, 1988: "Notes towards a description of social representations", in: *European Journal of Social Psychology*, 18: 211–250.

Moscovici, Serge, 2001: *Social representations: Explorations in social psychology* (New York: New York University Press).

Piaget, Jean, 1926: *La représentation du monde chez l'enfant* (Paris: Presses Universitaires de France).

Piaget, Jean, 1932: *Le jugement moral chez l'enfant* (Paris: Presses Universitaires de France).

Piaget, Jean, 1980: *Les formes élémentaires de la dialectique* (Paris: Gallimard).

Psaltis, Charis; Duveen, Gerard, 2006: "Social relations and cognitive development: The influence of conversation type and representation of gender", in: *European Journal of Social Psychology*, 36: 407–430.

Psaltis, Charis; Duveen, Gerard; Perret-Clermont, Anne-Nelly, 2009: "The social and the psychological: Structure and context in intellectual development", in: *Human Development*, 52: 291–312.

Psaltis, Charis; Zapiti, Anna, 2014: *Interaction, communication and development: Psychological development as a social process* (London: Routledge).

Rateau, Patrick; Lo Monaco, Gregory, 2016: "La théorie structural ou l'horlogerie des nuages", in: Lo Monaco, Gregory; Delouvée, Sylvain; Rateau, Patrick (Eds): *Les représentations sociales: théories, méthodes et applications* (Louvain-la-Neuve: De Boeck Supérieur): 113–130.

Salvia, Agustín; Donza, Eduardo, 2017: *Informe: pobreza y desigualdad por ingresos en la Argentina urbana (2010–2016)* (Buenos Aires: Observatorio de la Deuda Social Argentina, UCA).

Salvia, Agustín; Muratori, Marcela, 2017: *Buenos Aires: observatorio de la deuda social Argentina* (Buenos Aires: UCA).

Sammut, Gordon, 2015: "Attitudes, social reprsentations and points of view", in: Sammut, Gordon; Andreouli, Eleni; Gaskell, George; Valsiner, Jaan (Eds.): *The Cambridge handbook of social representations* (Cambridge: Cambridge University Press): 83–95.

Smetana, Judith; Villalobos, Myriam, 2009: "Social cognitive development in adolescence", in: Lerner, Richard; Steinberg, Laurence (Eds.): *Handbook of adolescent psychology, Vol. 1: Individual bases of adolescent psychology* (Hoboken, NJ: Wiley): 187–228.

Starklé, Christian, 2015: "Social order and political legitimacy", in: Sammut, Gordon; Andreouli, Eleni; Gaskell, George; Valsiner, Jaan (Eds.): *The Cambridge handbook of social representations* (Cambridge: Cambridge University Press): 280–294.

Valsiner, Jaan, 2014: *An invitation to cultural psychology* (London: Sage).

Verges, Pierre, 1999: *Ensemble de programmes permettant l'analyse des évocations* (Aix-en-Provence: LAMES-MMSH).

Wagner, Wolfgang, 1998: "Social representations and beyond: Brute facts, symbolic coping and domesticated worlds", in: *Culture & Psychology*, 4: 297-329.

Wagner, Wolfgang, 2015: "Representation in action", in: Sammut, Gordon; Andreouli, Eleni; Gaskell, George; Valsiner, Jaan (Eds.): *The Cambridge handbook of social representations* (Cambridge: Cambridge University Press): 12–28.

Wagner, Wagner; Duveen, Gerard; Farr, Rob; Jovchelovitch, Sandra; Lorenzi Cioldi, Fabio; Markovà, Ivava; Rose, Diana, 1999: "Theory and method of social representations", in: *Asian Journal of Social Psychology*, 2: 95–125.

Wagner, Wolfgang; Hayes, Nicky, 2005: *Everyday discourse and common sense: The theory of social representations* (New York: Palgrave Macmillan).

# Chapter 7
# Common Sense in Gramsci's and Moscovici's Writings: Inspiration, Subversion and Revolution in Sociopolitical and Scientific Fields

Suzzana Alice Lima Almeida

## Introductory Comment

**Maria de Lourdes Soares Ornellas**

*The author Suzzana Alice Lima Almeida consolidated her formation as a researcher in the field of social[1] representation through the postgraduate programme in Education and Contemporaneity at the University of Bahia State – Brazil (PPGEduC-UNEB, Brazil). In her field she has already contributed with the subject 'Education, SR and Subjectivity' offered, each term, to students of the masters and doctoral degree of PPGEduC; besides that, she is a member of the Study and Research Group on Psychoanalyses, Education and Social*

Suzzana Alice Lima Almeida is a professor at the University of Bahia State – UNEB, Campus VII, Brazil. She has a doctorate in Education and Contemporaneity. The preliminary translation of this chapter into English was carried out by Pascoal Eron Santos de Souza of the University of Bahia State in Brazil. Email: suzzanaalice@hotmail.com.

_____

[1]Maria de Lourdes Soares Ornellas is a full professor at the State University of Bahía (UNEB), where she leads the Study Group in Education Research and Social Representations (GEPPE-rs). She is a psychoanalyst and a member of the Psychoanalytic Association of Bahía. She collaborates in the International Centre for Studies in Social Representations and Subjectivity – Education (CIERS-ed) and the Study Group in Discourse Analysis at the Catholic University of Salvador (UCSAL). She also coordinates the International Research Centre in Social Representations (NEARS). Email: ornellas1@terra.com.br.

© Springer Nature Switzerland AG 2021
C. Prado de Sousa and S. E. Serrano Oswald (eds.),
*Social Representations for the Anthropocene: Latin American Perspectives*,
The Anthropocene: Politik—Economics—Society—Science 32,
https://doi.org/10.1007/978-3-030-67778-7_7

*Representations – Geppe-rs/CNPQ, integrating lines of research focused on social representations and education. Geppe-rs is linked to the International Centre for Studies on Social Representations (CIERS/UNESCO/Carlos Chagas Foundation/ FCC) and has biennially produced the collection entitled* Social Representations and Education: Imagetic Letters. *At the time of writing, the last issue (4) was published in December 2017, with the participation of the scholar Denise Jodelet. Geppe-rs is also responsible for organizing some events in Brazil (Salvador – Bahia): the State Symposium on Social Representations and Education (SERS) and the International Symposium on Education, Social Representation and Subjectivity (SIERS), always with the participation of nationally and internationally important and renowned researchers. The author coordinates the executive committee for the events and has participated in all of them.*

*The article entitled "Common sense in Gramsci's and Moscovici's writings: inspiration, subversion and revolution in the sociopolitical and scientific field" highlights common sense as a construct and shows its relationship with cognitive, affective and social aspects of individuals. It deals with a complex concept that does not correspond to notions of common sense as inferior, without scientificity, worthless, colourless, and characterised by intellectual indolence. These ideas give us the opportunity to discuss the constitution of its signifier and signified and, from this perspective, common sense refers to the ethical and political dimension, as well as practical knowledge.*

*Questions measured by Moscovici in relation to science and how it is amalgamated in popular culture bring about changes in the way one thinks about society and show how knowledge of common sense becomes knowledge. In his seminal work analysing social representations Moscovici (2003a) demonstrated that there is a correlation between popular knowledge and the common sense of particular groups regarding scientific knowledge. The author states that social representations are engendered in daily life; they are constituted of contemporary knowledge which comes from common sense and, by this movement, social subjects feel, assimilate, interpret, apprehend the reality and everyday practices in an integrated form. This everyday life of common sense, lived by subjects in their relationships with different groups, is registered in a way that is reminiscent of a camera that portrays reality in each socio-historical period, and, in this register, pieces of information are stored, allowing us to formulate the history of humanities.*

*In the research into and construction of the concept of common sense, the author invites the thinkers Moscovici and Gramsci to an intellectual banquet in order to weave their theoretical approximations and detachments, taking as reference simple logic, the choices, the political process and the implications of its subjective constitution. Gramsci's conception of common sense is seen as the view of the world which the subject designs through the relations of force that exist in society, which becomes a polysemous conception of various integrating elements: education, religion, science, fiction, etc.*

*Thus, Moscovici postulates that common sense can be described this way:*

*(...) as essence of our consensual universe; it includes the cultural and historical meanings of our experiences and activities. It is a rich knowledge, highly diversified and particular for each context. In the consensual universe, society becomes visible; it is innovative, has voice, acts before the world answering to it and prompting changes in the world. (2003a: 323)*

*For the author, it is a knowledge that owns speech, answers to the vicissitudes of the contemporary world and tries to demonstrate the need for changes. It is worth indicating that either Gramsci or Moscovici agree with the idea that thought and language are engendered in common sense; they are expressed through speech, through listening and through the exchanges an individual makes in everyday life. Gramsci and Moscovici emphasize that the usage of language provides subjects with some devices for a possible symbolic (re)construction of society.*

*The author's article demonstrates theoretical and epistemological consistency and it is relevant to the constitution of Social Representations Theory. It deepens a concept that is very important to us, because the ideas presented by Gramsci and Moscovici sometimes seek for the lost object, sometimes rediscover it, go through the presence/absence of it and when this mythical instant happens, there are clear nuances of its conceptual constructions on this object.*

*The article has social relevance because we live in uncertain times marked by a perverse political conjuncture where one can observe a crisis of paradigm in the field of knowledge in which affections of liking and disliking go across individual and collective subjects. In this sense, to deepen common sense in this scenario is to be uncertain of learning to create emancipatory common knowledge.*

*According to Santos (2001: 109) "The emancipation-knowledge, becoming common sense, does not despise the knowledge which produces technology, but understands that such knowledge must be translated into the wisdom of life". It is the life drive, not death drive, in Freudian terms, we are to urge from this society because common sense favours the action, which is not taught or even transmitted.*

*The path historically travelled from common sense knowledge to scientific knowledge may give rise to a new scenario in which there is a shift from scientific knowledge to common sense knowledge. It may be necessary to find a new direction and to attempt social rupture with the objective of facilitating the emergence of the desirable new common sense – emancipatory common sense.*

# References

Moscovici, Serge, 2003: *Representações sociais: investigações em psicologia social* (Rio de Janeiro: Vozes).

Santos, Boaventura de Souza, 2001: *A crítica da razão indolente: contra o desperdício da experiência* (São Paulo: Cortez).

**Abstract** The article entitled *Common sense in Gramsci's and Moscovici's writings: inspiration, subversion and revolution in sociopolitical and scientific fields* represents part of the results of theoretical investigations undertaken for my doctoral thesis, which also had the objectives of investigating convergences between Gramsci's and Moscovici's theories and trying to articulate, through these convergences, an educational proposal for praxis in order to help train people to become proactive, politically-minded members of their community who take action to improve living conditions for all. I undertook theoretical research using both authors' key books on the development of the theory, such as *The Prison Notebooks* (Gramsci 1975) and *Psychoanalysis: its Image and its Public* (Moscovici 1979), besides other sources by scholars who have studied the work of these authors. The methodological path I followed had approximations to paradigms that use change and historicity as guiding elements; therefore, I sought support from critical theory and the Gramscian philosophy of praxis to back up my data analysis. Results suggest that the main theoretical convergence between the authors is related to their approach to common sense, from which emerged subcategories that gave better visibility to such approximations between them. In this chapter I highlight the fecundity of the studies carried out so far and indicate that there is much more to be explored and deepened.

**Keywords** Social representations · Gramsci · Moscovici · Common sense

## 7.1  First Writings

Although there are some tendencies in American social psychology that apparently want to depoliticize this field of knowledge, with resonances also in Social Representations Theory, other lines of thought take the progressive and politicized approach over the deepening of their concepts and categories, especially in European and Latin American social psychology. This allows us, with more vitality, to elucidate and analyse the intertwining of ideas produced by different authors, initially with different theories from social and human sciences which demarcated the ways in which we understand the pattern of modern society and the position of common sense in this scenario. What emerges is the way that the epistemological horizons combine with dialogues which have conceptual adhesions to theories elaborated in different periods and contexts, inviting us to participate in an unprecedented intellectual banquet; to drink from the cup of two important thinkers and discover that they are closely aligned through the complementarity of concepts present in their work. Let us make a toast, then, to this deep and necessary meeting arranged here: Antonio Gramsci (1891–1937) and Serge Moscovici (1925–2014), face-to-face.

Through readings and analyses of the work of both authors, I selected the category that articulates and promotes the approximation of their ideas: *common sense*. This category provides material for thinking about limitations in the conceptual adhesion of the stated theories. From this perspective, I present some subcategories resulting from my analyses.

## 7.2  Theoretical Approximations between Gramsci and Moscovici Through Common Sense

When Gramsci (1999a) criticises Gentile[2] for failing to value common sense in his work, his critique implicitly outlines his own concept of the term, as well as the importance he himself places on common sense. Furthermore, the open-mindedness of the scholar is revealed when he places popular knowledge on the same level as scientific knowledge, demystifying the latter. In other words, the basis for Gramscian thought on common sense is combined with its valorization and importance in Social Representations Theory, although, unlike Moscovician studies, Gramsci's studies do not focus solely on common sense. Gramsci states:

> Gentile's philosophy, for example, is completely contrary to common sense, whether one understands it as the naive philosophy of the people, which rejects any kind of subjectivist idealism, or whether one understands it as being good sense and an attitude of disregard for the obscurity and abstruseness of certain forms of scientific and philosophical expositions. This flirtation of Gentile with common sense is quite bizarre. What we said previously does not mean that there are no truths in common sense. It means that common sense is an ambiguous, contradictory and multiform concept, and that to refer to common sense as a confirmation of truth is a nonsense. It is possible to state correctly that a certain truth has become part of common sense in order to indicate that it has spread beyond the circle of intellectual groups. (Gramsci 1999a: 117–118)

Common sense is not understood in a reductionist way by the author, representing a naïve view or a world-view that has been accepted by society. Gramsci understands common sense as a conception of the world that has been built by history and by the relationship between the powers that exist in society; such a conception is believed to be capable of legitimizing the domination that exists through the action it takes in people hearts and minds or, in an opposite way, it is able to disarticulate this situation.

About this, in *Notebook 11*, Gramsci explains:

> Maybe it is useful to distinguish in a "practical way" philosophy from common sense in order to indicate clearly what has been said: Philosophy specifically means a conception of a world with strong individual features; common sense, on the other hand, is the conception of a world that is diffuse in a particular historical period of a popular mass. If common sense is to be changed, new common sense must be created; this is why it is necessary to take care of the "simple". (1999a: 107)

Thus, it is from its heterogeneous composition that common sense envelopes the seeds for a new perception of world and the will for its realization. This is why it is an important category in both Gramscian theory and the theory of social representation.

Moscovici (2012: 41), inturn, emphatically talks about the discredit of common sense in scientific milieu:

---

[2]Giovanni Gentile, an Italian philosopher, was, with Benedetto Croce, one of the most important exponents of philosophical neo-idealism, and a prominent character in Italian fascism.

Well, our point of view is very clear: these representations are neither an "ancient" form nor a "primitive" form of thinking, nor a way to be in the world; they are normal in our society. Whatever the development of science is, it will always be transformed to become part of everyday life in human society.

The author goes on:

They go above of what is given in science and philosophy, beyond the given classification of facts and events (…). What is received, even in these zones, is submitted to an evolving work of transformation, in order to become the kind of knowledge that most of us use in everyday life. When this happens, the universe is populated by beings, behaviour becomes full of meanings, concepts are highlighted or become concrete (they objectificate, as it is said), enriching the structure of what is reality for each person. (Moscovici 2012: 47)

Moscovici's implicit attitude towards common sense is evinced through his speech, in which his approach is close to poetic. What we see is the delimitation and the affirmation of the place that common sense occupies and its importance in the everyday life of men and women. From that point of view, common sense is also knowledge – another kind of knowledge, beyond the one previously considered valid and dependable by some classical scientific conceptions.

When asked in an interview with Marková about what Marxists thought of the effect of science on common people, Moscovici answered:

So, during the war I started thinking about the impact of science in people culture, how it changes their minds and behaviour, because it becomes part of their system of beliefs, etc. As you can see, this is the kind of question Gramsci asked himself during his years in prison. At that time there were no clear positions on the problem. (2003b: 309)

This appears to provide evidence that Moscovici read Gramsci, especially his *Notebooks 10* and *11*, volumes in which the author discusses culture and science. Writings about knowledge, epistemology, common sense and ideology are also registered in these notebooks; that means they contain the theoretical and epistemological basis of Gramscian theory, whose foundations are also found in Moscovici. In these notebooks, Gramsci (1999b) also criticizes the determinism and mechanicism present in some circles that study Marxism when they discuss historical materialism.

This echoes the answer Gramsci himself gave when he wanted to know: *How are our philosophical choices made*? It is in trying to answer this question that Gramsci gets closer to what is *lived* in social practices, and therefore to what is felt by individuals. This can be explained by "experienced ideological relations" (Gramsci 1999a). It seems that the author, through this question and answer, provides greater clarification of the position that philosophers take when they are engaged in daily life and have the benefit of experience and knowledge. This happens when Gramsci describes, in his explanations of the choices we make – i.e. from our political stance as men and women – our initial engagement with daily social practices and the complexity that arises from this. In the works of Gramsci this engagement is manifested in common sense; even if, according to Gramsci, common sense is regarded as a less prestigious form of knowledge, we can identify in his work complex men and women who also contribute know-how derived from their culture. On this point, the alignment with Moscovici's outlook becomes closer, especially when Gramsci confers on common sense the privileged place of maintaining subjectivity and cultural relations which influence the choices we make and help to determine our way of living.

For Gramsci (1999a), while the foundation of philosophy is the individual and the exhortation "know yourself" – i.e. each person's critical analysis of himself or herself and his or her social reality – "common sense is the conception of world that is uncritically absorbed through several social and cultural environments in which the moral individuality of the average man is developed" (Gramsci 1999a: 113). Because common sense arises from uncritical absorption of various ideological elements from the past and the present, "its basic feature and most fundamental characteristic is being (also in the brain of one individual) a fragmentary, incoherent, inconsequential conception in conformity with the cultural and social position of those masses whose philosophy it is" (Gramsci 1999a: 114).

With regard to "philosophy" being interpreted as "non-philosophy", the heterogeneity which is present in common sense prevents it leading to unity, to an "intellectual order" that creates unity and coherence in the thought of social groups, therefore it could be termed non-philosophy. Cartesianism, order, would be consistent with that philosophy, as previously highlighted.

About this, Moscovici (2012: 73) states:

In fact, social representation summarizes a direct, diverse and diffuse collective rationality from which each participant is self-taught [...] This has some implications for the reciting, descriptive, arborescent style with repetitions, advances and retreats of the texts written through "thoughts that lead to immediate communication.

From this perspective, common sense is not understood as reductionism, or as representing a naive view that is accepted by the majority of society. When analysing Gramsci's thought about the function of common sense, Martins (2008) emphasizes that:

[...] common sense is no longer a simple naive idea, since it is strongly self-serving, that means, it is not ethically or politically naive, neutral or disinterested. In fact, more than merely being a naive knowledge about reality or an idea widely accepted by social groups and classes, in Gramsci, common sense become a plural conception of world, because it gathers different elements (religion, science, fiction, etc) [...]. (Martins 2008: 291–292)

This conception of common sense, with regard to its intention, which is also present in its composition, is pointed out by Moscovici (2003b) – and reinforced by Marková (2015) – when he discusses the "ethics of social representation" and addresses the characteristics of the reified universe and the consensual universe. So, the author says:

Common knowledge is the essence of our consensual universe; it involves the cultural and historical meaning of our experience and activity. It is rich, diverse and specific to different contexts. In the consensual universe, society turns itself visible; it is innovative, it reacts to and, at the same time, provokes changes in the world. (Moscovici 2003b: 323)

In Gramsci (1999a), common sense cannot be disconnected from economic relations, because it has a dialectic association with them, being determining and determined, legitimizer and reproducer, a structuring aspect of the system. Gramsci emphasizes that: "Common sense is not a single conception, identical in time and space: it is the 'folklore' of philosophy and, like folklore, it takes different forms [...]" (Gramsci 1999a: 113).

The richness of the concept of common sense is also apparent when the political dimension appears, occupying the creative space of daily social practices. Staccone emphasizes:

> The field of common sense in Gramsci is always open to new compositions, being converged by religious and pseudoscientific elements from the past which are linked to ideology of dominant classes, from the past and from the present, **but it is also marked by ideas and ethical impulses spontaneously generated in the core of subaltern classes, being the fruit of their experiences of resistance and struggle**. (1993: 82; my emphasis)

The ethical dimension that is present in common sense, sometimes called *good sense* by Gramsci (1999a), is also circulated in everyday knowledge; such a dimension is defended by Moscovici (2003a), as recently analysed by Marková (2015). The researcher says:

> Despite the superficial socialization with their fellows, the individual lives in a deep loneliness of spirit and "with sweet words and hugs, he/she colludes against life and fortune of friends and close people". We can infer that the barbarism of the reflection refers nowadays to an attempt to justify rationally some scientific theories which are themselves irrational like racism and Nazism. Those, according to Moscovici, "take place in colleges and universities, not on the streets" and, being that way, they were legitimated by intellectuals with a significant power of thought. (Moscovici/Marková 2000: 228; Marková 2015: 89)

This fecundity, then, is imbued with common sense when it is understood as a place where, especially in moments of crisis, the two contradictory faces of awareness of social groups are revealed. The superficial, explicit one is exhibited, most frequently, in a verbal way; it is declared and we can witness it routinely in social networks; it is almost always inherited from dominant ideology. The implicit one, on the other hand, results from struggle and solidarity, experiences which we are generally accustomed to neglecting.

## 7.2.1 Gramsci and Knowledge: Epistemological Principles and the Place of Common Sense

In the construction of this analysis, it is relevant to first design a theoretical view of Antonio Gramsci's conception of science and knowledge production in order to stimulate debate on his ideas and Serge Moscovici's, especially Social Representations Theory.

Antonio Gramsci developed his theory on the basis of principles that ruptured and questioned the positivist ideas which were in ascension in Italy; he was motivated, as well, by the adherence of this movement to the actions of the fascist government of Mussolini. The author dedicated Notebooks 10 and 11 to epistemological discussion through his writings on philosophy, including the deconstruction of the philosophy of Benedetto Croce[3] – a scholar to whom Gramsci showed some allegiance – and of Giovanni Gentile. The notebooks contain his ideas about science, scientific knowledge, and common sense, as well as some notes about the philosophy of praxis, highlighting the unity between action and thought, and the individual as being active and creative, and not just a passive object of history. Philosophy of praxis alters not only the common way of defining philosophy, but also its relationship to common sense; it conveys the epistemological bias which underlies his work.

Philosophy, according to Gramsci's understanding, is a view of the world; therefore, it is also a political condition, and, for this reason, all men are philosophers. For Gramsci (1978), the philosophy of praxis is a critical device for refining ancient ways of thinking, in which existing concrete thought, and the existing cultural universe, are important elements.

Gramsci states: "There is no philosophy – meaning conception of the world – without our awareness of historicity..." (1999a: 95). Therefore, he continues, "actually philosophy in general does not exist: there are philosophies or conceptions of the world, and one always makes a choice between them (...). The choice and the critique of a conception of the world are also political facts." (1999a: 95–96).

Praxis, understood as a dialectical unity between theory and practice, is not a merely mechanical factor, but a construct of historical development. This historical development needs to be understood in human logic as the expression of collectiveness and its transformative actions of itself and others, whose relations are of a social and historic nature. Thus, this unity between theory and practices (praxis) is a dialectical relation that places the historical being as a political being, expanding the view of both philosophy and politics, whereby philosophy is history in act, i.e. the existential condition itself (Gramsci 1978).

In *Notebook 11*, Gramsci states:

> In reality, it is possible to "scientifically" foresee only the struggle, but not its concrete moments, which cannot but be the results of opposing forces in continuous movement, which are hardly reducible to fixed quantities since within them quantity changes continually into quality. (Gramsci 1999a: 121)

This perspective of the processes and historical developments in the way we live and survive in the world displays a distinctive conception of the world, a new method of working and the horizon of a new epistemology in the historical context, and also reveals the expansion of the horizons within classic Marxism when it suggests anti-scientificism, i.e. other ways to see the world without any kind of determinism. Gramsci criticized the positivism that was inherent in various

---

[3]Benedetto Croce (1866–1952) was an Italian historian, writer, philosopher and politician. His writings cover a wide range of topics, mainly on aesthetics, philosophy, historiography and history. He is considered one of the most important personalities of the Italian liberalism of the twentieth century.

positions that masquerade as dialectic materialism and were based on a 'naive' conception of matter. He openly criticized the concept of scientific objectivity as well as the assertion that human inter-subjectivity only existed in history. The convergence of these two characteristics could not avoid becoming anti-determinist Marxism, in addition to criticizing "transcendental subjectivism". Semeraro (2000: 4) states: "In the basis of knowledge, as understood by Gramsci, there is no place for any kind of myth, whether it is of a rationalist, empiricist, or irrationalist character. And it criticizes the transcendental subjectivism that ascribes to a universal and abstract mind the assurance of truth."

The originality of his ideas is also anchored in the place he gives to common sense, positioning it close to philosophy, suggesting that elaborated knowledge constantly dialogues with the "simple", in contrast to the arrogance of modern scientificism which is characterized by underlying positivism whose neglect of popular knowledge is fundamental in current attitudes. Yet, in *Notebook 11*, Gramsci says:

> Maybe it is useful to make a "practical" distinction between philosophy and common sense in order to indicate the passage from one moment to the other more clearly. In philosophy, the characteristics of individual elaboration of thought are evident, whereas in common sense it is the diffuse and freed characteristics of a generic form of thought belonging to a specific period and a specific popular environment. But every philosophy tends to become the common sense of an environment, even if it is restricted to intellectuals. Therefore, it is a matter of elaborating a philosophy that – possessing a certain diffusion or the possibility of diffusion, being linked to practical life and implicit in it – becomes a renewed form of common sense with the coherence and vitality of individual philosophies. This cannot occur without feeling the permanent need for cultural contact with the "simple". (Gramsci 1999a: 100–101)

This perspective forms the basis of Santos's (2001) observation regarding "prudent knowledge for a decent life", prompting his first theoretical exhortation *against the waste of experience* and letting us know about the vanguardism of Gramsci's thought.

Analysing Gramsci's theory of knowledge, Semararo points out:

> In fact, Gramsci frequently speaks about a new type of philosopher, about the "democratic philosopher" who, being aware that "every master is always learner and every learner is always master" (Q 1330-2), sets a dialectic relation between science and life, acts to change the environment of which he is part and realizes that the environment itself, the objective reality, "works as a master", while it demands a permanent learning and an unstoppable overcoming of knowledge. (2000: 4)

When Gramsci (1999a), in prison, was feeling distant from the "molecular complexity of real life", he found that the lack of direct and personal contact with real interlocutors became a cognitive obstacle, almost preventing him from thinking and writing, due to the distance from "vivid, direct and immediate impressions of the life of Peter, of Paul, of John, of real people; without understanding it is not possible to understand what is generalized and universalized" (Gramsci 1999a: 222).

Yet, in *Notebook 11*, criticizing Croce's work, *The Popular Manual*, Gramsci (1999b) points out some epistemological fundaments which give support to his form of understanding the world and, therefore, leading his own way to produce knowledge. He affirms:

> But it is the concept itself of "science", as it emerges from the *Popular Manual*, which must be critically destroyed. It is taken root and branch from the natural sciences, as these were the only sciences, or the science par excellence, as decreed by positivism. But in the *Popular Manual* the term science is used in various meanings, some of them explicit, some implicit or barely mentioned. The explicit sense is the one that "science" has in physical research. At other times, however, it seems to indicate the method. But is there a method in general, and if it there is, then surely it can only mean philosophy? At other times, it could mean nothing more than the formal logic: but can that be called a method and a science? It has to be established that every piece of research has its own specific method and constructs its own specific science, and that the method has developed and been elaborated together with the development and elaboration of this specific science and research, forming with them a unique whole. Believing that it is possible to advance the progress of a scientific research project by applying a standard method, chosen because it has given good results in another field of research to which it was genuinely suited, is a strange delusion which has nothing in common with science. (1999a: 122)

Gramsci goes on to state his views about the methodological bias needed for every piece of research, regardless of the scientist's specialization, pointing out some "general criteria" and some necessary competences for researchers from different fields:

> However, there are also certain general criteria which could be held to constitute the critical consciousness of every scientist, regardless of "specialization", and such criteria must always be spontaneously active in his or her work. Thus, one can say that someone is not a scientist if he or she displays a lack of assurance in his/her particular criteria or does not have a complete understanding of the concepts he/she is using, or has scant information about and understanding of the previous state of the problems he/she is dealing with, or is not cautious of his/her assertions, or does not proceed methodically, but in an arbitrary and disconnected way, or cannot take into account the gaps which exist in knowledge already acquired, but rather ignores them, being satisfied with purely verbal solutions and connections instead of declaring that they are provisory positions which may have to be reconsidered and developed, etc. (each of these points can be developed with appropriate exemplifications. (1999a: 122–123)

It is worth noting the practicality, the accuracy and the theoretical and methodological relevance of his views with regard to the principles of qualitative research. Gramsci, the thinker who did not obtain any academic awards, such as a university degree – despite studying at the University of Turin and being an excellent student of literature without, however, completing the course due to the economic problems which led him to become a journalist – wrote in prison, with precision, about science, about scientific knowledge and about ethics in science; however, he did not regard science as the modern god, as was fashionable in the first decades of the twentieth century; instead he correlated scientific knowledge with "the simple" and elucidated the incompleteness of this same knowledge. Gramsci lives.

## 7.3   Theoretical Dialogues between Gramsci and Moscovici: Common Sense and Epistemology

As in Gramscian theory, Moscovici's attempts to map the forms of how we deal with the world evoke an epistemological framework: Social Representations Theory, based on common sense, is not just a kind of knowledge, but has prompted a theoretical debate in scientific circles.

### 7.3.1   Historical Processuality: The Movement of Common Sense

The approximation of these epistemological frameworks in the theories developed by Gramsci and Moscovici appears, initially, in the bias of the processuality that sets the foundations for their conceptions of the world, illustrated by the expression of the movement and dynamism of common knowledge that circulates in social groups, and it is referenced in the works of both authors. In this sense, historicity is something to be built rather than an absolute and immutable matter.

Moscovici (2003a: 12) states:

> Subject does not exist without system and system does not exist without subject. The role of shared representations is to ensure that this co-existence is possible. I mean, it is this state of things which turns the notion of conflict so essential in our theory, whether it is about cognitive transformations or about public communications. Without this notion, one cannot understand either the dynamism of society or the change of any part of it. Well, for reasons that are not strange, social sciences, and social psychology in particular, resist recognizing this role of conflict, of dissension, both in theory and in practice. This, consequently, sets a very static view of individuals and of society.

Society, as represented by Moscovici, is the society of his time – modern western society – and plays a central part in his theory. Therefore, the author is interested in both analysing the role of social representations in real situations of everyday life, with their dynamism and fluidity, and understanding the process of constructing social representations.

Moscovici gets closer to Gramsci when he evinces the movement, the dynamism and the contradiction that are present in his conception of the world. Regarding this, Gramsci (1999b), still criticizing Croce's work *Popular Manual*, says:

> Thinking about a philosophical statement as true in a particular historic period, i.e. as a necessary and indivisible expression of a particular historic action, of a particular praxis, but overcome and "emptied" in a subsequent period, without, however, falling into scepticism or moral and ideological relativism – in other words, seeing philosophy as historicity – is a hard and difficult mental operation. Rather, the author falls into dogmatism, therefore into a form, although a naive one, of metaphysics (…) Philosophy that is implicit in the *Popular Manual* could be called a positivistic Aristotelianism, an adaptation of the formal logic to the methods of physical and natural sciences. The law of causality and the search for regularity, normality and uniformity replaces the historical dialectic. (1999a: 119–120)

During the elaboration of his theory, Moscovici apprehended, in addition to the structural difference, the difference present in the dynamism of modern society compared with previous historic periods (Duveen 2003). The path of changes in previous societies is completely different to the speediness that characterizes modern social dynamism, which is much more accelerated and fluid. It was with the intention of distinguishing between and also keeping away from the integrating and positivistic conception of Durkheim, which is contained within the concept of "collective", that Moscovici chose to replace this concept with the term "social" (Guareschi/Jovchelovitch 2003).

Thus, the term "social" also refers to particular methodological and theoretical developments and temporal demarcations, thereby emphasizing the contrast between the various social groups and their representations. It is additionally necessary to understand the dynamism of the hegemonic conquering of particular representations, mainly through fast mass communication vehicles. The author says: "Knowing that representations are, at the same time, acquired and constructed, one can understand them without the static and predefined features of classic views. It is not the substratum, but interactions which must be considered." (Moscovici 2001: 62).

In this respect, the interaction favoured by communication – which occurs within the limits and possibilities of the cultural and historical context of a society – determines the place where representations are produced. So, the dynamism of the process by which social representations are formed reflects the game of influence, or battles, established between various social groups that united a country or a given society. This occurs, for example, when different groups represent, in diversified ways, the same object and try to bestow on it a hegemonic character which corresponds to their own particular representations (Duveen 2003). This is how the concept of 'reality' sets the scene for the structures and maintenance of power by particular groups. It is worth emphasizing that the confrontations and disputes which occur within social groups clearly represent the principles of Gramscian theory when the author points out the necessity of conflict as a contra-hegemonic principle and reveals the transgressive character of common sense and Social Representations Theory.

With regard to this, in *Notebook 11*, Gramsci (1999a: 93) states: "In acquiring one's conception of the world one always belongs to a particular group which is that of all the social elements which share the same mode of thinking and acting. We are conformists of some conformism; we are always men-in-the-mass or collective men."

## 7.3.2   Communication and its Relation with the Propagation of Common Sense

This approximation is even more apparent when Gramsci (1999b) highlights the role of communication in the production and propagation of knowledge derived from common sense. In *Notebook 11*, Note III, he emphasizes: "If it is true that every language keeps the elements of a conception of the world and of a culture, it will be equally true that it is possible, from the language of each individual, to assess the greater or lesser complexity of his/her conception of the world" (Gramsci 1999a: 94).

Gramsci also demonstrates this when he points out particular institutions which are opinion-makers and therefore influence subjects in the elaboration of their conception of world. The author states:

> On the other hand, it will not be difficult to realize when such proposals for discussion emerge from interested motives and not from scientific ones. Likewise, it is not impossible to think that individual initiatives could be disciplined and subject to an ordered procedure, so that they must pass through the inspection of academies or cultural institutes of various types to become public after undergoing a selection process. It would be interesting to study concretely, within a given country, the cultural organization that keeps the ideological world in movement in order to examine how it works in practice. A study of the numerical relation between the section of the population professionally engaged in active cultural work in the country in question and the population as a whole would also be useful, together with an approximate calculation of the unattached forces. The school, at all levels, and the Church, are the biggest cultural organizations in every country, in terms of the number of people they involve. Then there are newspapers, magazines and the book trade and private educational institutions, either those which are part of the state system, or cultural institutions like the Popular Universities. Other professions include among their specialized activities a fair proportion of cultural activity, such as doctors, army officers, and the judiciary. However, it must be noted that in all countries, though in different degrees, there is a great gap between the popular masses and the intellectual groups, even the largest ones, and those nearest to the national peripheries, teachers and priests. (1999a: 115)

It seems that Moscovici read *Notebook 11*, paying particular attention to the notes which discuss issues related to philosophy, science, and common sense, and then choosing, on the basis of these considerations, a group formed by various subjects/contributors for the thesis from which his theory was created and whose focus was, as previously mentioned, social representations of psychoanalysis in French society, particularly the Parisian population.

Moscovici (2012: 31) explains about the "surveyed population"

(...) Liberal population (L.P.) in which are included teachers, doctors, lawyers, technicians, and clergymen. Worker Population (W.P.), a group that engages both specialized and non-specialized workers, and foremen etc. Student Population (SP), students from the University of Paris. Population of the students from technological school (T.P) including students from courses of secretariat, optics, ceramics, etc., who are 18 to 22 years old.

Additionally, he analysed the contents of the press. The author states: "A large part of the press was studied from January 1952 up to March 1953" (Moscovici 2012: 33). It seems that such choices and parameters were no coincidence.

### 7.3.3  About Method or Methods: Critique of Modern Science and the Discredit of Common Sense

The approximation between Gramsci and Moscovici in this debate about science and knowledge production appears in these authors' thoughts about the dualism that exists in the scientific arena and the consequent dispute that emerges in it about methods and methodology, etc, in addition to the discrediting of common sense. As previously noted, Gramsci (1999a: 122) pointed out: "But is there a method in general? […] Believing that it is possible to advance the progress of a scientific research project by applying a standard method, chosen because it has given good results in another field of research to which it was genuinely suited, is a strange delusion which has nothing in common with science."

In the same way, Moscovici (2003b: 14) states that:

> There are no methods for a field of knowledge that has true intellectual content. The objective is to find truth. How to get there, nobody knows. Experimental methods, mathematical ones, the different techniques, they are not methods of work to find truth. A biologist or a physicist will never become creative because someone has told them: here you have some methods, try them with a new organism. This is done by someone who does not know what to give for students do. It is an admission of failure.

Such concern, marked by this epistemological bias which proposes ways to overcome dichotomies and search for a dialogue between the different currents that guide the production of knowledge, is present in the work of each author when discussing scientificity. The similarity is so evident that it prompts visions of a fictitious meeting between those two authors just to discuss the subject. Gramsci (1999a: 131) affirms:

> The question is closely related, which is understandable, to the question of the value of the so-called exact or physical sciences, as well as the position they have taken within the philosophy of praxis, a position almost of fetishism; in other words, a position of the only true philosophy or knowledge of the world.

Moscovici (2003b: 14), enthusiastically corroborating with Gramsci, replies:

> I have repeatedly written: I am fundamentally opposed to the tendency to fetishize a particular method. Using experimental methods, or other non-experimental methods, as a royal guarantee to reach knowledge is as pernicious as any fetishism.

Another discussion about the choice of appropriate methods is related to the strand of objectivity and rationality in the studies. Marková (2015), defending "common sense as an epistemology of social representations", notes:

> In one of his most recent works, Serge Moscovici affirms that he has always been intrigued by the fact that "most of the theories or discussions related to prejudice, stereotypes and group relations are expressed in terms of categories and the logic of facts" (Moscovici[4] 2011: 445), totally ignoring values and ethical choices. Opinion and attitude scales, questionnaires about attributions, stereotypes, prejudices and influences – all of them aim to examine facts and categories, information and citizens' rational thought, as if participants in such studies were rationalized machines that express their thought without any active engagement in such socially valued phenomena. (Marková 2015: 86)

Concerning that, in *Notebook 11*, still criticizing the theory of Croce, Gramsci (1999a: 319) turns to Missiroli's 1932 manuscripts to refer to the growth

---

[4]This text was published in *Social Science Information* (2011) under the title 'An essay on social representations and ethnic minorities', not available in Brazil.

tendency of the rationalist, experimental model based on the "absolute positivism" embraced by the fascist regime, where social practices and popular culture
– mouthpieces of common sense – are neglected.

> To understand Croce's atitude in the second post-war, it is useful to recall the answer
> sent by Mario Missiroli to an inquiry promoted by *Saggiatore* magazine and published
> in 1932 (it would be interesting to know all the answers to the inquiry). Missiroli wrote
> (cf. *Fascist Critique*, 15 May 1932): "I cannot see anything well delimited, but only the
> overall state of attitudes, tendencies, morals. It is hard to predict what the culture orien-
> tation will be; but, I do not hesitate to formulate a hypothesis that we are marching to an
> absolute positivism, which honours again science and rationalism in the ancient meaning
> of the word. Experimental research may be the glory of this generation that ignores and
> wants to ignore the verbalism of the most recent philosophy. It does not seem reckless to
> predict the resumption of anti-clericalism, which, personally, I am far from desiring.

This scenario allows us to reaffirm the approximation of the authors' ideas,
even though Moscovici did not register or give any direct indication of his views
on this dialogue or his theoretical affinity. The resistance of certain currents,
within social psychology, to legitimising the important role of social phenomena
related to their political dimension, which is characteristic of the known dual-
isms between psycho X social, can provide some clues about this reality. For a
psychologist, using Gramsci as a referent still seems to characterize an epistemological conflict within currents of psychology and sociology even in the current
context. Furthermore, Gramsci's detachment from and disenchantment with the
Communist Party, of which he was a member in his youth, seems to be a delim-
iter of his positioning. In the late 1940s and in the 1950s and 1960s, the release of
Gramsci's manuscripts – all thirty-three notebooks – was at the discretion of the
Communist Party, and they were initially edited for publication in a more tenden-
tious format; the thematic version served as party-political auto-promotion, despite
the shaken bonds and the detachment of Gramsci himself from the party during
the entire period he was writing his theory in prison.

Thus, it is important to note that the debate and correspondence of theories are
demonstrated when the authors discuss the place of common sense in the knowledge production process based on their conceptions of the world and the society
present in these categories. Other concepts, such as the Gramscian concept of ideology, widen the visibility of this theoretical similarity, using common sense as a
point of reference.

## 7.4   Theoretical Dialogues Between Gramsci
## and Moscovici: Common Sense, Social
## Representations and Ideology

Studies have already suggested that in scientific literature there are at least sixteen
different views about the concept ideology (Eagleton 1997), some of which are
compatible and others incompatible. It is the concept originating from Marxism

which is at the forefront of analyses in the current literature. In Gramsci's work the concept of ideology is used to signify the *theory of ideology in general*. Hence, it is inherent in any human society and is dissociated from particular interests; its role is to provide the cohesion of society through a set of ideas, values and concepts, and a shared view of world. The other perspective concerns the theory of specific ideologies whose main role is to ensure the domination of a class. Both perspectives – general and specific – were also developed by Althusser (1996).

In this direction Xavier (2002: 32) states: "Marx's concept of ideology is only one of the faces of ideology; preceding it, there is another view: the one of the ideas and representations of society in general, about which influences and manifestations of particular ideologies are established."

So, the political and social character of ideology is also linked to its role in the composition of individuals. Individuals' ideas and views of the world ground their practices; there is no separation; people embrace or reject roles on the basis of their ideology. Therefore, the power and strength of ideology – understood as general ideology – is its characteristic of defining and legitimizing conduct; general ideology contains the raw materials of particular or specific ideologies.

Gramsci rejected the negative notion of ideology, with the meaning of domination and alienation, in several parts of his notebooks. In his theory, he stressed the differences between arbitrary ideologies (general and spontaneous) and organic ideologies (specific, the ideologies of class). In his concept of ideology, individuals are autonomous. Thus, ideology is, for Gramsci "the highest meaning of a conception of the world and it manifests itself, implicitly, in art, in law, in economic activities, in all kinds of manifestations of collective and individual life." (1999a: 98–99).

Despite this inherent character of ideology, the author divides the concept into four levels: philosophy, religion, common sense and folklore, in a descending order of systematization and intellectual articulation. For him, all levels are in the realm of ideologies in general or arbitrary ones, and, in turn, correspond to what we understand *as social representations*. Even at the first and highest level – philosophy – Gramsci associates the intellectual with common sense when he affirms: "all of men are philosophers" (1999a: 92). That means that common sense also develops theories, in spite of not being systematically driven by organic or "pure" reason. This emphasizes even more the close approximation with the definition of social representations, taking into consideration the discursive bases in various categories in the author's studies, mainly because they are linked to knowledge that is spontaneously developed in everyday life and therefore generates common sense and orientates conduct.

In this scenario, the power of ideology is also the power which we attribute to social representations to guide the behaviour of men and women. Knowledge which circulates within social groups and is assimilated as a reference point for conduct and choices cannot be ignored. Nevertheless, Moscovici (2003b) several times refers to the concept of ideology in its strict sense, i.e. how *organic ideology* (specific ideologies) – in practice, the discursive register that appears in the *Prison Notebooks* written by Gramsci in the early twentieth century – approximates this category based on what he himself wrote about the concept of social representations in the middle of the same century, taking into account Gramsci's explanation of arbitrary ideologies (ideology in general).

It seems relevant to reaffirm Jodelet's definition of social representations. She states that it

> is a kind of knowledge socially shared and elaborated, with a practical objective, and contributes to the construction of a common reality for a social group. It is equally indicated as knowledge of common sense or even naive knowledge, natural; this kind of knowledge is differentiated, among others, from scientific knowledge. However, it is considered an object of study as legitimate as that one, due to its importance in social life and the elucidation that enables cognitive processes and social interactions. (Jodelet 2001: 22)

About this, Gramsci (1999b), in *Notebook 11*, in a didactical way, explains how this socially shared knowledge leave marks in the individual and guide his/her perception of the reality, and, therefore, in the guidance of his/her conduct.

> On what elements, then, is his philosophy founded? And, in particular, is his philosophy in a form that has great importance for him with regard to standards of conduct? The most important element, undoubtedly, is one of non-rational character: it is an element of faith. But faith in who, or in what? Mainly in the social group to which he belongs, so far as it also thinks as he does: the man of the people thinks that so many people cannot be wrong, as the arguing opponent would like him to believe; he thinks that he himself is not able to support and develop his reasons as the opponent does; but he thinks that in his group there is someone who could do that, and could certainly argue better than the referred opponent; and he remembers, in fact, having heard someone expounding, coherently, in a way to convince him, the reasons for his faith. He does not remember the precise reasons presented and he would not be able to repeat them, but he knows they exist, because he has heard them being expounded and was convinced by them. (Gramsci 1991)

For Gramsci, 'faith' refers to a secular dimension of the term. It is about believing in something motivated by "certain historical rationality", but of "impassioned finalism" (Gramsci 1999a: 106). Thus the subject of individual will meets the subject of reason. The subject of social psychology also appears when Gramsci conceptualizes the idea of "faith" which he constantly uses in his manuscripts. Moreover, he emphasizes the importance of common sense – referring in his theory to "its philosophy" – and its role in "standards of conduct".

Moscovici confirms that this aspect is also present in Social Representations Theory:

> It qualifies as a social representation because it is collectively elaborated and shared, which gives sense to behaviour. Shared social representations produce meanings which help individuals to understand, act, and orientate themselves in the social environment. (Moscovici 2003a: 65)

Moscovici also highlights the formation of social representations and their function in integrating the new into the social philosophy that directs conduct according to social reality.

> The particularity of social representation is not in the fact that it is formed by the structure of society, but in its role in the guidance of social conduct and in its cognitive function to integrate the new into social thinking. The importance of studying the social representations of individuals regarding certain aspects of reality lies in the social function of orientating behaviours and preparing actions, which is, in fact, considered more important than

the analysis of its social production, since, like ideology, science and myths, it is socially determined. (Moscovici 1978: 76)

In Notebook 11 Gramsci (1999b: 111) emphasizes the process of adhesion to the new – a new ideology – within social groups. Like Moscovici, he also gives visibility to the cognitive aspects – or "rationality" – involved, and warns against the "arbitrary" constructions which are also part of the process. However, Gramsci stresses that constructions which meet the needs of a certain historical moment tend to be incorporated easily.

> The adhesion or non-adhesion of the masses to an ideology is the way in which the real critique of rationality and the historicity of the modes of thinking are verified. Arbitrary constructions are pretty rapidly eliminated by historical competition – even if sometimes, because of a combination of immediately favourable circumstances, they enjoy some popularity – since constructions that respond to the demands of a complex and organic period of history always impose themselves and prevail. (Gramsci 1999a: 111)

The author adds: "It would be interesting to study concretely, within a given country, the cultural organization that keeps the ideological world in movement in order to examine how it works in practice" (Gramsci 1999a: 112). Thus, we could replace the term *ideological world* with the term *world of social representations*, and the sense would be the same, considering the approach of the authors: "It would be interesting to study concretely, within a given country, the cultural organization that keeps the world of social representations in movement in order to examine how it works in practice."

Brazil would be an excellent sample to study at the current political and historical time, since its cultural features are clearly demarcated; world-views, values and affections are heightened, and therefore offer the possibility of demonstrating the importance of studying such concrete realities through the theoretical approaches of Gramsci and Moscovici, whose works, whether from the perspective of arbitrary ideologies or social representations, allow us to understand a bit more about the common sense knowledge that filters through to the social environment. In the same way, common sense has received attention and respect in Brazilian research and has influenced the majority of analyses.

## 7.5 Theoretical Dialogue Between Gramsci and Moscovici: Common Sense and Popular Knowledge

The similarity between pieces of information that circulate about common sense in the works of Gramsci and Moscovici also appears in the space they both give to popular knowledge. The transgressive dimension of their work is much more visible in this category. Whether in political, cultural or scientific aspects, resistance to modern standards is manifested in various senses. Note the attention to language, in both theories, as the main provider of the meeting between common sense and popular knowledge.

### 7.5.1 Language and Popular Knowledge: Persuasion and Influence in the Production of Common Sense

In *Notebook 11* Gramsci refers to a popular saying that was common in Italy in the twentieth century to explain the breadth and importance of popular knowledge; it is also familiar nowadays in Brazil, with particular resonance in Bahia State, adapted to *"doing things with philosophy"*, or *"doing things with science"*. Gramsci asks:

> What is the popular image of philosophy? It can be reconstructed through expressions in common usage. One of the most widespread is "being philosophical about it", which, if it is analysed, is not to be entirely rejected. It is true that it contains an implicit invitation to resignation and patience, but it seems that the most important point is rather the invitation to people to reflect and to be aware that whatever happens is basically rational and must be confronted as such, and that one should apply one's power of rational concentration and not let oneself be carried away by instinctive and violent impulses. These popular expressions could be put together with similar expressions used by writers of a popular stamp (examples being drawn from a large dictionary) which contain the terms "philosophy" or "philosophically". Thus, one can see from these examples that the terms have quite a precise meaning: that of overcoming bestial and elemental passions through a conception of necessity which gives a conscious direction to one's actions. This is the healthy nucleus of "common sense", which could be called good sense and which deserves to be developed and turned into something more unitary and coherent. So, it becomes clear that it is not possible to separate what is known as "scientific" philosophy from "vulgar" and popular philosophy, which is only a fragmentary collection of ideas and opinions. (Gramsci 1999a: 97)

The first question followed by the answer Gramsci himself gives, "What is the popular image of philosophy? It can be reconstructed through expressions in common usage", leads us to associate his ideas with the Social Representations Theory of Moscovici. In this respect, the term *idea* can be replaced with the term "social representation of philosophy". Yet, according to the approach of Gramsci (1999a), the terms *idea* or *social representation* could be understood through expressions of common language that circulate in social groups, in the same way as defended by Moscovici (2003a) when he affirms that in studying common sense, or popular knowledge, we are also studying something that "connects society, or individuals, to their culture, their language, their familiar world" (Moscovici 2003a: 322).

It is also important to emphasize the conception of people that Gramsci highlights when he analyses the popular saying "Being philosophical about it": "the invitation to people to reflect and to be aware that whatever happens is basically rational and must be confronted as such, and that one should apply one's power of rational concentration and not let oneself be carried away by instinctive and violent impulses." The subject of social psychology is represented here: objectivity and subjectivity, reason and emotion/affection. The idea of a complex person – someone who goes beyond thinking, and doesn't just function on a purely cognitive level – is visibly presented by Gramsci as someone who is also moved by "instinctive and violent impulses". It is this person who develops the sense and perception that things have, and popularizes these ideas through particular forms of language that circulate within the group – popular expressions. Thus, the author demonstrates that "it is not possible to separate what is known as 'scientific' philosophy from 'vulgar' and popular philosophy."

With regard to this, Marková (2003) summarized a fragment of her conversation with Moscovici, in which he talked about language and popular knowledge:

> Thus, if I understood you correctly, in order to develop a social psychology of knowledge, it is necessary to start with questions related to popular knowledge and cultural knowledge, which social representations are part of, and through them, they develop. Its genesis is studied through conversation, advertisement, media and other means of communication based on language. Representations are inserted in the meaning of the words and, consequently, are recycled and perpetuated through public speech. And, of course, you mentioned before that culture plays an important role in the formation of social representations. (Marková 2003: 321)

Thought and language that are present in common sense, expressed through everyday speech, stress the need to always take into consideration – when we intend to approach social thought – the fact that the usage of language provides individuals with mechanisms for creating a symbolic (re)construction of reality, giving sense to facts that surround one's existence, as emphasized by Gramsci.

In *Notebook 11*, Gramsci (1999b) clarifies how knowledge becomes popular, or how "new conceptions of the world are spread" through language usage. The didactic explanation of the author reminds us about the process of objectification and anchoring in the production of social representations by people within their social groups. Gramsci asks:

> Why and how do new conceptions of the world spread, becoming popular?
>
> In this process of diffusion (which is simultaneously substitution of the old conception and, very often, combining old and new), do they influence (how and to what extent) the rational form in which the new conception is expounded and presented, the authority (in so far as this is recognized, appreciated, at least generically) of the expositor and of the thinkers and scientist from whom the expositor takes support, the participation in the same organization of that man who upholds the new conception (after taking part in the organization, but for reasons other than sharing the new conception)? In reality these elements vary according to social groups and the cultural level of the groups in question. But the research has a particular interest in relation to the popular masses, who seldom change their conceptions, or who never change them in the sense of accepting them in their "pure" form, but always and only as a more or less heterogeneous and bizarre combination. The rational and logically coherent form, the perfection of reasoning which never neglects any argument, positive or negative, of any significance, has a certain importance, but is far from being decisive. It can be decisive, but in a secondary way, when the person in question is already in a state of intellectual crisis, wavering between the old and the new, when he has lost his faith in the old and has not yet favoured the new, etc. The same can be stated about the authority of thinkers and scientists. It is very widespread among people. But, in fact, every conception has been favoured by scientists and thinkers, and authority is divided; besides this, it is possible, with every thinker, to cast doubt on whether he really has said such things exactly that way. (Gramsci 1999a: 108–109)

Thus, Gramsci presents a mental and collective scheme in the process of apprehension of a "new" conception of the world, in which some elements are highlighted, such as: diffusion, the rational form of the exposure, the authority of the expounder, participation in the group, popular masses, rejection of unqualified acceptance of the new, intellectual crises involving "the new and the old", and rejection of the rational form as decisive in the apprehension of the new. Gramsci reminds us about the need to correlate new information with popular knowledge, given that, within social groups, other knowledge is already anchored to form common sense and there will be conflict, which means a tendency to resist the process of incorporation, even if new pieces of information are linked to the knowledge of authorities – thinkers and scientists.

Confirming this theoretical adherence and the approximation of this category with Social Representations Theory, in her interview with Marková/Moscovici (2003) uses the example of the prevention campaign against AIDS in the 1980s in England, and its consequent inefficacy in some social circles, emphasizing that the language used did not dialogue with the popular knowledge that was already present within some social groups.

Marková (2003) highlights:

> I'd like to give an example to enhance the theme. What you have said about the study of attitudes in social psychology is also applied, in general, to the study of the thought of problems solutions, of concepts and of the formation of concepts. These subjects are also based in the ontological assumption that the object of study and the self are independent. This has important epistemological consequences for the theory of concept formation, solutions to syllogisms and anagrams, acquisition of meaning for words and so on. My personal critique about such a position was made more than twenty years ago, but to make the question clear, I will refer to a more recent example in the field of health education, relating to AIDS. The campaign, in England in the 80s, was conducted under the slogan "Do not die of ignorance". It was supposed, in that campaign, that the individual, in order to protect himself against HIV/AIDS, had to obtain technical knowledge. It totally ignored the existence of popular 'knowledge', in which there were representations of HIV/AIDS which were part of the culture and, for that reason, were part of the individual's mindset; this popular knowledge and these representations were anchored in sin, sexually transmitted disease, obscenities and other undesirable phenomena. These representations had a stronger influence on people's activities than neutral and objective knowledge about the virus, antivirus, infected needles, and condoms that were given to them through health campaigns. The representations of HIV/AIDS were threatening to the self: taking preventive action against the acquisition or transmission of HIV could, at the same time, constitute proof, to others, that the individual might be infected. This, meanwhile, could lead the individual to be rejected by others. In general, what these campaigns could have done was to take seriously the social and popular representational knowledge, its linguistic expression and the particular reasoning. (Marková 2003: 320–321)

Moscovici (2003b: 321) answered: "No, I cannot see differently."

In Note IV of *Notebook 11*, Gramsci reaffirms his critique of this detachment of scientific knowledge from popular knowledge and emphasizes the need for dialogue and socialization of discoveries in order to guide of the actions of social groups in a responsible way. He says:

> Creating a new culture does not only mean one's own individual "original" discoveries. It also, and overall, means the diffusion in a critical form of truths already discovered, their "socialization", as it were, and, therefore, changing them into the basis of vital action, an element of coordination and of intellectual and moral order. The fact that a crowd of men are led to think coherently and in a unitary way about the present reality is a

"philosophical" event far more important and "original" than the discovery by some philosophical "genius" of a truth which remains the property of small groups of intellectuals. (Gramsci 1999a: 95–96)

Thus the theories of both Gramsci and Moscovici recognize the diversity and richness of the knowledge of common sense as a guide to everyday reality. Their works demonstrate that common sense is marked by dialogical tensions and different perspectives. It is this movement which strengthens the production and renewal of common sense. The contradictory character of common sense is also apparent, mainly when it is linked to popular knowledge and varies according to circumstances and contexts. Thus, language is, for both Gramsci and Moscovici, a living expression of conceptions about life, of popular common sense and of culture in general.

## 7.6   (In)conclusive Considerations

When the writings of Gramsci and Moscovici were examined in a quest to identify elements which demonstrate their views on the dynamic of knowledge that circulates in the everyday lives of individuals, filters into the social environment, and influences the behaviour of groups, the main convergence in their works was found to be common sense. This signifier is illustrated by the quantity of categories presented in the results, and in the subcategories which organically emerged from them. These latter provided a framework and gave support to this analysis and these interpretations. In general terms, this analysis suggests that in both theories there is the recognition of common sense knowledge with all its diversity and richness, guiding individuals' conduct in everyday life.

Common sense can be accredited as a kind of knowledge different from scientific knowledge. When we study common sense, we are also studying a category that unites society – people – with their culture, their language and their familiar world. There is much more to be explored and investigated in this scenario and these approaches.

## References

Althusser, Louis, 1996: "Ideologia e aparelhos ideológicos de estado", in: Zizek, Slavoj (Ed.): *Um mapa da ideologia* (Rio de Janeiro: Contraponto).

Duveen, Gerard, 2003: "O poder das ideias", in: Moscovici, Serge: *Representações sociais: investigações em psicologia social* (Rio de Janeiro: Vozes).

Eagleton, Terry, 1997: *Ideologia: uma introdução* (São Paulo: Boitempo/UNESP).

Gramsci, Antonio, 1978: *Concepção dialética da História* (Rio de Janeiro: Civilização Brasileira).

Gramsci, Antonio, 1999a: *Introdução ao estudo da filosofia: a filosofia de Benedetto Croce* (Rio de Janeiro: Civilização Brasileira).

Gramsci, Antonio, 1999b: *Cadernos do cárcere* (Rio de Janeiro: Civilização Brasileira).

Guareschi, Pedrinho; Jovchelovitch, Sandra, 2003: *Textos em representações sociais* (Petrópolis: Vozes).

Jodelet, Denise (Ed.), 2001: *As representações sociais* (Rio de Janeiro: EdUERJ).

Marková, Ivana; Moscovici, Serge, 2003: "Ideias e o seu desenvolvimento: um diálogo entre Serge Moscovici e Ivana Marková", in: Moscovici, Serge: *Representações sociais: investigações em psicologia social* (Petrópolis: Vozes).

Marková, Ivana, 2015: "Ética na teoria das representações sociais", in: Jesuíno, Jorge Correia; Mendes, Felismina Rp; Lopes, Manuel José (Eds.): *As representações sociais nas sociedades em mudança* (Petrópolis: Vozes).

Martins, Marcos Francisco, 2008: *Marx, Gramsci e o conhecimento* (Campinas – SP: Autores Associados/Americana/Centro Universitário Salesiano de São Paulo).

Moscovici, Serge, 2001: "Das representações coletivas às representações sociais: elementos de uma história", in: Jodelet, Denise (Ed.): *As representações sociais* (Rio de Janeiro: Ed. UERJ).

Moscovici, Serge, 2003a: *Representações sociais: investigações em psicologia social* (Rio de Janeiro: Vozes).

Moscovici, Serge, 2003b: "Le premier article, *Journal des Psychologues*, n° hors série sur Serge Moscovici: le père de la théorie des représentations sociales: seize contributions pour mieux comprendre": 10–13.

Moscovici, Serge, 2012: *A psicanálise, sua imagem e seu público* (Petrópolis: Vozes).

Santos, Boaventura de Souza, 2001: *A crítica da razão indolente: contra o desperdício da experiência* (São Paulo: Cortez).

Staccone, Guiseppe, 1993: *Gramsci: 100 anos de revolução e política* (Petrópolis: Vozes).

Xavier, Roseane, 2002: "Representação social e ideologia: conceitos intercambiáveis?", in: *Psicologia & Sociedade*; 14,2 (July/Dec): 18–47.

## Other Literature

Semeraro, Giovanni, 2000: "*Para uma teoria do conhecimento em* Gramsci", at: http://www. acessa.com/gramsci/?id=284&page=visualizar. (May 19 2020).

# Chapter 8
# Diffusion, Propaganda and Propagation: The Actuality of the Construct

Claudomilson Fernandes Braga

## Introductory Comment

Pedro Humberto Faria  Campos

*In the introduction to the second part of Volume II of*[1] Attitudes et Représentations Sociales *(La Psychologie Sociale collection, compiled by Jean-Léon Beauvois 1996), Dechamps and Beauvois (1996) make a curious statement that the researchers of cocial cognition (also known as 'cognitive social psychology') and those engaged in the field of Social Representations Theory work on the same platform:*

> *Today, without truly communicating, without truly ignoring each other, the two traditions coexist and work on two very similar platforms. The study of attitudes, on the one hand, is mainly active in Anglo-Saxon literature; whether you like it or not, it integrates, through concepts of beliefs and knowledge, the elements that refer to social representations. The study of representations, on the other hand, is markedly active in literatures in Latin languages; and it cannot be exempted from the study of evaluative processes which are, as we have seen, characteristic of attitudes, to the point of being the common denominator of their definitions (Dechamps/Beauvois 1996: 133).*

---

Claudomilson Fernandes Braga gained his PhD in psychology at the Pontifical Catholic University of Goiás (PUC Goiás). He has a post-doctoral degree in psychology from PUC Goiás, and is Professor of the Postgraduate Programme in Communication (PPGCOM) in the Faculty of Information and Communication at the Federal University of Goiás (FIC/UFG). Email: milsonprof@gmail.com.

---

[1]Pedro Humberto Faria Campos is full professor at Salgado Filho University. He obtained his undergraduate degree in psychology at the Catholic University of Goiás (1988), his Master's in education at the Federal University of Goiás (1994), his Master's in social psychology at the Université de Provence (1995), and his PhD in psychology at the University of Provence in France (1998). His areas of interest are social representations, the structural school of social representations, social exclusion, social practices and violence. Email: pedrohumbertosbp@terra.com.br.

© Springer Nature Switzerland AG 2021
C. Prado de Sousa and S. E. Serrano Oswald (eds.),
*Social Representations for the Anthropocene: Latin American Perspectives*,
The Anthropocene: Politik—Economics—Society—Science 32,
https://doi.org/10.1007/978-3-030-67778-7_8

*The difference is that the first theory (SRT) primarily focuses on the processes of communication, i.e. interactions 'in the making'. It is not a big deal for those who know of Moscovici's claim to have always been associated with the "cognition train" (Moscovici, 1986). Not by chance did he act decisively and become the first president of the European Association of Experimental Social Psychology; not by coincidence was the experimental study of SR based on game theory among the first two lines of research in the laboratory he created at the École des Hautes Études en Sciences Sociales (EHESS). However, we should not allow ourselves to make the mistake of believing that SRT is just another theory in the field of social cognition (such as Cognitive Dissonance Theory, or Attributions). Placing communication at the centre of the theory is highly significant, since communicative practices cannot be studied exclusively from either a cognitive or a sociological point of view.*

> I was always asked what I understood by sharing a representation, by shared representations. It is characterized not by the fact that it comes from itself, that it is common, but rather from the fact that its elements have been formatted by communication and are in relation to communication (...) I speak of shared representations to indicate that the forms of our thinking and our language are compatible with the forms of communication and the limits imposed by this (Moscovici 2013: 111).

*As social-cognitive phenomena, social representations are organically embedded in social interactions, from both a more objective perspective (considering institutions, power relations, social positions, institutional roles, media and forms of communication in mass societies), and a more 'subjective' (symbolic and interactional) perspective of interactions. Additionally, SRT is constructed in part with a constructivist epistemological basis and a stance against all radical objectivisms and subjectivisms. Again, not by chance, Moscovici (1987) uses his theory of social representations to clarify more dynamic and interacting sociological visions like those of Weber, Simmel, Goffman, and Cicourel.*

*Let's return to the importance of studying communicative processes in SRT. A central postulate is the establishment of the processes of objectification and anchoring as the basic processes (founding or structuring) of the functioning of SR. Both are endless processes, always present – although generally salient, more visible and more intense during the genesis of each social representation (Campos 2017, Moliner 2001) – in the interactions that involve large affective loads, strong references to collective memory and/or group identity, and the perception of autonomy on the part of the subjects (Abric 1987, 1996; Campos/ Rouquette 2003); or even in the face of radical changes in objective reality (Guimelli 1989). Both objectification and anchoring are processes that operate through communication: through intense intra-group exchanges; in the selection (reception, filtering, evaluation) of the information that comes from the 'outside world' to the group, particularly via the media (printed, TV or digital); in the production of images (face figurative of SR); in the experimentation of practices, and the adoption or rejection of these; or in the suggestive argumentation (the production, maintenance or transformation of meanings), or the normative dynamics of influence, whether of the majorities or the minorities. Finally, the affirmation of a basic and continuous operation in the objectification and the anchorage, which are mechanisms that organize the interaction between the group and the infinite variability of the situations, reinforces the need to study SR as processes, not only as products; and for this, the study of forms of communication, the study of communicative interactions becomes unavoidable.*

*The considerations made so far lead us to two other findings: the paucity of empirical research in the field of SRT that is dedicated to the study of communicative processes; and the need to revisit some constructs used by Moscovici in another historical period to discuss communicative practices, coupled with the need for a theoretical investment in the relationship between the field of communication and that of SRT, investment in an updating of the forms, interactional modes or 'communication systems' acting in the construction of common sense at the present time.*

> *... it was necessary to rethink representation as a network of concepts and images that interact and whose contents evolve continuously over time in a specific environment. The way in which the network evolves depends on the size and speed of communications, but also on media communication. And their social character is determined by the interactions between individuals and/or groups, the effect that one has on others by virtue of the bond that unites them (Moscovici 1987: 135).*

*The need to intensify communication studies in the field of SRT is demonstrated by the text presented here by Braga, which discusses once again the communication models studied in* La psychanalyse, son image, son public *(Moscovici 1976) through diffusion, propaganda and propagation 'systems'.*

*It is in this area that Claudomilson Fernandes Braga develops what can be called an authorial research line, above all interrogating the forms of social communication and social interaction that underpin the formation of common sense in the present time. In particular, their interrogations and findings attentively observe the impact of technological advances on forms of communication, and the way in which the media has produced changes in values – and, indeed, has facilitated the intensification and acceleration of conflicts, and created new scenarios for the field of struggles, new rhythms and spaces for dissension and the emergence of 'consensus' (Mosovici/Doise 1992). In summation, the media has accelerated the forms of social interaction and the historical processes of change; in other words, on the basis of the social relations that modernity has created, we can observe the enormous impact of new communication resources. The symbolic dimension of social reality has been tilted (not to say run over) by a radical transformation of the objectivity of communication processes.*

*Claudomilson Fernandes Braga is a consolidated researcher, with the breadth of interdisciplinary richness proper to the field of social and ingenious communication (in the sense attributed by Marková 2017). Along the way, he completed a Bachelor's degree in communication at the Fernando Pessoa University, in the city of Porto, Portugal, and a PhD and a postdoctoral degree in social psychology at the Pontifical Catholic University of Goiás (PUC-Goiás, Brazil). He is currently Professor of the Postgraduate Programme in Communication (PPGCOM) in the Faculty of Information and Communication at the Federal University of Goiás*

*(UFG, Goiás, Brazil) and a researcher and director of the Laboratory Research Group Media Image and Citizenship. Its main lines of research are: a) modes of communicative interaction and relations with culture; b) media and the State: processes of hegemony and social representations as a form of resistance; c) social groups, social communication and identity processes; d) communication and social invisibility.*

*His research works and reflections seek to advance the knowledge of the role of communicative practices in the functioning of social representations (genesis, maintenance and transformation), and also of the role of representations in social dynamics in the social communication scenario. These interests are always anchored in a concern, an axis of thought, which assures them continuity and coherence: the construction of citizenship in Brazil, and the impact of the processes of social exclusion on the identities of marginalized youths and indigenous populations, using basic education teachers as examples.*

*Four of his main works that highlight his constant interrogation of citizenship are the articles "Invisible and subordinates: the social representations of indigenous" (2013) and "Social representations, communication and identity: indigenous in the print media" (2013), and the books* Social representations, potentially communicative and conflict situations: the case of the Raposa Serra do Sol Indigenous Reserve *(2012) and* Social representations and communication: the social image of the teacher in the media and its reflexes in (Re) Identity Significance *(2016).*

# References

Abric, Jean Claude, 1996: *Exclusion sociale, insertion, prevention* (Saint Agne: Eres).
Abric, Jean Claude, 1987: C*ooperation, compétititon et representations sociales* (Cousset/Suisse: Editions Del Val).
Braga, Claudomilson Fernandes, 2013: "Representações sociais, comunicação e identidade: o índio na mídia impressa", in: *Comunicação e Sociedade*, 16,2: 107–122.
Braga, Claudomilson Fernandes; Campos, Pedro Humberto Faria, 2013: "Invisíveis e subalternos", in: *Psicologia e Sociedade*, 24,3: 499–506.
Braga, Claudomilson Fernandes; Campos, Pedro Humberto Faria, 2012: *Representações sociais, situações potencialmente comunicativas e conflito: o caso da Reserva Indígena Raposa Serra* (Curitiba: Appris).
Braga, Claudomilson Fernandes; Campos, Pedro Humberto Faria, 2016: *Representações sociais comunicação: a imagem social do professor e seus sreflexos na ressiginificação identitária* (Goiânia: Kelps).
Campos, Pedro Humberto Faria, 2017: "O estudo da ancoragem das representações sociais e o campo da educação", in: *Revista de Educação Pública*, 26,63: 775–797.
Campos, Pedro Humberto Faria; Rouquette, Michel-Louis, 2003: "Abrodagem estrutural e componente afetivo das representações sociais", in: *Reflexão e Crítica*, 16,3: 435–445.
Deschamps, Jean Claude; Beauvois, Jean-Leon, 1996: "Attitudes et representations sociales: présentation de la deuxième partie", in: Deschamps, Jean Claude; Beauvois, Jean-Leon (Eds.): *Des attitudes aux attributions: sur la construction sociale de la réalité* (Grenoble: Presses Universitaires de Genoble: 101–133).
Guimelli, Christian, 1989: *Chasse et nature au Languedoc* (Languedoc: L'Harmattan).

Moliner, Pascal (Ed.), 2001: *La dynamique des representations sociales* (Grenoble: Presses Universitaires de Grenoble).

Moscovici, Serge, 1986: "L'ère des representations sociales", in: Doise, Willem; Palmonari, Augusto (Eds.): *L'étude des representations sociales* (Neuchâtel: Dalchaux & Niestle): 34–80.

Moscovici, Serge, 1987: "Answers and questions", in: *Journal for the Theory of Social Behavior*, 17,4: 513–519.

Moscovici, Serge, 2013: "L'histoire et l'actualité des representations sociales", in: Kalampalikis, Nikos (Ed.): *Le scandale de la pensée sociale* (Paris: Éditions de l'École des Hautes Études en Sciences Sociales): 65–117.

Moscovici, Serge; Doise, Willem, 1992: *Dissensions et consensus: une théorie générale des decisions collectives* (Paris: PUF).

**Abstract** Throughout his work, Moscovici gave a preponderant role to social communication. Communication has a fundamental role in the maintenance and dissemination of social representations. Moscovici (1961, 2012) characterizes communication systems, or as Vala (2010) calls it, a typology of communicative acts of social representations: diffusion, propagation and advertising. Five decades after the seminal work of Serge Moscovici was released, we have an exciting theoretical legacy. However, the legacy left by the author in relation to the role of communication in the process of disseminating social representations and, above all, communication systems has been neglected over the years. This chapter seeks to discuss the actuality of social communication and its three core constructs: diffusion, propaganda and propagation.

**Keywords** Social representations · Communication · Diffusion · Propaganda · Propagation

## 8.1 The Genesis of Research in Social Representations and Communication

The concept of social representations (SR) which prevails today stems from the primary work of Serge Moscovici, *Psychanalyse: son image et son public*. It has evolved into an important theoretical tool for understanding various aspects of the social world, such as cognitive, cultural, ideological and symbolic. In the second part of his work, the author dedicates a privileged place to communication, in order to understand how these aspects of the social world influence the diffusion of the concepts of social representations of psychoanalysis in and by the French press.

According to Moscovici (2003), it was believed that human comprehension was the most common faculty and that it was first and foremost stimulated by the exogenous world and human understanding born, in fact, from social communication – i.e. Moscovici (2003) attributes to communication a fundamental role in the maintenance and dissemination of social representations.

The term 'social communication' is used to exemplify the multiple forms of human interaction resulting from a social construction of reality in such way that all – or at least most – of the information we have is distorted, to a higher or lower degree, by its representation. The aim of this distortion is to make objects conventional, naturalized and appropriate to categories which already exist in society.

The behaviour of the social man and the way he allots, shares and transforms knowledge into practice seems to be influenced by the media, since the media, with its dynamics and symbologies, persuades society to construct a social reality, and consequently influences representations about the world. After extensive analysis, Moscovici (1961, 2012) characterizes in his studies what he called communication systems, and what Vala (2010) calls typology of communicative acts of social representations: diffusion, propagation and advertising. More than five decades after the inspiring work of Serge Moscovici was released, we have an exciting theoretical legacy, predominantly in areas involving psychology, education and health, and an important theoretical and methodological legacy, which includes approaches such as the Structural Theory developed by the researcher Jean Claude Abric (1998) and the Societal Theory elaborated by the researcher Willem Doise (2002).

However, the legacy left by Moscovici in relation to the role of communication in the process of disseminating social representations and, above all, the communication systems developed by Moscovici (1961, 2012), have been neglected over the years, that is to say, little research in the last five decades has been undertaken from this perspective. Strictly speaking, most research on social representations has been dedicated to identifying the representations of certain groups in relation to certain social phenomena, and when they involve social media, they privilege the social representation that circulates via the vehicle without necessarily identifying the communication system that is present in this process.

It was Michel-Louis Rouquette who, in the 1970s, developed several pieces of research from this perspective, retaking the Moscovician prism for studies of social representations from the media with particular emphasis on communication systems in *Les communications de masse* (1973, 1984, 1998), *Diffusion, Foules, Opinion, Propagation, Propagation, Psychology of masses, Psychologie politique, Rumeurs* (1991) and *La communication sociale* (1998).

In Brazil since 1991, the Laboratory of Social Psychology of Communication and Cognition (LACCOS) at the Federal University of Santa Catarina has been developing research activities within the scope of undergraduate and graduate studies related to social representations and communication.

In an unpublished survey, Simoneau and Oliveria (2014) present an overview of the research involving social representations and communications which was published in scientific journals indexed in Brazil between the years 2000 and 2011.

The researchers identified fifty-four scientific productions in the decade studied, twenty-three of which were considered pertinent to the inclusion criterion: complete texts in Portuguese based on Social Representations Theory and destined for analysis by the media and/or social communication. Simoneau and Oliveira concluded that there is a consensus on the importance of the analysis of communications and representations through the media, even though the frequency of studies has not been great compared with the general output inspired by the theoretical framework adopted (Simoneau/Oliveira 2014: 1). That is, despite the amount of research and publications that involve communication and social representations, strictly speaking, few or almost none were dedicated to studies originally undertaken by Moscovici.

Despite this observation, the importance of studies on social representations and communication was emphasized by Moscovici when he affirmed that studies involving the press make it possible to uncover the simple but fundamental lines of the social models that use it as a starting point (Moscovici 2012: 284). Their verification allows us to better understand the object of study and situate the ideological groups in relation to this object (Moscovici 2012), especially with regard to the ideological, symbolic and cultural aspects.

## 8.2   Communication Systems and Social Representations: The Moscovian Legacy

When studying the processes of communication it is important to observe and consider the multiplicity of possibilities and relationships involved: the relationship between the sender and the receiver; the relationship between the published content and the understanding of this content; the organization of content and the conduct of the recipient, and so on – this complexity needs to be observed and understood.

It is in this tangle of possibilities that Moscovici constructs his research and elaborates on what the author called "communication systems". Defined as diffusion, propaganda and propagation, the communication system outlined by the author explains how psychoanalysis is treated and portrayed by print vehicles in France in the 1960s.

When analysing several publications, previously defined as vehicles intended for specific audiences – the population in general; Catholics and Christian communists – Moscovici (1961–2012) seeks ultimately to identify which image of psychoanalysis was conveyed by the media in relation to the readers for whom those publications were intended.

It is in this binomial public image that the author defines the functionality and role of each communication system. Diffusion, the first of the systems, seeks to establish a relationship of equality between the vehicle and the reader; it seeks to form a kind of unity and, at the same time, a kind of differentiation. Distancing is another feature of the diffusion system. With regard to the sender and the receiver, the vehicle in this system, according to Moscovici (2012), has the role of mediator, and by assuming this function it automatically occupies the place of the receiver – in other words, at the same time that it emits, it receives. So diffusion is not intended for a specific group; it is, in fact, intended for the masses. Although it is made of distinct subjects, there are certain connections between them – above all, of a social nature – which enable it to maintain uniformity and diversity while allowing the reader/receiver a certain margin for making decisions about his or her own conduct. What the system intends to create is unitary behaviour; the desire to talk about and speak of the featured topic (Moscovici 2012).

The second system of communication, propagation, was identified during the analysis of the vehicles of communication targeted at Catholics. Unlike diffusion, propagation-type communication is purely instrumental and therefore seeks to express clearly the object of communication to a defined group. In this relationship there is no reciprocity – i.e. unlike diffusion, non-media propagation communicates.

Because it is endowed with authority and a certain degree of autonomy, the propagation of the sender to the receiver occurs in normative contexts in which cognitive and social rules prevail, so that communication in this context becomes hierarchical and authoritarian, aiming to reach a conception already existent on the one hand, and, on the other, to orientate itself in relation to this conception, generating appropriate attitudes. Propagation is therefore systematic and can guide the members of the group and control the conduct of these members according to the foundational and fundamental norms of the group, where norms and conduct provoke a global conduct, as opposed to the more autonomous mindset associated with diffusion.

The last system, advertising, is a similar form of communication to propagation: "same symbolic systems, same implication, same relations with readers" (Moscovici 2012: 361). However, unlike propagation, which seeks to renew meanings, advertising's goal is to create them, to reinforce them. The dichotomy is exhibited in a controlled and evidenced form. It is in the antagonism of communication and communication that propaganda is built.

By polarizing the theme, the conflict, the phenomenon, propaganda regulates and organizes the representation of which the communicative act speaks, strengthening opposition and establishing itself as a group identity, and in this way creates a representation of the phenomenon.

## 8.3   Communication Systems and Social Representations: An Attempt to Update the Construct

Taking as a reference the Moscovian perspective that the human comprehension of the world necessarily passes through communication, and considering the changes that have occurred, especially in the last two decades, in relation to communication, it seems naive to believe that the communication systems originally identified by Moscovici remain or, at best, remain unchanged.

In other words, if communication is the basic support of SR (Vala 2010), given that so many changes have taken place in the sphere of communication with the appearance of Information and Communication Technologies (ICTs), it is impossible not to reflect on the phenomenon of SR in this new digital environment (Mazzotti/Campos n.d.).

According to Castells (2000), this new setting of globalized communication, networked and fed by the collaboration of millions of users, is not only a tool at our disposal; it has transformed and keeps transforming interpersonal communication and has changed the way we live, because we are not the same person after we are inserted into social networks. The internet as a social space in the contemporary world presents itself as a hybrid place, built in the interface between direct and mediated experience through communication and, above all, through internet communication. Traditional vehicles, especially the printed ones, have also reinvented themselves and now the same printed version is available in the virtual space. The copydesk technique takes care of supplying this multiplicity of platforms.

The same work by Moscovici on communication systems that has inspired researchers has now become a cause of concern: do communication systems remain? Moscovici analysed print media vehicles with pre-defined social brands, which represented to some extent organized institutions or social groups wishing to assert and even impose their world-views. Are there vehicles today with this objective? Do contemporary media vehicles advertise social representations, stereotypes, or exactly what? And what is the role of the reading public: do we speak of a passive or active audience? What about the role of information and communication technologies in this trans-mediatized space? Communication occurs within this new framework of, for example, printed media, television, radio, digital platforms (Facebook, Twitter, Instagram, weblogs, etc.), such that media vehicles are now part of a network of communicative possibilities with multiple platforms. McLuhan's (1964) suggestion that "the medium is the message" prevails. That is, the importance of the media is detrimental to the messages, in that the contents need to fit the vehicles. As a consequence, audiences have multiplied and now have divergent attitudes towards the same content. From the point of view of audience studies, it seems impossible to characterize a single audience for a particular vehicle.

In other words, the same medium, given the multiplicity of broadcasting possibilities (print, radio, television, digital) takes on new languages, new shapes and, as a fundamental consequence, reaches new audiences, new crowds and new publics. Thompson (2008), for example, while studying the influence of the media on the formation of modern societies, developed the concept of what he called mediated quasi-interaction to explain that, with the advance of technology – including the printing press – a new type of interaction emerges whereby symbolic forms are no longer transmitted to a group of subjects. Now the receivers are unlimited, expanding the people's field of view, so that space and time no longer matter. Now everyone can see everything – and see it, moreover, at the same time. From a group perspective, it seems that everyone can be a part of all groups. Geographical and cultural barriers are senseless and no longer prevent entry. This is what Castells (2000) called the network.

The Arab Spring, as social uprising tool built essentially by the network, is an emblematic example of this new social configuration that has been created because of the new media. Distinct groups with diverse interests seem to commune with the same libertarian ideal shared on the network and disseminated on a large scale by the media. The absence of a definite social marker in relation to these groups becomes irrelevant. Could we talk about a new way to construct SR? Using the Thompson perspective, because the new technologies provided this new experience to the social actors, at the same time the vehicles also assumed another role: the role of actors. Actor-network theory explains that in contemporary culture the so-called non-human actors (intelligent devices – computers, sensors, servers, etc.) and humans act mutually, interfere and influence each other's behaviour, forging connections between other non-humans, changing the order of human life, dictating the rhythm of thinking and acting, acting as a mediator. While this establishes human interaction at all social levels (Latour 2011) and spreads beliefs and values on a global scale, it weakens the classical notion of groups.

When defining the so-called global village in the 1960s, McLuhan (1964) asserted that the effects of communication would be felt throughout the entire world and the speed of communication would eliminate all distances and lead us to a process of retribalization, whereby cultural, ethnic and geographic barriers, among others, would be relativized, leading to sociocultural homogenization and a scenario which the author called "world society".

If we take as a reference the assumption of SR as the representation of a group related to a certain phenomenon, in the scenarios described by McLuhan (1964) and advocated by the actor-network theory, this does not happen … everyone seems to be part of a large group, now globalized.

It is in this context that, due to information and communication technologies, all the traditional vehicles (print, radio and television) and digital media converge with each other from unifying platforms that on the one hand assume a definitive role in relation to the contents, and on the other hand make it possible to receive new behaviours, new postures and also roles: to consume information, to share and, above all, to create content.

Terra (2012) explains that these are the media users of contemporaneity: groups of audiences that can be understood as those who not only consume content and pass it on, but also interact with comments on other people's initiatives and produce content quite frequently. This is what Jenkins (2008) calls "audience empowerment". That is, it is with the substantial change of media and with the active participation of audiences that communication has reinvented itself and assumed new roles. Contrary to the reality experienced by Moscovici at the time of the identification of communication systems, in the present day communication is multiple, comprehensive, divergent, and, at least in theory, unifying; the reader is now an active and participatory user. In this new environment, communication takes on new contours and is increasingly being guided by the audience, and with this new format, classic communication theories have a direct impact on their assumptions. The Agenda-setting theory, for example, that asserted in the 1970s that it is the media that determines what is discussed in the social world (McCombs/Shaw 1993), no longer holds up completely. The now-empowered media user determines on a large scale which subjects will be part of the media schedule.

In this way, the classical notion of communication systems – diffusion, propaganda and propagation – which inspired this study, and which also seemed troubling, takes on a new format. Nowadays, with the reorganization of traditional media and the growth of new media, there is a new *modus operandi* in the maintenance and dissemination of social representations.

In surveys which focused on the social representations of Brazilian indigenous people in mainstream media, Braga and Campos (2012) found that, in relation to the contents conveyed, the traditional media still resists setting themes of interest that are inclusive, exhibiting instead a strong tendency towards hegemonic discourse. The phenomena reported are biased and certain groups are favoured over others. The social reality is not accurately described and it cannot be said that there is an accurate social representation that circulates in the published material. Media content is loaded with particular opinions and usually represents the interests of specific groups. This study indicates that there is a clear partiality in reinforcing the invisibility of the indigenous.

In the same vein, when studying the social representations of the Unified Health System (SUS), Menezes and Braga (2014) identified a disconnection between the vehicle and the reality experienced by the users of the system. There is a certain lack of awareness among journalists about the agenda. It would be impossible to affirm that there is a social representation of SUS in the media if the contents ruled do not represent the SUS and are described by a group that does not know – because it does not live – the experience. In addition, users do not recognize the system itself and therefore they do not have a social representation of it either. One cannot speak of a social representation by a structured group in relation to a certain phenomenon if the group does not 'know' the phenomenon. In this case neither can the media.

When studying the social representations of violence in young people and the media, Mendes (2017) identified a disconnection between the representation that circulates and the representation that the group has of itself in relation to the discourse conveyed. According to Mendes, if, on the one hand, the young people do not assimilate the media content without criticizing it, associating the analysed newspaper with characteristics of sensationalism, violence, and superficial news that underestimates the capacity of interpretation of its reading public, on the other hand they have a representation which is different from the one that circulates. The result indicates that, despite experiencing violence in different ways, young people were unanimous in stating that the media is not impartial or neutral in portraying their daily lives. In this sense, the social representations that circulate do not reflect reality. On the contrary, they appear partial, biased and do not disseminate a fair representation. Rather, it is an opinion, a version, a vision. Although we should recognize that the socio-historical moment of the original research developed by Moscovici was different, it is important to emphasize that when studying social representations in the media we must not ignore the fact that the advance of ICTs has changed communication processes.

## 8.4    Final Considerations

To return to the initial concerns that inspired this chapter – that the original studies of communication systems represented vehicles with defined social markers and this condition influences contents. We see this perspective still remains, communication vehicles still have cultural markers. Even if they are spread across several platforms, the contemporary media still maintain a certain homogeneity, a certain cultural formation. Based on the notion that communication, understood here as media, has changed in form and content, the notion of originally identified mediation has undergone substantial changes, and vehicles have become mediators not only of content over audiences. To put it another way, the vehicles of yesterday mediated content in relation to public readers; the vehicles of today mediate content in relation to the public and in relation to other vehicles and are at the same time mediated by the public as they too (the readers) are producers of content.

In this approach, the definitions of mediation (diffusion), instrumentalization (propagation) and polarization (propaganda) do not, in my view, support the current reality. However, there is still a fundamental question: because of the advent of ICTs, what exactly do contemporary media vehicles spread? Social representations, stereotypes or something else? If we start with the notion that there is, in social networks, an intense interaction between digital vehicles and audiences and that much of it is reflected in traditional media, it can be affirmed that there is a process of sharing and consequently of orientations and social representations, a preparation for action, which directs behaviour, remodels and reconstitutes the elements of the environment in which behaviour must take place, integrating a network of relationships (Moscovici 2012); and if the media process operates nowadays, it is fair to say that social representations circulate in the media, even in digital media. However, the dynamism of communication and the speed with which the content circulates between recently formed groups pose a new challenge. Can online groups be considered groups? I believe that Harré's (1989) positions are convincing and that, just like offline groups, online groups discuss and share beliefs and opinions, circulating virtual representations of the real world. However, the speed of information and communication that seems to have no limit provides daily challenges in this regard. The opportunities to study social representations in the age of digital media justify research in this area. However, as this process has no end – quite the opposite – it seems to be only at the beginning. What can we say, for example, about Fake News, the latest 'invention' circulating in the media?

## References

Abric, Jean Claude, 1998: "A abordagem estrutural das representações sociais", in: Moreira, Antonia Silva Paredes; Oliveira, Denize Cristina de (Eds.): *Estudos interdisciplinares de representação social* (Goiânia: Ed. AB).

Braga, Claudomilson F.; Campos Pedro Humberto Faria, 2012: *Representações sociais, situações potencialmente comunicativas e conflito: o caso da reserva indígena Raposa Serra do Sol (2005 – 2009)* (Curitiba: Apris).

Braga, Claudomilson F.; Campos Pedro Humberto Faria, 2016: *Representações sociais e comunicação: a imagem social do professor na mídia e seus reflexos na (re) significação identitária* (Goiânia: Kelps).

Castells, Manuel, 2000: *A sociedade em rede* (São Paulo: Paz e Terra).

Codol, Jen-Paul, 1988: "Vint ans de cognition sociale", in: *Bulletin de Psychologie*, XLII: 472–491.

Doise, Willem, 2002: "Da Psicologia Social à Psicologia Societal", in: *Psicologia: Teoria e Pesquisa*, 18,1: 27–35.

Harré, Rom, 1989: "Grammaire et lexiques, vecteurs des représentations sociales", in: Jodelet, Denise (Ed.): *Représentations sociales: un domaine en expansion* (Paris: PUF).

Jenkins, Henry, 2008: *Cultura da convergência* (São Paulo: Aleph).

Latour, Bruno, 2011: "Networks, societies, spheres – reflections of an actor-network theorist", in: *International Journal of Communication*, 5 (special issue): 796–810.

Mazzotti, Alda Judith Alves; Campos, Pedro Humberto Faria, 2011: "*Cibercultura*: uma nova 'era das representações sociais'?", in: Almeida, Angela; Santos, Maria de Fátima; Trindade, Zeidi Araujo (Eds.): *Teoria das Representações Sociais: 50 anos*, (Brasilia:Technopolitik): 457–488.

McCombs, Maxwell; Shaw, Donald, 1993: "A evolução da pesquisa sobre o agendamento: vinte e cinco anos no mercado de idéias", in: Tranquina, Nelson: *O poder do jornalismo: análise e textos da teoria do agendamento* (Coimbra: Minerva 2000).

Mendes, Gardene Leão de Castro, 2017: "Representações sociais de jovens de Goiânia: a negociação de sentidos em relação aos discursos midiáticos a respeito de si" (PhD thesis, Universidade Federal de Goiás, Faculdade de Ciências Sociais).

McLuhan, Marshall, 1964: *Understanding media* (London: Routledge).

Menezes, Kalyne; Braga, Claudomilson F., 2014: "As representações sociais do SUS" (MSc dissertation, Universidade Federal de Goiás, Faculdade de Informação e Comunicação).

Moscovici, Serge, 2012 [1961]: *A psicanálise, sua imagem e seu público* (Petrópolis: Vozes).

Moscovici, Serge, 2003: *Representações sociais: investigações em psicologia social* (Petrópolis, RJ: Vozes).

Moscovici, Serge, 1976: *La psychanalyse, son image, son public* (Paris: PUF).

Noelle-Neumann, Elisabeth, 1993: *The spiral of silence: public opinion – our social skin* (Chicago: University of Chicago Press).

Rouquette, Michel-Louis, 1984: "Les communications de masse", in: Moscovici, Serge (Ed.): *Psychologie sociale* (Paris: Presses Universitaires de France): 495–512.

Rouquette, Michel-Louis, 1991: "Diffusion"; "Foules"; "Opinion"; "Propagande"; "Propagation"; "Psychologie des masses"; "Psychologie politique"; "Rumeurs", in: *Grand dictionnaire de la psychologie* (Paris: Larousse).

Rouquette Michel-Louis, 1998: *La communication sociale* (Paris: Dunod).

Simoneua, A. Sancho; Oliveira, Denize Cristina de, 2014: "Representações sociais e meios de comunicação: produção do conhecimento científico em periódicos brasileiros", in: *Psicologia e Saber Social*, 3,2: 281–300.

Thompson, John B., 2008: "A nova visibilidade", in: *Matrizes*, 1 (April).

Terra, Carolina Frazon, 2012: "Usuário-mídia: o formador de opinião online no ambiente das mídias sociais", Paper for the VI Congresso Brasileiro Científico de Comunicação Organizacional e de Relações Públicas – VI Abrapcorp, São Luiz, MA, Brasil.

Vala, Jorge, 2010: *Psicologia social* (Fundação Calouste Gulbenkian: Lisboa).

# Chapter 9
# The Figurative Core of Social Representations and Figures of Thought

Rita de Cássia Pereira Lima

## Introductory Comment

Tarso  Mazzotti

*I[1] met Professor Rita Lima in a technical meeting at Centro Internacional de Estudos e Representações Sociais e Subjetividade – Educação (CIERS-ed), at Fundação Carlos Chagas, in São Paulo, when she worked in Ribeirão Preto (SP). Later I met her again on the Post Graduate Programme in Education at University Estacio de Sá (PPGE-UNESA), in Rio de Janeiro, where I started working in 2005. Since then I have noticed that Professor Rita Lima has all the desirable attributes of a professor and researcher: she is not afraid to ask questions, is a good listener, knows how to appreciate her colleagues' and students' contributions, and is fully committed to her research and teaching activities. While I was working at PPGE-UNESA I had the pleasure of collaborating with Professor Rita Lima on the definition and development of research involving her Master's and doctoral students. The work she is presenting now shows our contribution to the development of conceptual frameworks in order to better assimilate the social representation phenomenon.*

---

Rita de Cássia Pereira Lima obtained her doctorate in Educational Science at Université René Descartes in Paris, France. She is a professor on the Postgraduate Programme in Education and the Pedagogy Course at the University Estácio de Sá (UNESA-RJ). Email: ritaplima2008@gmail.com.

---

[1]Tarso Mazzotti is a full professor at the Federal University of Rio de Janeiro (UFRJ), and an associate researcher at both the Carlos Chagas Foundation and the Estácio de Sá University. He has a degree in Pedagogy from the State University Paulista Júlio de Mesquita Filho (1972), a Master's degree in Education from the Federal Universidade of São Carlos (1978) and a PhD in Education from the University of São Paulo (USP, 1987). He has extensive experience in the fields of education, philosophy of education and social representations, with significant contributions to the study of the subject. His main areas of research are: social representations theory, the philosophy of education, epistemology, and social representations. Email: tarso@mazzotti.pro.br.

© Springer Nature Switzerland AG 2021
C. Prado de Sousa and S. E. Serrano Oswald (eds.),
*Social Representations for the Anthropocene: Latin American Perspectives*,
The Anthropocene: Politik—Economics—Society—Science 32,
https://doi.org/10.1007/978-3-030-67778-7_9

**Abstract** This paper proposes reflections on the 'figurative core' of social representation by highlighting the roles of figures of thought. The starting point is the expression 'figurative model' in the seminal work of Serge Moscovici, *Psychoanalysis: Its Image and Its Public,* and later theoretical proposals of the 'core', such as Jean Claude Abric's Central Core Theory and, more recently, Pascal Moliner's Matrix Nucleus Theory. The notion of 'image' runs through these reflections. Starting from these referential concepts, Tarso Mazzotti's studies on the relationship between 'figurative core' and figures of thought are prioritized, especially when the author proposes an 'argumentative core' for analysing speeches. An example of some research in which analysis uses rhetorical devices is presented. Overall, the work highlights the presence of a core in social representations which approximates to the approach of rhetorical social representations, and seeks ways to both identify and analyse it.

**Keywords** Social representations · Figurative core · Figures of thought, social representations core

## 9.1 Introduction

In Moscovici´s approach (Moscovici 1976), common aspects approximate social representations to figures of speech, according to the rhetorical perspective: both have a cognitive function, express forms of knowledge, relate thought and action, and offer guidance on looking at and interpreting social objects. Based on this background, both can be described as social constructions with epistemological roots in theories of knowledge, one in Social Representations Theory (SRT), embedded in the social psychology of knowledge, and the other in rhetoric.

With a psychosocial look and seeking to approximate these two theories, this paper intends to show that the 'figurative core' of social representation consists of figures of thought, an idea already presented by Mazzotti (1998) twenty years ago. He proposes rhetorical devices for analysing such figures in studies of social representation, mainly when seeking the 'figurative core'.

Figures of thought are here considered according to the concept of Mazzotti (2008, 2015, 2016), who also refers to them as 'rhetorical figures', 'argumentative figures' and 'argumentative schemes'. Such figures can be, for example, metaphors, metonymies, synecdoche, and irony. Mazzotti seeks to evidence the way the figurative dimension operates in discursive processes. Thus, for him (2008, 2015, 2016), figures of thought: (a) consist of sources of arguments about reality; (b) activate classifications in the world; (c) express the attitudes of members of a social group; (d) coordinate and condense implicit meanings in discourses; (e) explain reasons for disputes and conflicts when opposers, who are expressing their opinions, are identified; (f) make it possible to establish the meanings of arguments elaborated through comparisons. Mazzotti (2016) proposes that social representations studies, which involve analysis of 'situated word', should be based on argumentative schemes defined by rhetoric and logic.

As stated by Mazzotti and Alves-Mazzotti (2011), social representations are produced in conversational contexts in which subjects/groups interact using arguments to persuade each other. They can be comprehended through rhetorical analysis, which shows what is 'preferable' and 'not preferable' for each participant. The subject of 'rhetoric' emerged in Ancient Greece and was systematized by Aristotle in the fifth century BC to enable people to argue and counterargue in assemblies and courts, mainly due to the absence of both lawyers and a representative political system. Its objective was to convince or persuade the audience. Alves-Mazzotti and Mazzotti assert that rhetorical analysis explains meanings shared by a group and indicates discourses situated in specific contexts, which makes it possible to identify similarities in social representations.

> […] social representations are always situated, constructed through negotiations of meanings which imply taking certain positions, just like when deliberating in a rhetorical situation (Mazzotti/Alves-Mazzotti 2011: 74).

These reflections draw attention to the meanings shared by subjects/groups and constructed in communication processes in which the social situation and cognitive system are articulated from the position taken towards a social object. It is a central idea in the perception of the figurative core of social representation proposed by Moscovici (1976) as an explanatory scheme for the psycho organization of the participants in his research on the social representation of psychoanalysis, which was conducted in France with several groups and involved organizing and ranking subjects. Thus, the figurative core would have an imagistic dimension, or scheme, which presents the subjects' selective appropriations according to values within the group, contributing to the social construction of reality.

Moliner (1996, 2016) develops the idea that the mental image which is expressed in a 'figurative scheme' reveals a privileged moment in the representation process. It is a social image because it is orientated by collective factors and shared by the group. This 'social image' also reveals the beliefs and values of the subjects/groups who construct them. According to Moliner, when abstract notions are associated with concrete objects, they can be organized in a 'figurative scheme', making the object of representation comprehensible. Thus the mental image leads to the construction of the social representation.

I believe that this 'social image' which evidences a figurative scheme can contain figures of thought, as proposed by Mazzotti (1998, 2002a, 2002b, 2003, 2008), whose theorisations enhance advances in the research on the core of social representation. Consequently, this paper is divided into three parts: (a) notes about the figurative core of social representation, using Serge Moscovici's reflections as a starting point, followed by authors who approach the core, especially the development presented by Pascal Moliner through the notion of 'social image'; (b) Tarso Mazzotti's contributions to the analysis of the figurative core in the field of social representations based on figures of thought; (c) the presentation of a case study grounded in rhetorical devices to propose a social representation figurative core. With the foundation presented, the main goal of the paper is to reflect on the figurative core in social representation research, seeking to deepen knowledge of the theme.

## 9.2   The Figurative Core in Social Representations Theory

Although not very often, the theme of 'figurative core' has previously been addressed in social representation studies. Without wishing to be exhaustive, we can cite, besides the seminal work of Moscovici (1976), Herzlich (1969), Chombart de Lawe (1972), Jodelet (1989b, 2014), and Duveen (1995). In Brazil those investigating the topic include Alves-Mazzotti (1994) and, more recently Alves-Mazzotti (2017), Lima and Soares (2017), Morais and Lima (2017), and Lima and Campos (2018).

The work in which Moscovici opens the field of social representations (1976; first edition 1961) presents a figurative model (later called 'figurative core') of the social representation of psychoanalysis by several groups in France, based on data collected mainly through devices which he called 'questionnaire-notebooks'. Figure 9.1 expresses this model.

In this proposal, Moscovici (2003: 71–72) shows the process of the objectification of the social representation of psychoanalysis by different groups, and asserts that objectifying is discovering the iconic quality of an imprecise idea; it is reproducing a concept as an idea. On the other hand, comparing is representing, filling with substance what is empty. When elaborating a knowledge of common sense, or a social representation of psychoanalysis, groups expressed their ideas using figure 9.1. As Moscovici states (2003: 72), "the images which were selected, due to their capacity of being represented, mix, or better yet, are integrated into what I have called a *figurative core* model, a complex of images which visibly reproduces a complex of ideas."

Jodelet (1984, 2014: 374–375) explained the elements of the figurative model in Figure 9.1 by using three stages of objectification:

1. Selection and decontextualization of elements of the theory: cultural and normative values led to the streamlining of information about psychoanalysis circulating at that time. They were separated from the scientific field by the public in order to be apprehended and dominated, and thus projected in their own universe.
2. Formation of a figurative core in which an imagistic structure reproduced a conceptual structure. Key notions like 'conscious' (desire, apparent, achievable) and 'unconscious' (involuntary, hidden, and possible) are visualized in the core, above and below a line of tension where conflict is expressed.

**Fig. 9.1** Figurative model of the social representation of psychoanalysis. *Source* Moscovici (1976: 116)

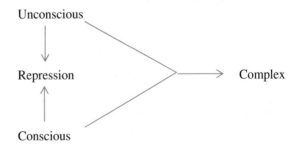

Contradiction in the form of repressed pressure – repression – produces 'complex'. Sexuality was hidden, and libido, the essential element of Psychoanalytic Theory, was subtracted from the scheme.

3. Naturalization: the figurative model enabled the elaboration and coordination of elements which became creatures of nature: "the conscious is restless"; "the complexes are agressive"; "the conscious and unconscious of the individual are in conflict". The elements of thought presented in the figurative core show the integration of elements of science with the reality of common sense.

According to Moscovici (2003), when a society accepts a particular figurative core, it becomes easier to talk about it, and the words start to be used more often in different social situations through formulas and clichés which summarize it. The author states:

> Although we all know that a complex is a notion whose equivalent objective is highly vague, we still think and behave as though it were something that really existed the moment we judge someone and relate them to it. It doesn't symbolize their personality, or their behaviour, but actually it represents them, and *it starts to be* their 'complexed' personality and their way of behaving (Moscovici 2003: 73).

Moscovici's study (2003: 68) showed the multiplicity of terms derived from a model, giving us a glimpse of a psychological symptom, for example 'shyness complex', 'power complex', which are no longer psychoanalytical terms, but words to define them. The author states: "Hence, those who speak, and those who are spoken of are forced into an identity matrix they have not chosen and have no control over."

According to Jodelet (2014: 376), the terms 'conscious' and 'unconscious' in the figurative core have existential ressonance because they express an intimate conflict experience with an imaginary and mythical dimension through a battle between 'powers' and 'antagonistic forces'. Cultural elements which are present in the "mental universe of subjects and groups can be applied in the activity or restructuring and can be relevant to illustrate ideological references or cultural models" (Jodelet 2014: 376).

Thus, the figurative core draws attention to an important notion approached in social representation studies: the 'image', which, according to Moliner (2016a), has a polysemic character and designates a wide variety of phenomena. As Jodelet states (2014: 369), the notion of image is not used in studies of social representations with the idea of 'authentic copy', 'mental sensation' or 'real reflex of the outside environment'. The term, in this field, is used with a different meaning, such as 'figure' or 'figurative group'. For the author, it means a constellation of concrete features, with interpretations which mobilize the intervention of the individual imagery or the social imagery or the imagination.

According to Moscovici (1976), two movements converge in the figurative core: (a) from theory to image: the scientific conception is confronted with the values of the groups, where apprehensions are selectively made; (b) from image to social construction of reality; the social representation is not only a 'double' of science, but also a 'profane theory' used to classify people and behaviours by category.

Pascal Moliner is the author who has conducted the most studies on the relationship between image and social representations, including the discussion of a core within it. According to Moliner, Rateau and Cohen-Scali (2002), social representations are an "objectified knowledge", consisting of the substitution of the concept for the precept. For the authors, in the objectification process some knowledge is related to the object by the individuals, who turn it into a mental image. Thus, the individuals who express a representation describe the environment as it is or as it is perceived by others.

Moliner (1996) grounds his view in the objectification process to establish a relationship between images and social representations, stating that the naturalization of an abstract concept transformed into object of the real world emerges through an image. That is, thinking about abstract notions may evoke in people images that approximate them to the social world. The analogy which comprises the engine of the naive thought is based on similitudes between different objects, and the image is crucial to establish analogies inspired by perceptive similarity.

According to Moliner, in essence, representations are disseminated visual images which are integrated in the form of a collective opinion, or 'social image'. Three things should be noted:

1. Social image can be defined as a set of features and characteristics which individuals give to a certain object;
2. The emergence of a social image depends on two conditions: individuals need to have information about the object and experiences which are comparable to such information, besides using previous common knowledge to interpret and refer to the information;
3. Social images have a dual role: they reveal the way certain objects exist in the cognitive environment of individuals, and also allow individuals to judge such objects; that is to say, the images have an evaluation objective (Moliner 1996).

According to Moliner (1996), the mental image expressed in a figurative scheme has shared values as a point of reference. In this sense, the notion of 'figurative scheme' is very close to the notion of 'social image' because its elaboration emerges through collective factors. SRT explains the mechanisms of image formation and describes its structure, which consists of two elements: (a) a descriptive one, involving activities of description; (b) an evaluation one, related to values attributed to knowledge which can ground the elaborated judgements of the object.

Moliner and Matos (2005: 3.2) state that Core Theory (Abric 1976) relates to the figurative model proposed by Moscovici (1976), composed of an imagistic structure and key notions. The change in the conceptual notion from the figurative model to the central core would correspond to the intention of changing from the process to the product, the figurative model being the embryo of the central core. For the authors, the latter sustains a genetical function of generating meanings, but privileges the contents of stable representations. Central Core Theory proposes that the structure of a social representation is organized according to a dual system of information, opinions and beliefs: the central core and the peripheral elements. The representation is proposed as a hierarchical set of beliefs involving the peripheral elements, organized around a core. According to Moliner (2016b) the main function of this dual system is maintaining the stability of the representation or the meanings the group attributes to the object of the representation.

In a more recent work, Moliner (2016b: 3.9–3.10), ponders some contradictions in Central Core Theory and proposes an alternative which he calls Matrix Core Theory, but without opposing Central Core Theory. Matrix Core Theory proposes specifying the functions generally attributed to the core. Thus, instead of 'meanings', 'organization' and 'stabilization', the author proposes 'denotation', 'aggregation' and 'federation' ('association').

'Denotation': is based on symbolic characteristics of the central elements – signs that allow individuals to indicate in what universe of opinions their discourse is located; alternatively, a central element can indicate that the object of representation itself would be excluded from the discourse.

'Aggregation' is connected to the strong semantic potential of central elements which allow individuals to combine experiences under a single term, for example, the association "work/salary". The central elements would be categories of language.

'Federation' emerges from precedent functions. When scattered elements of definition are offered to the group, the core offers a 'common matrix', allowing the individual to evoke the object of representation, but with existence of different personal experiences. Not all members of the group are expected to adopt all elements of the core of a representation.

Moliner (2016b: 3.11) critically revisits Central Core Theory, which tends not to favour the notion that the structure is a set of beliefs and individuals who build bonds with such beliefs. Matrix Core Theory reactivates this reflection, suggesting that opinions and beliefs can also be bonds with individuals. Thus, it is not a case of individuals building bonds with different beliefs, but of beliefs building bonds with people. The author proposes the use of metaphorical language which does not demand that two people dominate all the words of the language to communicate with each other; their respective records are enough for each to understand that they are speaking the same language. According to Moliner,

> Social representations are languages which social groups create to think about society and to communicate with it. In these languages, 'matrix cores' play a specific role because the elements which they are composed of denote a social object which builds bridges between individuals (Moliner 2016b: 3.11).

In this more recent work the author's great interest in language can be seen. In a previous text, Moliner (1996) referred to image as a metaphor of a discourse, or as a mental or figurative consequence of this discourse. He had already stated that the notion of 'image' plays a preponderant role in social reality, mainly in the communication between social actors, drawing attention to the metaphors. This leads to the question: would the matrix core also bear images which might be accessed through metaphors?

Moscovici (2003: 77–78) verifies the union of language and representation: "Each case implies a social representation which transforms words into flesh, ideas into natural powers, notions or human languages into a language of things."

According to him, representations are more consolidated when mediated through analogies, implicit descriptions, explanations of phenomena and personalities, along with the necessary categories for the understanding of behaviours:

> Thus the tendency to turn verbs into nouns or the contrary, through gramatical categories of words is a sure sign that grammar is being subjectified, that words do not merely represent things but create them and invest them with their own properties." (Moscovici 2003: 77).

The reflections discussed in this section about the figurative core of social representation lead us to Tarso Mazzotti's works, presented below, which, in my opinion, have advanced theories about the relationship between the representation core and figures of thought, incorporating a notion of 'image'.

## 9.3   The Figurative Core and Figures of Thought: Tarzo Mazzotti's Contributions

This section addresses areas of Tarso Mazzotti's work which focus on social representations, especially the topic of 'figurative core'. It is important to state that, apart from those relating to SRT, Mazzotti's vast bibliography of works in the field of rhetoric will not be discussed.

With regard to the core of social representation, it is possible to notice two phases in texts in which Mazzotti likens SRT to rhetoric: (a) the one which starts in 1998 and continues until around 2007, when the central reflection emphasizes the figurative core of social representation, with particular reference to the metaphor as a figure of thought; (b) from 2007 onwards, the author is more committed to tackling other figures of thought and the relationships between them in the discursive processes, and mainly from 2015 he starts to mention what he calls the 'argumentative core', which also covers other phenomena apart from social representations. This separation is based on didactic criteria, with no intention of linearity, since there are texts in which these two movements interlace.

When proposing that the figurative cores can be investigated as metaphors, Mazzotti (1998, 2002a) refers to the work of Moscovici and Vignaux (1994), authors who propose comprehension of social representations based on the *themata concept*, in which the 'themes', 'ideas-sources and 'concepts-images' can be considered elements of the figurative core. Sustained by this idea, Mazzotti (1998) seeks to demonstrate that the figurative cores are composed of figures of thought.

To develop his reflection, Mazzotti returns to Holton's affirmation (1982) that *topoi* (places) are *themata* involving antithetical pairs (for example 'reductionism/holism', 'hierarchy/unit') which play an important role in the organization of scientific discourses as they show the conditions for accepting statements and evoke contrasting arguments, the 'counter-themes'. According to Mazzotti (1998, 2002a), such arguments can include not only *topoi*, but rhetorical figures (metaphors, metonymies, synecdoches) not privileged by Holton.

Mazzotti (2002a) claims that *topois* are neither organizers of scientific theories nor social representations, for the antithetical pairs may receive different meanings according to the metaphor used in the discourse. According to Mazzotti, Holton confuses *topoi* with metaphors, does not distinguish between them, and fails to consider the organizing role of metaphors in communications. Furthermore, he does not understand that *themata* are controlled by metaphors, therefore he does not recognize the disputatious value of figures or rethorical schemes. As to Moscovici and Vignaux (1994), Mazzotti (2002a) emphasizes the affirmation that "thematization is predication" and questions how predicates (or categories) emerge:

> Predication emerges through an approximation between what is familiar and what is unfamiliar. The familiar offers the predicates, or categories, which can assimilate the unfamiliar through negotiation between people. But the theme is not a predicate; it is a general scheme from which one starts to argue (Mazzotti 2002a: 107–108).

Mazzotti (2002a: 108) diverges from Moscovici and Vignaux (1994) because the old themes, or primitive notions referred to by them are what Aristotle regarded as "common places", which are not predicates or qualities of something. In fact they are "argumentative lines coordinated by metaphors which establish the predicates". Mazzotti states that Moscovici and Vignaux (1994) also fail to recognize the role of metaphors in the coordination of meanings, and he reiterates that they are central to every argument. That is, for him, in order to understand the figurative core, it is essential to identify metaphors, not the *themata*, as proposed by Moscovici and Vignaux (1994). Hence, he states:

> cognitive metaphors organize representations, determining the role of places or topoi. If that is so, then themata – which Holton and later Moscovici and Vignaux consider to be central in representations – are subordinated to the metaphors which operate them (Mazzotti 1998: 3).

When stating the argumentative effectiveness of obvious and hidden metaphors, Mazzotti (2002a: 109) defends the necessary identification of metaphors in studies of social representation. He states: "The operation of figures or rhetorical schemes – mainly metaphors – in the core of representations is clear to me." In this first phase of Mazzotti's work the noticeable emphasis on metaphors and rhetorical and linguistic procedures can be identified through markers (for example, 'X is like Y', or comparatives such as 'like', 'the same way', 'similar to', 'it looks like', 'it is about', 'in other words', 'putting it another way'). Another characteristic of the rhetorical reading, for Mazzotti (2002b), is orientation based on questions aimed to the speaker: Who? When? Against what? Why? How? The author states that the dialogue with the text allows the researcher to ask to whom the discourse is addressed, placing him or her face to face with the real audience being addressed by the speaker.

According to Mazzotti (2002a: 112), in the study of social representations it is necessary to analyse the discursive *corpus* of the subjects when seeking to understand the negotiation of meanings between the speaker and the audience, apprehend the semantic operators and identify the central metaphor in a representation or argument. According to the author: "metaphors condense and coordinate meanings, thus they activate the cores of social representations, once they have established and managed the predicates and common places (*topoí koiná*)".

According to Mazzoti (1998), metaphors are ways to assimilate the new, allowing an unfamiliar object to become familiar to the members of a social group, mainly because they play an organizing role in the practices of this group.

> A metaphor is, at the same time, a product, the result of a process, and the process through which the 'new' is assimilated in the previous representations. The metaphorization process emerges from the transformation of the 'object' into something presented as an 'image', materializing it into an intelligible form for the social group, which is the the the point of support or anchor for the meanings placed in the metaphor (Mazzotti 1988: 4).

Thus, the author establishes some similarities between metaphor and social representations: (a) they result from the same deviation process (tropos) of meanings, named by Jodelet (1989a) 'subtraction', 'distortion and 'supplementation'; (b) both mean something to a particular social group, although they may be 'translated' by other groups, since they share the same semantic field and are in the same social enviroment (Mazzoti 1998: 5). The imagistic dimension can be added, which would allow comparisons with Moliner's previously mentioned studies.

The author continues by reinforcing his view that social representations benefit from the argumentation and rhetorical theories, mainly because argumentative practices are carried out by members of social groups. Thus, he reiterates the ideas of Wagner (1998), who takes an interest in the sociogenesis of the public discourse of "reflexive" groups whose members recognize and categorize themselves as members of this particular social unit (Mazzotti 2002b, 2003).

According to Mazzotti (2002b, 2003), Wagner notes that social representations are produced in a social context with sociogenesis based on consensus and participation. They are produced in groups and by groups who think ("reflexive"), and coordinated and harmonic interaction is needed. Otherwise, the group and its representations cannot be built. Thus, for a representation to be produced, one condition is essential: the coordination of the actions among the social group. For Wagner, it is necessary to consider unstructured external attributes, which constitute knowledge of the representation and its relationship with the group. The author draws attention to the "externalities", mainly because no representation is shared by 100% of members of the group. In fact, there is a functional consensus which makes the group categorize themselves and be maintained as a reflexive social unit. Wagner also states that something unfamiliar which results from changes in the life conditions of a group initiates collective communication processes and thereby becomes familiar; but in many cases this can provoke conflicts among social groups. The author highlights the role of discourses (arguments and contradictions) in these contexts. Inspired by Wagner's proposal (1998), Mazzotti (2003: 9) states:

> Once reflexive groups have developed their representations through conversations, intending to assimilate and accommodate new objects which have been presented to them, it is necessary to analyse the argumentative process carried out in the groups and, at the same time, evaluate social representations through argumentative figures.

According to this, therefore, prioritizing metaphors is an essential element in discursive coordination and plays a persuasive and efficient organizing role in the cohesion of a group. Mazzotti (2002b, 2003) opposes two aspects proposed by Wagner when he states: (a) it is not necessary to consider what Wagner denominates "externalities" of social representations; (b) forming social representations is not the objective of groups. According to Mazzotti, social representations are produced in and by groups, based on their material and dialogic practices, and when analysed through figures of thought, they expose the practices which cause them. According to the author, if the "reflexive groups" proposed by Wagner formed their representations through conversations (to assimilate and accommodate new objects) it would be necessary to analyse the argumentative process of these groups by investigating the social representations through argumentative figures.

Although he also refers to metonymy and synecdoche as schemes or cognitive forms, Mazzotti (2002b, 2003) in this period emphasises the study of metaphors, understood as condensations of meanings based on analogies, a definition which allows them to be regarded as condensed analogies. Therefore, they are at the centre of social representations. Mazzotti (2002b) agrees with the cognitive linguists Lakoff and Johnson (1980) when they say that metaphors are cognitive models with embodiment, structuring thought, and forming categories.

Metaphors have the following characteristics, according to Mazzotti (2002b): (a) they are linked to actions from which cognitive or conceptual schemes emerge; (b) recognition of them is associated with the type of emotional and cognitive relationship established in the group, which reinforces their meaning; (c) they are centred in argumentative processes, particularly since they condense comparisons between heterogeneous factors to establish the intelligibility of something; (d) they reveal the operational process of forming a representation and the social genesis of what is valued in analogies; (e) they rely on schemes which allow persuasion to be used; (f) they express and coordinate the social practices of groups when using particular questions as conversation subjects; (g) they operate in a semantic, cognitive, expressive and praxeological way; (h) they state that something is a fact, and that this fact creates an attitude, and that this attitude is preferable to another.

By way of illustration, Mazzotti (1998, 2002b) cites the "natural selection" of species, when Charles Darwin establishes an analogy between, on the one hand, the selection of animals and plants carried out by men to produce species which interest them, and on the other hand, the emergence of new species only by the intervention of particular conditions. That is, the metaphor "natural selection" originates from the comparison of "selection produced by men". Nevertheless, "natural selection" is not intentional. Thus the meaning of the original metaphor has undergone a modification, since the concept of "natural selection" suppresses the crucial factor of intentionality which is present in the selection of animals and plants carried out by men. Thus, if the intentional aspect is returned to the concept of "natural selection", according to Mazzotti, Darwinian Theory becomes creationist and finalist. In this example, the analogical relationship is: the production of new organisms controlled by men is related to intentional selection as the emergence of new species in natural conditions is related to the accidental or natural selection. According to the author, A/B:C/D, which would be the general form of analogical relationships, presenting metaphor as an analogy which condenses meanings, constituting a predication. According to the author (Mazzotti 2002b), predication is a process which uses the distinctive qualities of each thing, as established by the social group. When something new arises, it is necessary to find predicates which characterize what is familiar and which can be used to assimilate the new element to the cognitive and emotional repertoire of the group.

When explaining operators of the social representation phenomenon, Mazzotti (2002b) enphasizes that the analysis of metaphors demands conflicts of discursive productions of fully antagonistic or completely opposing groups. He considers that social representations emerge from conversations between members of a group and express the negotiation of differences to apprehend something placed as an object. This ype of negotiation of differences condensed in schemes or in cognitive and emotional figures is a characteristic of rhethoric. According to Mazzotti, if research on social representations identifies metaphors utilized by social actors, taking into consideration the context of the discourse in which opponents express analogical non-negotiable elements would make it possible to analyse their sociogenesis and show them more accurately when accessing a more controlled interpretation of their arguments.

The idea that social representations constitute the argumentative or conversational practices of groups in the process of interaction can be seen in other texts by Mazzotti. According to Duarte/Mazzotti (2004: 83): "It is when we use discourse (argumentative practices) that we present the meanings which define things we talk about." They believe that figures of thought express and condense the movement of the subject to accommodate himself or herself to what is new when trying to communicate it or, when there is nothing new, the movement of accommodation to new audiences. The authors refer to Fahnestock's (1999) definition of language figures, in which they are "the epitomes of argument lines or thinking. In other words, figures are common strategies of thinking which summarize an argument" (Duarte/Mazzottii 2004: 87). The proposal of "summarizing an argument" suggests the figurative meaning of social representations.

Although already mentioned in previous texts such as the one cited above (Duarte/Mazzotti 2004), in publications from 2007 onwards regarding the approximation between SRT and rhetoric made by Mazzotti, there is greater consolidation of what started to be referred to as "rhetorical change of philosophy". "Rhetorical change" values argumentative practices and actions of men in the construction of knowledge, unlike the previous hermeneutical approaches grounded on romantic sense. Based mainly on the work of Perelman and Olbrechts-Tyteca (1958, 1996), the "new rhetoric" proposes approximation between the speaker and the audience – the construction of identity of feelings, negotiation and the sharing of meanings – resuming the rhetorical method elaborated by Aristotle.

In this second movement of texts by Tarzo Mazzotti, the fundamentals of "new rhetoric" seem to be more accentuated, favouring relationships between figures of thought within the approach of the figurative meaning of social representation (no longer emphasis on metaphors, although they remain significant) and later the proposal of an "argumentative core" of discourse.

When approaching the relationship between figurative core and different figures of thought, Mazzotti (2008: 98–99) refers to Serge Moscovici:

> But what does Moscovici mean by figure? "The word figure expresses more than imagery, that is, it is not only a reflex, a reproduction, but also it is an expression, a production of the subject". (Moscovici 1976: 63, note 1). Thus, figures can be understood as figures of thought: metaphors, metonymy, (synecdoche) and irony. Not by means of arbitrary movement of approximations between language theories and rhetoric, but by means of characterization of the core of social representation established by Moscovici (1976: 127): "once the social representation core is naturalized, it is still necessary to demarcate and set the personal practices and put them in order so that they are in accord with such a core. This is the task of classificatory thinking."

According to Mazzotti (2008: 99), this classifying mindset would systematize the core of social representation, drawing attention to the way that classification (organization, hierarchization) of things in the world emerges and that meanings result from negotiations between people about the new, supported by previous figures/schemes. This implies that categorization is used to establish equalities and inequalities.

Mazzotti notes the usefulness of metaphors and metonymies, figures which show the similarities used by the subjects when defending what they consider preferable. Synecdoches (the substitution of a term, evoking enlargement or reduction of its meaning) would be included in both. The comparison of gender and species between different beings produces metaphors. And the comparison between related beings originates metonymies. Both are basic schemes or figures of classification. For Mazzotti (2008: 99), "if the core of social representations is figurative, resulting from classifying processes of things in the world, then these schemes can only be metaphors and metonymies." According to him:

> Metaphors and metonymies coordinate and condense actions of speech, organize human actions. But still: in conflictual situations one resorts to irony to reject representation of others, which allows us to understand representation as what supports and what is told about the other's representations (Mazzotti 2008: 104).

Mazzotti (2008) states that "saying is knowing", that actions of speech only carry meanings when situated in social groups which maintain constant relationships of interlocution, creating an identity through what they say. These aspects, he says, show the confluence of sociolinguistic, pragmatic rhetoric and social psychology approaches dedicated to studying social representations.

In this context Mazzotti takes an interest in "the situated word", understood as a social situation which reveals meanings selected and preferred by subjects, who believe that others are making the same choice. According to him, what they are willing to do is express their beliefs in order to influence actions so that they can achieve what they wish.

Mazzotti (2015: 3) refers to implicit argumentative schemes, not declared, but hidden, yet intelligible to both the speaker and the audience. Once the speaker knows the unspoken outlook of the audience he or she can use this as a base on which to start building a persuasive argument. It is thus possible to make explicit what is implicit. Mazzotti emphasizes once again the relevance of rhetorical analysis, which "has as a regulating framework the explicit awareness of social relationship between the speakers (*ēthos*) and the audience (*páthos*) which is produced in the situated word (*lógos*)."

According to Mazzotti (2015: 8, adhesion to a specific narrative makes it possible to identify the audience's preferences. For him, "it is like saying what their values or 'argumentative core' are". The author states that the audience can be described by the speaker they support. That is, it would be necessary to know the arguments accepted by the audience, which can support or not the arguments presented by the speaker. According to the author, identifying figures of thought, metaphor and metonymy makes it possible:

> to understand position-taking when apprehending the nucleus of the comparisons that make them up, and to comprehend the causes of endless conflicts and disputes that cannot be resolved through scientific theories. The reason for this is the attitudes of the social actors who select the figures of thought. [Figures of thought] are the origin of the arguments about what is real and at the same time the expression of the attitudes of the members of the social group (Mazzotti 2015: 11).

The author develops this thinking in another article (Mazzotti 2016: 1), reinforcing the idea that social representation reflects the socially situated world which anchors or sustains itself on implict elements. He states: "Researchers whose conceptual view is influenced by Social Representations Theory face a key problem: the implicit." According to the author, the "implicit" can be identified through figures of thought (for example, metaphors and metonymies) and also through argumentative mistakes or fallacies used in conflictual situations. These figures help to reveal the meanings of the premises which sustain a particular argument.

The author also refers to what audiences consider preferable, emphasizing the arguments of opposers and the explanation of meanings of figures/schemes applied by speakers, which can be described according to variations of the "argumentative core" of discourses. The author states: "As arguments are always against others, it is necessary to identify opposers to better comprehend what is being told about something" (Mazzotti 2016: 11). The predominance of a scheme in the discourse makes it possible to understand what is being classified or characterized by the speaker:

> *Categorizing* is differentiating without assuming any of the hierarchy which is used in classification. Classification can use *dissociation* to split an existing notion into two *new notions*, or terms. The first term provides peripheral details; these complement the information provided by the second term, which presents the qualities considered fundamental and more important; hence the first term reveals information missing from the second term (and vice versa) (Mazzotti 2016: 7).

The idea presented above indicates that, when identifying a dissociation of notions, it is possible to comprehend the "argumentative core", something which is not just confined to social representation discourses. This would be an argumentative scheme in which a notion which is traditionally considered to consist of one component is divided into two and compared with the first version. It usually appears first in the discourse and has its meanings controlled by the second one, which expresses the chief and the best qualities affirmed by the speaker (Mazzotti 2011).

In Mazzotti's articles the idea of a "core" in the process of the construction of knowledge is still present in this second movement. It is an "argumentative core" organizer of the discourse, constituted by a set of arguments containing cognitive and emotional operations which might contain the "figurative core" of social representation, with schemes/figures of thought forming the meanings of the premises.

Mazzotti (2016) reaffirms the relevance of studies of social representation, based on the situated word, to resolve argumentative schemes proposed by rhetoric. The next section shows the results of a study which aimed to utilize rhetorical analysis devices. I consider this a helpful way to reflect on social representation cores.

## 9.4  The Figurative Core Based on Rhetorical Devices: A Research Example

The research conducted by Lemos (2016) shows how syllogism partly helps to explain the figurative core of a social representation. The study aimed to investigate the social representations of school by artisanal fishermen of Atafona, a district of São João da Barra, in Rio de Janeiro state. All the content discussed in this section was taken from Lemos's research.

The study was conducted in a specific social environment, where there was a territorial conflict situation. Atafona is a community with the characteristics of a village. Artisanal fishing is the main economic activity of its inhabitants. Most local families derive their only income from fishing, characterizing it as a family business which passes from generation to generation and constitutes their life histories and their social and cultural identities. In general, these fishermen did not finish basic education, largely due to the difficulty of balancing their fishing work with school organizational rules, especially with regard to school hours.

In 2007, the local peaceful life was disturbed by the arrival of LLX Company, a member of RBX Group, which was simultaneously developing activities in sectors such as petroleum extraction, mining, offshore shipbuilding and port logistics. The arrival of the company on the one hand evoked expectations of development and job opportunities in the area. On the other hand, it caused environmental impacts and invaded the community's fishing territory. The construction of Porto do Açu and of a pier in the best shrimp fishery site expropriated fishermen from their workplace, bringing unhappiness, rebellion, and tension over the possibility of an end to the fishing activities.

In this conflictual environment, the company developed strategic mitigation and service-delivery programmes in an attempt to minimize unfavourable reactions by fishermen to the operations at the site. The company's financial investment offered the north and north-west regions of Rio de Janeiro the opportunity to establish vocational courses in different areas, such as port logistics, shipbuilding and engineering.

As the company was legally obliged to invest in social responsibility actions, schooling and literacy programmes were offered, such as: "Alfabetização e Letramento" (literacy) and "Rede do Saber" (knowledge network). The companyy formed partnerships with institutions such as *Serviço Nacional da Indústria* (SESI) and *Instituto Federal de Educação Ciência e Tecnologia Fluminense* (IF Fluminense). The company invested in renovation, restructure and equipment for a school located in the same fishermen's land.

The arrival of the company also increased employment opportunities in the sea, though not in fishery. For some jobs, training courses were offered by the Merchant Navy, but admission required at least six years of basic education. The proposal of education upgrades and certificate programmes through literacy forced the fishermen to attend school.

The possibility of improving life conditions (regular salaries, labour rights, housing, the acquisition of assets) made the fishermen and their families curious about the role of the school offered by the company, which could guarantee basic education and, as a consequence, another job. Attending school and starting another type of work meant abandoning fishery and building another life history to guarantee subsistence.

Social Representations Theory was quite helpful when investigating school representation by fishermen in an area of territorial conflict at a moment in their lives when they felt they belonged to artisanal fishery and did not imagine being in an educational institution. The company responsible for Porto do Açu, to a certain extent, compelled them to attend school. Hence, a dilemma was created between on the one hand starting or returning to school, and on the other hand struggling to maintain artisanal fishery in their area.

As fishing was an exhausting and risky activity and kept the fishermen away from home for several days, they felt it was incompatible with school life, and this belief discouraged them from attending school. Despite realizing how valuable society considered school, and the possibility it offered to improve their children's lives through jobs in Porto do Açu, the fishermen resisted abandoning the jobs which constituted their life histories and guaranteed their subsistence.

In order to immerse herself in the field of research, Lemos (2016) adopted ethnography, which for two years guided her contact with the locals. The observation was constant during this time, along with interviews and focus groups with the fishermen, document analyses and records with photographs. Emphasis was given to discourses produced by the community, especially interviews with twenty-five fishermen and a focus group with nine participants. Three articulated themes were noticed in the discourses: 'fishery', 'the port' and 'the school'. The analyses of their categories favoured a conclusion/syllogism which contributed to the proposal that fishery formed the participants' figurative core of social representation.

According to Mazzotti (2007), it was Aristotle who proposed syllogism as a means of linking arguments: from two propositions which are placed together,

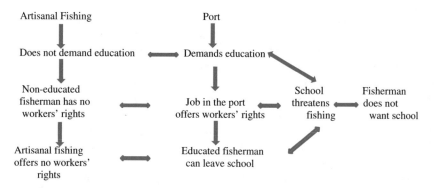

**Fig. 9.2** Deduction/syllogism based on discourses of artisanal fishermen. *Source* The author

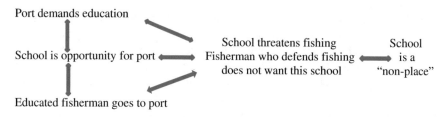

**Fig. 9.3** Figurative core of the social representation of fishermen of Atafona regarding school. *Source* The author

one can reach a third one resulting from both. For Mazzotti (2011), syllogism is a discourse in which a conclusion is drawn from some premises. Such premises are propositions with a subject and a predicate, and the conclusion results from the transfer of meanings placed by the predicate. Figure 9.2 proposes the idea of syllogism.

Although one cannot say it is a complete syllogism, well constructed, with implications and direct and necessary relationships, it was this conclusion/syllogism that allowed the proposition of a "figurative core" of social representation of this artisanal fishermen group regarding school, illustrated in Figure 9.3.

According to Moscovici (2012: 114), the figurative model (or 'core') "is not only a way to classify information, but the result of coordination which characterizes each term of representation." He states that the core is "a complex of images which clearly reproduces a complex of ideas" (Moscovici 2012: 72).

The subjects' discourses seem to aim at 'non-place' related to school, implying the central 'place' that fishing occupies in their lives. Some meaningful lines can be quoted: "My dream was be a fisherman"; "On the boat the illiterate ain't disturb nobody"; "Look, I be sincere: school ain't mean nothing"; "I think school ain't a place for me"; "Many people didn't care for study, they wanted work, fish"; "I'm fishing since I was 10. At that time we didn't get many choices. We had to choose work or study. We couldn't do two things"; "We went fishing morning, came back in afternoon, sometimes we was tired 'cause we got home two or three in after-noon and in evening we couldn't study, then was study or fishing".

The testimonies regarding school life show the views people built, some of them converging towards common traits of the social group. The 'non-place' of school in some of the fishermen's life histories, which was considered a "waste of time", something "which is not worth", started to gain importance in the research. It is important to evidence that fishing, the central activity in that culture, full of values and symbology, reveals the being and defines the existence of the group. The port, in the context which was studied, is something which invaded the space, disrupting it and imposing a new direction to something which appeared to be definitive and grounded. This situation seemed to support rejection of the school which was being offered. It seemed to be an opportunity for better life conditions, mainly for the older fishermen's children, but at the same time it was a threat to fishing activities.

Fishermen strongly affirmed "fishermen do not want this school" which with-draws them from their work activities and social identity. The "non-place for us" seems to join with the group's feeling of exclusion, deeply present in certain social situations, such as: the need for subsistance guaranteed through fishing; little attraction to the school they were offered; the difficulty of balancing the tiredness caused by fishing activities with the sacrifice which attendance at school demands; and social and workers' rights which have little effect on their lives. The "non-place" is something associated with environments which make them feel excluded, since this is the way they established an analogy with school.

Rhethoric devices used by Lemos (2016) were essential for understanding the genesis of the social representation of school produced by the group, illustrated in the figurative core. The threat the port posed to fishermen meant the end of fishing, and the school supported such a situation. This interfered not only with survival, but with the loss of the group's social identity. Invasion of their fishing community would expel fishermen not only from their workplace but also from their lives, destroying their origins, social identity, and the knowledge which was built all through their lives and did not depend on school education. Arguments expressed by the fishermen showed that resistance to school is not related to the values which society as a whole gives to this institution, but to the threat it represents to the survival of the social fishing group in that specific social environment.

## 9.5   Final Remarks

This chapter has shown approximations between SRT and rhetoric, but was not intended to epistemologically discuss their contribution as fields of knowledge, their differences or their similarities. The major point was to demonstrate the existence of a social representation "core", which can approximate both SRT and rhetoric from a psychosocial perspective.

Considering discourse essential when analysing the figurative core of social representation, one point can be seen approximating SRT and rhetoric: every discourse is situated, involving context, consent and dissent in groups. This is a way to develop future studies which focus mainly on the role of influence. In the field of social psychology, which supports social representation studies, one of the objects is influence. With regard to rhetoric, influencing means persuading.

This is related to group interactions in situations when speakers try to persuade others, convince through words, or, in conflictual environments, control disputes caused by opposition. The identification of figures of thought in minor or major group discourses can also contribute to future studies seeking to objectify and anchor social representations.

Other aspects can be mentioned, such as figures of speech which facilitate comparisons between different social groups and cultures and value the knowledge of researchers in the field being investigated. From an empirical point of view, in such a context it would be relevant to assess figures of thought in the groups, in order to verify if they can recognize them in their discousers.

This chapter also sought to show the need for perfecting analysis for the identification of the core of social representation, which would require favouring the devices offered by rhethoric. As Mazzotti states (2008: 105), "this will allow the perfecting of explanatory and analytical devices, increasing awareness of Social Representations Theory and its investigation techniques."

Finally, I believe that theoretical and methodological refinement of the "argumentative core" proposed by Mazzotti (2015, 2016) can increase the use of figures of thought in research about social objects investigated according to SRT. An important aspect which grounds his works is emphasis on conflict, disputes and social differences when knowledge of an object is being constructed. This is an aspect to be more valued in the field of social representation, in which several research studies tend to privilege consent and the stability of knowledge elaborated by groups.

# References

Abric, Jean-Claude, 1976: "Jeux, conflits et représentations sociales" (PhD dissertation, Université de Provence).

Alves-Mazzotti, Alda Judith, 1994: "Do trabalho à rua: uma análise das representações sociais produzidas por meninos trabalhadores e meninos de rua", in: Neto, Fausto (Ed.): *Tecendo saberes* (Rio de Janeiro: Diadorim/UFRJ): 9–45.

Alves-Mazzotti, Alda Judith, 2017: "Usando conjuntos de representações sociais na análise de trajetórias", in: *Educação e Cultura Contemporânea*, 14,37: 78–92.

Chombart de Lawe, Marie-José, 1972: *Un monde autre: l'enfance. De ses représentations à son mythe* (Paris: Payot).

Duarte, Mônica; Mazzotti, Tarso Bonilha, 2004: "Análise retórica do discurso como proposta metodológica para as pesquisas em representação social", in: *Educação e Cultura Contemporânea*, 1,2: 81–108.

Duveen, Gerard, 1995: "Crianças enquanto atores sociais: as representações sociais em desenvolvimento", in: Jovchelovitch, Sandra; Guareschi, Pedrinho (Ed.). *Textos em representações sociais* (Rio de Janeiro: Vozes): 261–292.

Fahnestock, Jeanne, 1999: *Rhetorical figures in science* (New York/Oxford: Oxford University Press).

Herzlich, Claudine, 1969: *Santé et maladie: analyse d´une représentation sociale* (Paris: Mouton).

Holton, Gerald, 1982: *Ensayos sobre el pensamiento científico en la época de Einstein* (Madrid: Alianza Editorial).

Jodelet, Denise, 1989a: "Représentations sociales: un domaine en expansion", in: Jodelet, Denise (Ed.): *Les représentations sociales* (Paris: PUF): 31–61.

Jodelet, Denise, 1989b: *Folies et représentations sociales* (Paris: Presses Universitaires de France).

Jodelet, Denise, 2014: "Représentation sociale: phenomènes, concept et théorie", in: Moscovici, Serge (Ed.): *Psychologie Sociale* (Paris: PUF, Quadrige): 363–384.

Lakoff, George; Johnson, Mark, 1980: *Metaphor we live by* (Chicago – London: The University of Chicago Press).

Lemos, Suely Fernandes Coelho, 2016: *"Pescadô num qué ir pra essa escola, não!": representações sociais dos pescadores de Atafona* (Curitiba: Appris).

Lima, Rita de Cássia Pereira; Santos, Ivan Soares, 2017: "Representações sociais e práticas em escolas do Ensino Fundamental: efeitos de Unidades de Polícia Pacificadora (UPP) no Rio de Janeiro", in: *Psicologia e Saber Social*, 6,1: 67–86.

Lima, Rita de Cássia Pereira; Campos, Pedro Humberto Faria, 2018: *"Núcleo figurativo da representação social: contribuições para a educação"*, August (unpublished).

Mazzotti, Tarso Bonilha, 1998: "Investigando os núcleos figurativos como metáforas", Paper for the 1st Jornada Internacional sobre Representações Sociais, Natal-RN, Brazil.

Mazzotti, Tarso Bonilha, 2002a: "Núcleo figurativo: themata ou metáforas?", in: *Psicologia da Educação*, 14,15: 105–114.

Mazzotti, Tarso Bonilha, 2002b: "L'analyse des métaphores: une approche pour la recheche sur les représentations sociales", in: Garnier, Catherine; Doise, Willem (Eds.): *Les représentations sociales, balisage du domaine d'études* (Montreal: Édition Nouvelles): 207–226.

Mazzotti, Tarso Bonilha, 2003: "Metáfora: figura argumentativa central na coordenação discursiva das representações sociais", in: Campos, Pedro Humberto Faria; Loureiro, Marcos Corrêa da Silva (Eds.): *Representações sociais e práticas educativas.* (Goiânia: UCG): 89–102.

Mazzotti, Tarso Bonilha, 2007: "A Virada Retórica", in: *Educação & Cultura Contemporânea*, 4, 8: 77–104.

Mazzotti, Tarso Bonilha, 2008: "Confluências teóricas: Representações sociais, sociolinguística, pragmática e retórica", in: *Revista Múltiplas Leituras*, 1,1: 90–106.

Mazzotti, Tarso Bonilha, 2011: "Análise retórica, por que e como fazer?", Paper for the VII Jornada Internacional sobre Representações Sociais/ V Conferência Brasileira sobre Representações Sociais, Vitória-ES, Brazil.

Mazzotti, Tarso Bonilha, 2015: "A palavra situada", August (unpublished).

Mazzotti, Tarso Bonilha, 2016: "A exposição do implícito nas representações sociais", Paper for the VI Simpósio Estadual de Representações Sociais e Educação/I Simpósio Internacional de Representações Sociais, Educação e Subjetividade, Salvador-BA, Brazil.

Mazzotti, Tarso Bonilha; Alves-Mazzotti, Alda Judith, 2011: "El análisis retórico en la investigación sobre representaciones sociales", in: Seidmann, Susana; Sousa, Clarilza Prado (Eds.): *Hacia una psicologia social de la educación* (Buenos Aires: Teseo): 67–91.

Moliner, Pascal, 1996: *Images et représentations sociales: de la théorie des représentations à l'étude des images sociales* (Grenoble: PUG).

Moliner, Pascal, 2016a: *Psychologie sociale de l'image* (Grenoble: PUG).

Moliner, Pascal, 2016b: "De la théorie du noyau central à la théorie du noyau matrice", in: *Papers on Social Representations,* 26,2: 3.1–3.13.

Moliner, Pascal; Martos, Anaïs, 2005: "La fonction generatrice du noyau des représentations sociales: une remise en cause?", in: *Papers on Social Representations,* 14: 3.1–3.14.

Moliner, Pascal; Rateau, Patrick; Cohen-Scali, Valérie, 2002: *Les représentations sociales: pratique des études de terrain* (Rennes; PUR).

Morais, Carlos Fernandes; Lima, Rita de Cássia Pereira, 2017: "Representações sociais de professores do Ensino Fundamental sobre afetividade na prática docente", in: *Educação e Cultura Contemporânea*, 14,37: 213–243.

Moscovici, Serge, 1976: *La psychanalyse: son image et son publique* (Paris: PUF).

Moscovici, Serge, 2003: *Representações sociais: investigações em psicologia social* (Petrópolis: Vozes).

Moscovici, Serge; Vignaux, Georges, 1994: "Le concept de themata", in: Guimelli, Christian (Ed.): *Structures et transformation de réprésetations sociales* (Lausanne: Delachaux et Niestlé): 25–71.

Nascimento-Schulze, Clélia Maria, 1994: "O núcleo figurativo das representações de saúde e doença", in: *Temas em Psicologia*, 2,2: 213–219.

Perelman, Chaïm; Olbrechts-Tyteca, Lucie, 1996: *Tratado da argumentação: a nova retórica* (São Paulo: Martins Fontes).

Wagner, Wolfgang, 1998: "Sócio-gênese e características das representações sociais", in: Moreira, Antonia; Oliveira, Denize (Eds.): *Estudos Interdisciplinares de Representação Social* (Goiânia: AB Editora): 3–25.

# Chapter 10
# "The Tradition Must Carry On": Representations and Social Practices of Gender and Ethnicity among Members of a Gypsy Group in a Brazilian Region

Mariana Bonomo and Sabrine Mantuan dos Santos Coutinho

## Introductory Comment

### Zeidi Araujo Trinidade

*First, I[1] must acknowledge that it was not easy to select the two professors that I present, given the excellent level of other alumni of the Graduate Programme in Psychology at the Universidade Federal do Espírito Santo (PPGP/UFES) who also work with Social Representations Theory.*

---

Mariana Bonomo is a professor at the Espírito Santo Federal University (UFES) and collaborates at the University of Bologna (UNIBO), Italy. She has an undergraduate degree from the Espírito Santo Federal University (UFES) and a PhD in Psychology from the Espírito Santo Federal University (2010) with a sandwich internship at the University of Bologna. Email: marianadalbo@gmail.com.

---

Sabrine Mantuan dos Santos Coutinho is a researcher at the Fluminense Federal University (UFF) and visiting professor at the Espírito Santo Federal University (UFES). She completed her undergraduate, postgraduate, doctoral and postdoctoral degrees in psychology at the Espírito Santo Federal University. Email: sabrinems@hotmail.com.

---

[1]Dr Zeidi Araujo Trinidade is Full Professor of Social Psychology at the Universidade Federal do Espírito Santo, linked to the Department of Social and Developmental Psychology and to the graduate programme in psychology. She received her PhD and has a postdoctoral research post in psychology at the University of São Paulo (USP). She is a member of the directory of the National Association in Postgraduate Programmes and Research in Psychology (ANPEPP), an evaluator for the Coordination for the Improvement of Higher Education Personnel (CAPES), and Coordinator of the Social Psychology Network (RedePso). Her areas of expertise are: social representations, social practices, culture, gender, fatherhood and motherhood, reproductive health and youth. Email: zeidi.trindade@gmail.com.

*I would like to present the academic background of the professors that I now introduce, Mariana Bonomo and Sabrine Mantuan dos Santos Coutinho.*

*Social Representations Theory (SRT) has been one of the most strongly represented theoretical bases in the area of social psychology in the Graduate Programme in Psychology at the Universidade Federal do Espírito Santo since the introduction of the programme in 1992.*

*Originally used in isolated dissertations, it started to be used more systematically with the implementation, coordinated by me, of the research group Rede de Psicologia Social – Redepso (Social Psychology Network) in 2000, whose work continues to be predominantly guided by SRT.*

*Mariana Bonomo arrived first, in 2000, as an undergraduate student participating in the implementation and consolidation of the group. She defended the thesis "Social identity and social representation of rural and urban in a rural community context: field of antinomies" in 2010, advised by Professor Lídio de Souza. During her PhD degree, she conducted a sandwich internship at the Università di Bologna in Italy, advised by Professor Augusto Palmonari. Her PhD internship was developed at the PPGP/UFES, also supervised by Professor Lídio de Souza, between 2008 and 2009, during which period she worked with social representations of gypsies among the non-gypsy population.*

*She is currently a professor at the Department of Social and Developmental Psychology and at the Graduate Programme at UFES, and has worked as an advisor of students who have already completed their Master's degrees. She is currently advising Master's and doctoral students with research in progress.*

*She is also a collaborating researcher at the Università di Bologna (UNIBO) in Italy and a member of the Scientific Committee of the Centro di Ricerca sull'Educazione e la Formazione Esperienziale e Outdoor (CEFEO), coordinated by Professor Giannino Melotti and Professor Alessandra Gigli. She coordinates the agreement between UFES and UNIBO, consolidating the exchange programme between the two universities. In this context, technical visits are being carried out and two research projects are being developed in collaboration with the Italian professors, one with Brazilian funding and another with Italian funding. Two articles on such a partnership have already been published.*

*Her subjects of interest include: identity processes; social representations; psychosocial intervention; culture of violence; and the ethnic-racial issue.*

*Sabrine Mantuan dos Santos Coutinho arrived at RedePso in 2004 as a graduate student. She defended the thesis "The owner of everything: what it means to be a woman, mother and wife according to the social representations of women of two generations" in 2008. Her thesis was published as a book entitled "The owner of everything: an intergenerational study on social representations of mother and wife" (Mantuan 2009). She was advised by Professor Paulo Rogério Meira Menandro, who also supervised her postdoctoral internship, conducted at the PPGP/UFES between 2009 and 2011.*

*Since 2013 she has been a professor in the Department of Psychology at the Universidade Federal Fluminense in Campos de Goytacazes, Rio de Janeiro, and a collaborating professor on the graduate programme in Psychology at UFES, and has worked as an advisor of Master's degrees already completed and in progress.*

*Her subjects of interest include: gender; social representations; family and marital relations; maternity and paternity; children and adolescents in situations of vulnerability; and psychosocial intervention.*

*The two researchers are members of the ANPEPP WG Memory, Identity and Social Representations and the research group registered at CNPq, Sociocultural Representations, Identities and Practices.*

## References

Mantuan, Sabrine, 2009: *The owner of everything: what it means to be a woman, mother and wife according to the social representations of women of two generations* (Vitória: GM Publishing).

**Abstract** This study investigates the representations and social practices related to ethnic and gender identities among members of a gypsy community in a Brazilian territory. The analysis was conducted in the light of the theoretical-conceptual contribution of Social Representations Theory.

**Keywords** Gender · Gypsies · Brazil · Social representations

## 10.1  Introduction

In the gypsy community where the study was conducted, there were few women who were born in gypsy tents, or were daughters of gypsy parents. Many of the women who lived there did not have the same origin and only became part of the group because they had married gypsies. Among the children present in the visited territory, the boys were not only the majority, but also almost the entire child population. These initial observations, resulting from an exploratory and ethnographic study, led us to question the absence of girls in the territory and, consequently, the configuration of gender relations within the group, a scenario that implies the need to consider gender and ethnicity as interdependent dimensions in the analysis of the identity phenomenon in this sociocultural context (Bonomo et al. 2008; Mendes 2000; Pizzinato 2007, 2009). It is important to emphasize, however, that this phenomenon cannot be interpreted as a characteristic of the gypsy culture, but rather was verified in the dynamics of social relations established within the specific investigated group.

Considering these reflections, the study aimed to investigate the representations and social practices related to ethnic and gender identities among members of a gypsy community in a Brazilian territory. The analysis was conducted in the light of the theoretical-conceptual contribution of Social Representations Theory.

## 10.2   Gender and Ethnicity in the Gypsy Socio-cultural Context

Although Brazil is a country recognized for its ethnic diversity, the investigations that focus on the articulation of gender and identity, in general, establish urban societies as a privileged stage, and therefore, in comparative terms, there are few studies on this subject which focus on non-hegemonic groups (Milhomen 2011; Sifuentes/Oliveira 2010). Studies on indigenous communities, quilombolas and gypsies, for example, are still scarce in the field of national gender discussions, a gap that needs greater attention from researchers in the humanities and social sciences. Sifuentes and Oliveira (2010) mention as a possible reason for this issue the difficulty among scholars of balancing the discussions on gender and cultural processes.

More recent studies on gender in the field of psychology seem to follow a tendency to understand the concept as a social, historical and relational construction, which must be considered in a contextualized way and interrelated with several other social dimensions, including ethnicity, race, generation, sexual orientation, and socio-economic level, among others (Louro 2004). But it was not always so.

Revisiting the way in which the relationship between men and women has been scientifically and socially understood for centuries – a belief in a biological determinism that hierarchizes the sexes and naturalizes such differentiations – it is necessary to highlight the important advances that have been made since the initial studies of gender, especially regarding changes in women's roles. The recognition that the gender differences are socially, culturally and historically produced was an important step towards giving visibility to inequalities and asymmetries previously understood as natural.

Feminist movements[2] played a key role in questioning the essentialism of the relations between men and women in Western societies. They emerged in Europe in the nineteenth century, in the context of industrialization and the consolidation of capitalism, gaining greater expression in the second half of the twentieth century (when they also gained ground in Brazil), constituting feminism as a movement advocating reform, political and academic in character. Such social movements fought for equal rights between men and women, questioning the idea of women's subordination, and rejecting explanations for inequalities anchored in biological differences. Thus, it could be said that it was in the wake of feminist movements that the concept of gender (which replaced the term social sex) began to be widely used, emphasizing the social quality of the distinctions between the sexes, and questioning the explanations based on the essentialism and biological determinism that prevailed until then (Galinkin et al. 2010; Galinkin/Ismael 2011; Scott 1995).

---

[2]We recognize the existence of several feminist theoretical currents and the discussion about the so-called great 'waves' of feminism. However, it is not the purpose here to discuss them in detail.

The historian Joan Scott brought important contributions and advances supporting the idea of gender as an analytical category very useful for understanding how social relations are constructed, defining it as a "constitutive element of social relations founded on the perceived differences between the sexes [...] a first way of signifying relations of power" (Scott 1995: 41). Contrary to the sexual binarism and biological determinism that even feminist movements of the period (despite trying to deny them) ended up reinforcing in defending the equality/difference opposition, Scott proposed that the male-female difference discourse was inadequate, because it concealed the differences between the women themselves and the men themselves, as if they were equal (Toneli 2012). In Scott's view, equality resides in difference – there is no antagonism – which led her to defend the thesis of "multiple rather than binary difference, understanding that women differ among themselves as to the origin of class, race/ethnicity, generation, behaviour, character, desire, subjectivity, sexuality, historical experience.

Another important name in this debate on gender is that of the philosopher Judith Butler (2003), who starts from a critical conception of the concept, understanding it as "cultural fiction and as an effect of repetitive discursive acts" (Galinkin et al. 2010: 25), which produce the categories man and woman. That is, for Butler (2003), gender does not imply "a cultural inscription (signification) on a previously given sex. Rather, body and sex are interpreted and instituted by gender" (Toneli 2012: 150). The tendency led by her problematizes several other concepts, such as subject and identity, and criticizes propositions that are understood as essentialist and apolitical (Galinkin et al. 2010).

The discussions on gender have gained space in the agenda of social psychology and it is currently being discussed from a dialogical and constructional perspective, emphasizing the importance of social interactions, the symbolic dimension of meaning production, and the historical contextualization (Galinkin et al. 2010). In this perspective, it is understood as "a system of meanings that is constructed and organized in interactions and that governs access to power and resources. It is not, therefore, an individual attribute, but a way of giving meaning to transactions: it does not exist in people, but in relationships" (Nogueira 2001: 21).

Recognizing advances in the understanding of gender issues in the academic-scientific field does not in any way mean that, in practical terms, there have been changes with the same intensity. Despite the undeniable changes in social terms that have occurred in the last few decades, which helped to reduce gender inequalities in different ways, and have contributed to the growing dissemination of a discourse that emphasizes equality between men and women, it is still evident, in the twenty-first century, that a social dynamic that maintains meanings about hegemonic masculinity and femininity, stuck to old stereotypes. In fact, what is identified is a coexistence of more egalitarian and traditional relations in the field of gender relations.

One aspect that should be highlighted when we think about this issue, and which has already been pointed out here, is the understanding that gender relations are intertwined with several other social dimensions, and socio-cultural insertion is one of them. In this sense, it is not possible to think of women's and men's models in ideal, generic terms, applicable to all groups, since the social, historical and cultural context is essential for understanding these relationships (Louro 2004). This is because it is from the interactions and relations established since birth that the individual, inserted in his or her socio-cultural context, establishes links with social groups and understands meanings about things and about the world – in the case of gender, about what belongs to the masculine and the feminine – reproducing or questioning them, in a process of permanent constitution and, at the same time, of differentiation.

In the gypsy socio-cultural context, gender relations can be considered as one of the central axes for the organization of group life, being fundamental to the regulation of family and kinship relations within the group itself or in agreements and contacts established with other gypsy communities (Mendes 2000; 2008). Thus, the family sphere occupies the function of central unit of regulation of the daily practices of the individuals, ensuring that norms are met according to cultural alignment. As Fonseca reports,

> These beliefs are the rigid taboos and formulas that protect against contamination – of the group, the person, and the reputation. They constitute the *romipen* – "the gypsyness" – and they are the key of the rare capacity that all the gypsies have to endure persecutions and drastic changes of many kinds, remaining always gypsy. Relationships between the *gadjé* and them are heavily regulated and restricted, as are the relations between men and women – and the burden of maintaining these customs falls mostly on women (1996: 65).

In many groups and in different nationalities, gypsy women are still deprived of education, exercise of a profession, and freedom of choice of spouse, among other factors that have limited the processes of autonomy of these women, in addition to the secular discrimination faced in the contact with non-gypsy contexts and societies (Ceneda 2002; Jovanović et al. 2015). If the image of the gypsies is directly associated with the figure of the gypsy woman, representing this ethnicity in general, women are also the main target of the prejudice directed at this group, a process of exclusion that intersects misogyny and antiziganism (Magano/Mendes 2014; Mendes 2015).

## 10.3 Social Representations and Identity Processes

Social Representations Theory stands out in its important role in the second half of the last century for the redirection of social psychology, since it sought to break with the predominant dichotomous model within the discipline, presenting an integrative proposal regarding the understanding of the relationship between subjective processes and social aspects, contributing to the production/expansion of a psycho-sociological perspective. It was a theory that emerged in Europe, but which has gained influence in different parts of the world, being widely used in Brazil today.

Moscovici (2012, 2003), who acknowledged the influence of theorists in the social sciences and psychology in his work, proposed studying the knowledge produced by individuals in their daily social interactions, understanding that this type of knowledge, socially established, has its own functions, allowing people to deal with things, with others and with the world, in a process of signification and production of meaning, which facilitates relations and social communication. This kind of knowledge produced and shared by ordinary people has been called social representations.

It is important to emphasize that social representations refer, simultaneously, to a theory and to the phenomenon/object studied by it (Jovchelovitch 2011; Sá 1998; Santos 2005). As a theory, it refers to "a set of articulated concepts that aim to explain how social knowledge is produced and transformed into processes of communication and social interaction" (Jovchelovitch 2011: 87). As a phenomenon, "it refers to a set of empirical regularities comprising the ideas, values and practices of human communities about specific social objects, as well as about the social and communicative processes that produce and reproduce them" (Jovchelovitch 2011: 87).

From the perspective inaugurated by Moscovici some theoretical-methodological developments focused on different aspects of social representations. The present study is in line with the approach that emphasizes historical and cultural dimensions in the understanding of social representations and is concerned with how they are produced and maintained by individuals and groups – the procedural or culturalist perspective (Jodelet 2001, 2005; Moscovici 2012, 2003). In this perspective, they are "treated at the same time as the product and process of an activity of appropriation of the social world by the thought and the psychological and social elaboration of that reality" (Chaves/Silva 2011: 306).

In relation to the construction of social representations, Moscovici (2012, 1978, 2003) emphasizes that this is based on a primary aim: to familiarize the unfamiliar. In this sense, "in its entirety, the dynamics of relationships is a dynamics of familiarization, where objects, people and events are perceived and understood in relation to previous encounters and paradigms" (Moscovici 2003: 55). That is, it is a dynamic in which prior knowledge acts, allowing the incorporation of what is unknown, new, into a known system, in order to make it understandable. This dynamics of familiarization are constituted by the action of two main sociocognitive processes that are the basis for the construction of social representations: objectification and anchoring (Moscovici 2003). These mechanisms, intrinsically related, "transform the unfamiliar into familiar, first transferring it into our own sphere, where we are able to compare and interpret it; and then reproducing it among things that we can see and touch, and consequently control" (Moscovici 2003: 61).

Objectification aims to make something abstract concrete, palpable, material; allows the iconic quality of an idea to be unravelled in an image (Moscovici 2003). It corresponds to "how the constituent elements of representation are organized and how these elements acquire materiality and form expressions of a reality seen as natural" (Vala 1997: 360). This course comprises three stages: (1) selective construction/selection and decontextualization of information (determined by cultural and normative criteria); (2) structuring schematization (concept of schema or figurative nucleus – the basic notions that form a representation are organized in such a way as to constitute a pattern of structured relations); and (3) naturalization (the elements of thought acquire concreteness, assuming the place of evidence in reality and in the field of common sense, the abstract becomes a full reality) (Nóbrega 2003; Moscovici 2003; Vala 1997). In other words, "it is a matter of privileging certain information over others, simplifying them, dissociating them from their original context of production and associating them with the context of the imaginary knowledge of the subject and the group" (Trindade/Santos/Almeida 2011: 109–110).

Anchoring, in turn, enables new, unknown, and threatening ideas to be integrated into familiar, known categories, i.e. classified and named based on a previous network of meanings (Moscovici 2003). For the new to be familiar and mastered, established systems of thought tend to predominate through the mechanisms of classification, comparison, and categorization of the foreign object being judged (Nóbrega 2003). It is an anchorage that allows the subject "to integrate the object of representation into a system of values of its own, naming and classifying it in function of the bonds that this object maintains with its social insertion" (Trindade/Santos/Almeida 2011: 110), instrumentalizing the individuals for their daily practices.

The relationship between representations and social practices can be further evidenced in the functions they play in the dynamics of everyday interactions (Chaves/Silva 2011), which, according to Abric (2000), include functions of: (i) *guidance*, guiding behaviours and social practices; (ii) *knowledge*, enabling individuals to understand and explain reality; (iii) *justifying*, allowing, *a posteriori*, the justification of positions and behaviours; and (iv) *identity*, allowing the protection of the specificity of the groups and the construction of identity.

In these dynamics of the functioning and construction of social representations, Jodelet (2001) emphasizes that the group's interpretive system has significant interference, so that the construction of a shared reality and the production of a reference system for directly related social practices related to cultural outlines and group identity is evidenced. That is, it presupposes an integration of cognitive/psychological processes with sociocultural membership and participation. It is this system of preliminary conventions, socio-historically constituted, prior to us, that allows the reading of social interactions, guides the actions of individuals and groups, and, consequently, leads to the production of identities.

For Deschamps and Moliner the relationship between social representations and identity processes is evident in the formulation of these phenomena, since representations define and, at the same time, are defined by identity: "Identity processes allow individuals to elaborate and maintain knowledge of themselves and others, of the different groups to which they belong and with which they are interacting" (2009: 81). Likewise, social representations interfere in the way in which their interactions with other people and groups occur, and in the knowledge they have of themselves and of others, intervening in the construction of their social identity.

In the same perspective of analysis, De Rosa and Mormino (2000) emphasize some dimensions of convergence between identity and representations: the sense of belonging guides the construction of the group imaginary (representational field and community memory); the social representations (mainly, the hegemonic ones) establish spaces of communication/ appropriation and exchanges within the scope of social life of the groups and individuals; and group identity maintains the unity and cohesion of the community through the principles of positive self-image, social distinction, and historical continuity sustained by effectiveness and investment in maintaining group values. Such processes can be verified in different works (Breakwell 1993), reinforcing the importance of the theoretical field of social representations in the study of the identity processes.

## 10.4 Methodological Strategies

### 10.4.1 Participants and Investigation Context

At the time the study was conducted, the community consisted of thirty families, linked by kinship ties to the paternal lineage. For more than two decades, the group abandoned their nomadic life and established housing in a semi-rural territory, in the countryside of a Brazilian municipality. The so-called gypsy trips, which in the past were carried out by travelling groups with troops of donkeys and mules, are nowadays experienced periodically and associated with the establishment of commercial or matrimonial agreements for the reproduction of the social fabric of the group. When they want to travel, they hire trucks to take their belongings and follow the fleet with their own car. They go to unfamiliar places and settle with their tents, repeating the tradition of gypsy journeys, now facilitated by new resources acquired by families.

The present study was carried out with nine members of this community. Among the participants, three men, aged between twenty-eight and sixty years, and six women, aged between twenty and forty-nine, were interviewed. All participants were married to gypsy men, but their belonging to the community was considered different. Among the women in this community there are females called *gypsy women*, who are daughters of gypsy men, and *dwellers*, non-gypsy women who have married gypsy men. The term *dwellers* means that these women come from the 'people who live in houses', as opposed to gypsies, who characteristically have the image of being nomadic and living in tents, even when they are fixed in some municipality for a longer period of time, as in the case of the group under analysis. All the men who were surveyed were married.

According to the group's customs, children are given a non-gypsy name at birth, a practice introduced into the groups due to the need to be registered and have economic, social and religious representation in their interactions with non-gypsies, and a gypsy name, which is secret and known only by the community

**Table 10.1** Characterization of participants. *Source*: The authors

|  | Fictitious name | Age | Origin and additional information |
|---|---|---|---|
| Women | Carmem | 20 years | They were born in groups from the same family as their spouses |
|  | Catarina | 40 years |  |
|  | Clara | 49 years |  |
| Men | Carlos | 28 years | They were born in the reference group where the study was performed |
|  | Caetano | 28 years |  |
|  | Casimiro | 60 years |  |
| Dwellers | Manoela | 23 years | Has been in the group for three years, wears gypsy clothes and knows the gypsy language |
|  | Mercedes | 26 years | Has been in the group for three years, wears gypsy clothes and knows the gypsy language |
|  | Miranda | 30 years | Has been in the group for six years, does not wear gypsy clothes and does not know the gypsy language |

itself. The fictitious names, presented in Table 10.1, are therefore intended to ensure the anonymity of the respondents regarding non-gypsy names known to the researchers. The fictitious names beginning with the letter C refer to (i) the gypsy women (Carmem, Catarina and Clara) and (ii) the gypsy men (Carlos, Caetano and Casimiro), and the names beginning with the letter M refer to (iii) the *dwellers* (Manoela, Mercedes and Miranda).

Regarding the origin of the women of the group, the gypsies who married members of this community generally belonged to the same lineage as the spouse's family or came from groups with which their family had been linked for some time. The entrance of the *dwellers* in this group occurred during the 'gypsy journeys', when single men chose their partners from the villages which they passed through (mostly in rural areas).

### 10.4.2 Data Collection Procedure

In order to create a more favourable interview environment for data collection, we used a thematic interview script with invisible structuring (Nicolaci-Da-Costa 1989). Under trees or in the gypsy tents themselves, we conducted the interviews individually with each of the participants, according to the availability of the members of the group.

The semi-structured interview script focused on the following information cores: (1) the personal data of the participant; and (2) questions regarding the socialization of gypsy girls and boys, the rules and gender roles in the group, and the views on what it means to be a gypsy woman or a gypsy man. As agreed, all interviews were recorded and later transcribed in full to enable the narratives to be analysed.

### 10.4.3   Procedure of Data Organization

The treatment of the data corpus was carried out using the methodological resources of Content Analysis (Bardin 2002, 2003), a strategy that allows reseachers to capture the "semantic or syntactic repertoire of a particular social group" (Oliveira 2008: 570) in relation to the social objects investigated and the themes associated with them. The organization of the information obtained was based on the identification of units of meanings related to thematic nuclei.

## 10.5   Results

The thematic units of meanings, identified in the discourse of the members of the gypsy community studied, are presented in three sections, interdependent as to their dynamics and content: (1) Gypsy Law: "the main thing is that the gypsy man has freedom and the gypsy woman does not"; (2) social representations of gender and ethnicity: being a gypsy woman and being a gypsy man; and (3) tensions between the Gypsy law and real life. To contextualize the analysis performed, the description of the thematic categories will be accompanied by fragments of the narratives of the participants, identified by their respective respondents using the fictitious names previously presented.

### 10.5.1   Thematic Axis 1: Gypsy Law: "The Main Thing is that the Gypsy Man has Freedom and the Gypsy Woman does not"

As in traditional non-gypsy societies in general, the process of socialization in this group maintains the classic model, with the woman focused on the house/ tent, while the man has knowledge and the mastery of the public space. Female demureness and men's unconditional freedom are conditions learned from childhood and reinforced throughout the entire life of the members of the group, with clear gender differences.

> ...After eight years old, we already start to separate: girls will only hang out with women; they can't be around men, because sooner or later, there may be comments. If the father or the family knows about it, or a relative, they will be very embarrassed (Casimiro).
>    Girls are taught how to take care of the house; and men learn how to work and provide the things for the household, you know? The boys, I mean, they don't bring any trouble for the family. The girl has to be raised inside the tent... (Clara).

The *dwellers* observe that submission is an extremely important value to be learned by the girls, associated with strict attention to the maintenance of virginity, giving them an education of permanent vigilance and distance from the masculine space.

> *Girls have to learn from an early age to be submissive to men. They observe how we relate daily, so they learn. This is their instruction. The girls grow up listening to this since they are little... what they can do and what they can't* (Mercedes).

For single young girls, the main prohibition is dating, be it with a gypsy or not. Carmem compares the rules directed at gypsy women in relation to non-gypsy women:

> *They* [Brazilian women] *can date; do whatever they please with their lives. We, in turn, are forbidden by our father, and this is our law. If gypsy people, I, a gypsy woman, if I date someone who is not a gypsy, our fathers won't allow. We are committed since we are little; our fathers arrange and make commitments; we go on and get married... we don't talk to the groom, we don't date... Gypsy can only marry if they are virgins... the bride has to be a virgin.*

For married women in the community, reserved behaviour, in order to preserve the image of the husband within gypsy society, is the main rule which, if not met, may lead to punishment. Episodes of conflicts are not uncommon. Thus, according to the norms of the group, if a woman dishonours her husband's family, she may be banned from the group.

> *She doesn't belong to the group anymore. If she makes a mistake, she will be banned from the group. The man has nothing to do with it. Men betray women, there is no problem* (Clara).

Gypsy women do not question male infidelity, and it is even naturalized. Among the *dwellers*, however, fidelity is required.

> *That's why gypsy women say it is OK to betray, but it isn't. They are resigned because when the gypsy woman gets married, it is for life, no matter what happens, it is for life. But not in our case. If my husband has a wife, I won't accept him living with a woman and living with me. But gypsy women, no! Because they don't want to lose or separate, so as not to be badly spoken of, they end up accepting* (Miranda).

### 10.5.2   Thematic Axis 2: Gender and Ethnical Social Representations: Being a Gypsy Woman and Being a Gypsy Man

The identification of the meanings associated with being a gypsy woman and being a gypsy man was conducted with the purpose of obtaining a more detailed characterization of the objects under analysis as a strategy to understand the processes that organize the field of social identification among the members of the community. In Table 10.2, we project the semantic units that characterize the masculine and feminine identities according to gypsy men, gypsy women and *dwellers*.

Among the portrayals by the subgroups of respondents of what it means to be a gypsy woman, what stands out is their submission to the set of rules dictated

**Table 10.2**  Field of significance for being a gypsy woman and being a gypsy man for women, men and dwellers of the group

|  | Gypsy women | Gypsy men | Dwellers |
|---|---|---|---|
| **Being a gypsy woman…** | Regarding the way of being:<br>• May dishonour the family;<br>• Embarrass men;<br>• Does not work (supported by the husband);<br>• Inferior to men;<br>• Quiet/reserved.<br>Regarding the rules they must obey:<br>• Does not date before marriage;<br>• Cannot betray husband;<br>• Cannot go to parties;<br>• Cannot drink alcoholic beverages.<br>Regarding the activities that she must perform:<br>• Takes care of the tent;<br>• Takes care of the husband (does everything he demands);<br>• Takes care of the children. | Regarding the way of being:<br>• Disrespects the men /family;<br>• Stays in the household;<br>• Unreliable;<br>• Quiet/reserved;<br>• Submissive to men (father and husband).<br>Regarding the rules they must obey:<br>• Does not date before marriage;<br>• Cannot betray the husband;<br>• Must always remain in the tent;<br>• No freedom.<br>Regarding the activities that she must perform:<br>• Takes care of the tent;<br>• Takes care of the husband (does everything he demands);<br>• Takes care of the children. | Regarding the way of being:<br>• Resigned/does not claim anything;<br>• Discriminated against by the gypsy men;<br>• Emotional;<br>• Dressed up/beautiful;<br>• Reads the hand;<br>• Does not care if husband is unfaithful;<br>• Marriages are arranged, not based on love;<br>• Wears colourful clothes;<br>• Lives under tight rules.<br>Regarding the rules they must obey:<br>• Should be submissive to men.<br>Regarding the activities that she must perform:<br>• Takes care of the tent;<br>• Takes care of the husband (does everything he demands);<br>• Takes care of the children. |
| **Being a gypsy man…** | Regarding the way of being:<br>• Likes alcoholic beverages;<br>• Gives comfort to women;<br>• Prefers boys to girls;<br>• Superior to women;<br>• Is more respected;<br>• Betrays the wife;<br>• Lives "partying" and in "streets".<br>Regarding the activities that he must perform:<br>• Supports the family. | Regarding the way of being:<br>• Gets embarrassed/irritated easily (which can be provoked by the wife and/or daughter[s]);<br>• Likes to eat early;<br>• Honoured;<br>• Likes dating/flirting;<br>• Does not like to be humiliated (which can be done by the wife and/or daughter[s]);<br>• Is more respected;<br>• Superior to women;<br>• Temperamental/nervous;<br>• Totally free;<br>• Works with smartness;<br>• Betrays the wife.<br>Regarding the activities that he must perform:<br>Supports the family. | Regarding the way of being:<br>• Sexist;<br>• Prefers boys to girls;<br>• Protects women;<br>• Feels superior to women;<br>• Betrays the wife.<br>Regarding the activities that he must perform:<br>Supports the family. |

Note: For the composition of the table of meanings associated with the objects in question, we identified descriptors present in all the interviews that composed the corpus of data analysed in this study. The treatment of the data set was based on the semantic criterion, disregarding the record of frequency of the elements identified according to the sample size and the shared nature of the information obtained by the gypsy culture. In addition to the classic role of the woman as "tent owner" (responsible for the family's tent, the husband and the children), the woman is considered inferior to the man, subject to the Gypsy Law (without freedom) and the carrier of male disgrace, so she should be reserved as a sign of obedience and virtue, an image that is ratified by gypsy men and women, as can be seen in the following fragments.

> For a gypsy man, anything a gypsy woman does is disrespectful. So, because of this, the gypsy woman is quieter, more reserved (Carlos).
>     The gypsy man betrays the woman and does not dirty the family name. Now, the woman, she does. The woman gives trouble to the family... she has to be more reserved (Clara).

The *dwellers*, however, question the condition of submission of gypsy women to the masculine universe. According to Manoela:

> Gypsy women have a life in which everything is OK for them, even if it's not. They don't claim anything. Life for them is: went to bed, got up; cooked, ate; washed clothes, ironed...

The social representations of being a gypsy woman, supported by negatively valued elements, according to the *dwellers'* discourse, are not a matter that affects them, since they do not feel they are gypsies (although they can be identified as such outside the gypsy camp, due to their clothes and typical props).

> I don't consider myself a gypsy and I have no desire for anyone to come and say "you are a true gypsy". I don't. I wear my clothes, I make my dresses, I walk like a gypsy, but my heart is a dweller's... [I'm] no gypsy! (Mercedes).

As for the depictions of being a gypsy man for men themselves, they reinforce their *status* of superiority – free from rules; betray wives, superior to women, and work smartly.

> ... The man in our tradition, for example, he is free, he can turn, as ancients would say, upside down, and it won't make a difference... but now, in the case of gypsy women, it is complicated (Casimiro).

The *dwellers* question the place of superiority of the men in the group, but emphasize the security and protection that they offer to women, a condition also emphasized by gypsy women who report the advantages of not having to work because men in the group support them.

> The man is always superior to women. And for them is it always like this... He has to be always two, three degrees above the woman (Miranda).
>     It is hard for the gypsy woman to work; men are the ones who provide things to women. Men work and support them. They work to give us things. Gypsy is the best thing there is, gypsy men. They give everything women want and need. Gypsy women only live comfortably, with the gold rings, necklace, and the lot (Catarina).

It also seems important to highlight, in this set of results, the proximity between the meanings presented by the different groups of respondents, demonstrating the effectiveness of the socialization process and the construction of the endogroup representations in line with the Gypsy Law.

### 10.5.3   Thematic Axis 3: Tensions Between the Gypsy Law and Real Life

Identified only in a group of the numerous territories visited (that is, this phenomenon cannot be generalized to other gypsy groups and segments of the region), an event emphasized by women refers to the donation of female babies soon after delivery.

> It is very serious if they betray the husband and dirty the family name. That's why we get rid of them, so it won't happen… We give the girl away to prevent this. If you marry a girl who is going to harm the family, the intrigue will remain forever for that family. They think that a boy will give less work and the girl gives more trouble to the family (Clara).

This practice seems to be based on the belief that gypsy women have the power to dishonour the gypsies and, consequently, all their descendants, if they betray the Gypsy Law. As a preventative measure, in the investigated group, parents gave their daughters away, removing the threat of dishonour.

> … If I find my daughter outside the house, kissing, that's it! We do not marry her anymore; she is no longer considered a damsel. She can talk with a guy, around us. He can ask her for water, "give me some water", something like this. But a guy alone, talking to her in the tent, hugging, kissing?! No, we don't accept it at all! If she does that, she is no longer a damsel for us. We stop loving her, father and mother, we no longer love her. Gypsy is like this. We don't accept it under any circumstances! The tradition must carry on… out law can't be betrayed (Caetano).

The *dwellers* view the practice of giving away girls as negative, as they identify themselves as possible mothers, but they also interpret it as a chance for their daughters to have a better life, escaping the rigour of the male hierarchy.

> To this day there are baby girls who are given away. There are many who are not raised at all. They find it unnecessary, inconvenient… what for? (Mercedes).
>
> I think that there are many women who have little girls and give them away, and sometimes they even thank God because it is very difficult. It is very hard to create a girl in here, in the gypsy environment. It is way harder… [they are] very discriminated against (Manoela).

It is also important to register the suffering of both father and mother in this process of giving away newborn baby girls. Although the negotiation occurs between the couple, women report having the final say. However, guided by the image of the possible destination of these children in case they are identified as traitors of the male honour, they generally choose to donate their newborn girls.

> The mother suffers more when giving away. The fathers and brothers give them away, afraid to have trouble in the future. But, mothers, they suffer more because a mother wouldn't want to give her child away, right?! But there are many gypsy girls who were given away in this world! (Catarina).
>
> Men do not force. If the woman doesn't want to give the baby away "ah, go on and give this girl away… I don't want her", this doesn't exist. Who decides is the man. But he does not push anything. He feels… as not to create trouble for the other children, because she will… Baby girls will cause trouble, so we already prevent it and give them away. But they are not forced to give them away. If the mother says, "I won't give her away", they won't. OK! (Clara).

The girls who are not donated go through the same process of socialization, which operates as a genuine attempt to control their behaviour and prevent anything which would be potentially capable of dishonouring the social image of gypsy men. One of the strategies is the promise of marriage between cousins while they are still children. According to the women, the confusion that might be caused by women after marriage can be better solved when the relations are within the same family group.

> I'm from the same group. I married a cousin, my father's nephew (Carmem).
>    In the families that keep the girls and don't give them away, if these girls get married, they only marry relatives... Then you see if the husband is treating her right... If he's not, then we split them and she goes back to the normal life she had as a child (Clara).

The belief that *gypsyness* is transmitted only through paternal blood ties ensures that new generations also rise through marriage with the *dwellers*, renewing the gypsy social fabric and keeping alive the ancestry of the group.

## 10.6  Social Representations and Identity Propositions: A Procedural Analysis

Based on the so-called Gypsy Law, as elaborated and experienced by the studied group, giving it coherence and symbolic support, some processes could be identified in the analysis of the results: (1) 'gypsyness', or 'being gypsy', being passed to descendants only through the blood of the gypsy man. Thus, the children of the *dwellers* are considered to be gypsies, although the mother has no blood relationship with the gypsy people, but the sons and daughters of the gypsies who left the group and married non-gypsies are not considered part of this ethnic group; and (2) women have the power to dishonour men, a process that unfolds in the public space of the community, where individuals seek the social recognition of others and are at the same time at the mercy of their constant evaluation and judgement.

The social representations of gender and ethnicity (being a gypsy woman or a gypsy man) identified in the study showed an apparent polarization between these spaces of belonging, while revealing the interdependence between these universes (Jodelet 1998; Marková 2006). According to the data presented in Table 10.2, women are assigned negatively valued meanings (without freedom, inferior to men, living under rigid rules, submissive to men, among others), while men are represented positively (they have freedom, are superior to women, are not subject to the rules, and are protective of women, among others). It seems interesting to note that the two cores of meanings are shared by men and women, and materialized in practices such as legitimizing male infidelity, controlling female behaviour, and giving gypsy girls away soon after birth, as well as the association of space: man-street and woman-tent. It should be noted, however, that this form of relationship, in which the man occupies a place of power and domination, and the woman a place of alleged subalternity and control, is anchored (Jodelet 2005) in a classic view of gender differences which is still in force in non-gypsy societies, although it may coexist with less traditional models.

Considering that individuals seek to maintain a positive social image, a condition mediated by the social group of belonging (Souza 2004; Tajfel 1982, 1983), the question arises: how would gypsy women (represented by non-gypsies in a negative way and in the endogroup context assuming inferior status), maintain their positive social identity?

Although this reality experienced by women may conflict with the identity dynamics (which is expected, as the identity processes are the conflicting dynamics of social relations), according to the principle of maintenance of the desired positive values, the meanings of being a man and being a woman are taken on by the individuals of the analysed group according to their ethnicity, as a kind of priority identity, seeking to maintain and protect the gypsy identity or 'gypsyness' through gender relations. Therefore, it is not possible to talk about gender without mentioning ethnicity in this group context. We are dealing with a dual dialogue, masculine/feminine and gender/ethnicity, which allows us to think of two propositions, interconnected in their genesis and functioning (see Figure 10.1).

In the gender proposal (I), according to the data presented, if the woman honours the man, it would be him, however, who confers on her virtue and comfort, attributes of positive belonging of the social category *gypsy woman*. Although the of lack of freedom of these women may inspire a negative reading of this

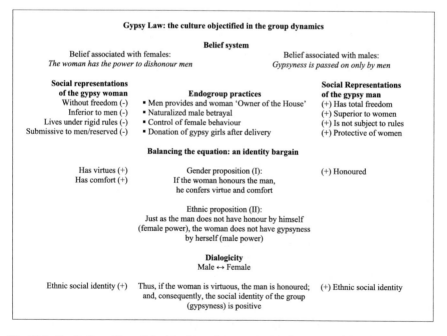

**Fig. 10.1** Synthetic outline of the identity and representational processes identified in the context under analysis

condition, gypsy women make use of this restriction to differentiate themselves from non-gypsy women by posing as virtuous because they do not date, drink, or walk outside without their husbands.

Following the second proposition formulated, the ethnic (II) one, just as the man does not possess honour by himself (feminine power), the woman does not contain gypsyness by herself (masculine power), conditions that are concretized in the relation of interdependence between the two universes. Thus, in the gypsy group investigated, the man (the one who bears the gypsyness) is the great "protector" symbol of group identity, although it is the woman who sustains the image of gypsy culture and tradition in the non-gypsy hegemonic social imaginary. Therefore, ethnic identity must be overvalued by all members of the group. For gypsies, therefore, the more they "honour" their men, the greater will be the attribution of positive values to their ethnicity, that is, they will be valuing more intensely their ethnic group and, consequently, their social identity.

Therefore (see Figure 10.1), if the woman is virtuous, the man is honoured, and, consequently, the positive social identity of the group (gypsyness) is achieved. The pact between the individual and the group is, therefore, balanced: the first representing his group in a positive way and the second offering social security, affective stability and, above all, a social identity to those who feel they belong to it. And in this sense, the social representations of being a gypsy woman and of being a gypsy man seem to contribute to the affirmation of the group, working towards its cohesion and integrity (Deschamps/Moliner 2009; Jodelet 2005; Moscovici 2003). Social representations act (Abric 2001; Jodelet 1998), thus orientating the construction of the identity spaces integrated by gender and ethnicity, producing significances that give meaning to the social life of the community and bringing together oppositions and ambiguities, which sometimes cause conflict and sometimes harmonize in favour of the protection and continuity of the group.

## 10.7  Final Considerations

Based on the analysis of the identity and representational processes, this study dealt with the representational dynamics and endogroup practices related to ethnic and gender identities among members of a gypsy group. Additionally, it sought to understand how members of this community understand the place they occupy within their group from the perspective of the socio-cultural standards of their social category, materialized in the so-called Gypsy Law.

The units of meanings identified in the discourse of community members revealed representations and practices related to the socialization process with clear gender differences: the Gypsy Law and the normative dynamics of the endogroup governed by men to be carefully followed by women; the misogyny strengthened by the community's concern to protect ethnic and male honour; and the gypsy identity or the gypsyness transmitted by the paternal side. Together, these results seem to demonstrate that the classic models of male hierarchy are strengthened in the ethnic identity sustained by the image of the man as the holder of gypsyness, which should be unconditionally protected by all the members of this social group.

The principle of conservation of the structuring and sustaining elements of a given social category, established in culture as a stabilizing force that organizes the reality of groups and individuals, cannot, however, be totally rigid. According to the results presented, the Gypsy Law, naturalized in daily practices of the endogroup, could be relativized by community members in situations where the affective dimension contrasted with the normative universe, such as the refusal of some parents to give away newborn gypsy girls. In this dimension, control and care are mixed. The different experiences of norms and rules that organize the life of the group, even in a culture with ethnocentric functioning, therefore demonstrate the complexity and dynamicity of the relationship between individuals and their socio-cultural context of insertion.

In this context of analysis, it is important to emphasize that the demonstration of examples revealing power asymmetries in gender relations within the analysed group cannot serve to reinforce prejudice against gypsies, nor can it be interpreted as a universal and characteristic process of this ethnic segment. In different groups and cultures other than gypsies, similar processes can be widely verified.

Far from exhausting the complexity of the phenomenon addressed in this study, among the possible questions arising from the analysis of the results remains the questioning about the process of social identification experienced by the *dwellers* in connection with the gypsy group. This question, which would form in an apparent identity paradox, is based on two complementary empirical evidences: (a) unlike the gypsies, the *dwellers* were not socialized under the reference of the norms, values and beliefs of the group; and (b) the double devaluation of gypsy women (by the non-gypsy society and their endogroup context) would tend to be characterized as a framework contrary to the adhesion of these women to the community. Given such observations, it would be interesting to develop new studies that investigate the process of social identification between non-gypsy women married to gypsy men, focusing on the process of identity re-signification.

Finally, we reiterate the importance of conducting studies in line with Jodelet (1998: 66) regarding the analysis of the "destructions that the negativity of social representations entails" in human experience. This is a permanent task that challenges us at different levels of psychosocial analysis.

# References

Abric, Jean-Claude, 2000: "A abordagem estrutural das representações sociais", in: Moreira, Antonia Silva Paredes; Oliveira, Denize Cristina de (Eds.): *Estudos interdisciplinares em representação social* (Goiânia: AB): 39–46.

Bardin, Laurence, 2002: *Análise de conteúdo* (Lisboa: Edições 70).

Bardin, Laurence, 2003: "L'analyse de contenu et de la forme des communications", in: Moscovici, Serge; Buschini, Fabrice (Eds.), *Les méthodes des sciences humaines* (Paris: Puf Fondamental) 243–270.

Bonomo, Mariana; Trindade, Zeidi Araújo; Souza, Lídio de; Coutinho, Sabrine Mantuan dos Santos, 2008: "Representações sociais e identidade em grupos de mulheres ciganas e rurais", in: *Revista Portuguesa de Psicologia*, 22: 151–178.
Breakwell, Glynis M., 1993: "Integrating paradigms, methodological implications", in: Breakwell, Glynis; Canter, David V. (Eds.): *Empirical Approaches to Social Representations* (London: Clarendon Press-Oxford): 180–201.
Butler, Judith, 2003: *Problemas de genero – feminismo e subversão da identidade* (Rio de Janeiro: Civilização Brasileira).
Cabecinhas, Rosa, 2004, "Representações sociais, relações intergrupais e cognição social", in: *Paidéia*, 14,28: 125–137.
Ceneda, Sophia, 2002: *Romani women from Central and Eastern Europe: a 'fourth world', or experience of multiple discrimination* (Czech Republic/Poland/Romania: Asylum Aid).
Chaves, A.M.; Silva, P.L., 2011: "Representações sociais", in: Camino, L.; Torres, A.R.; Lima, M.E.O.; Pereira, E.M.: *Psicologia Social: temas e teorias* (Brasília: TechnoPolitik): 299–349.
De Rosa, Annamaria Silvana; Mormino, C., 2000: "Memoria sociale, identità nazionale e rappresentazioni sociali: construtti convergenti: guardando all'Unione Europea e i suoi stati membri con uno sguardo verso il passato", in: Bellelli, Guglielmo; Bakhurst, David; Rosa Rivero, Alberto (Eds.): *Tracce: studi sulla memoria collettiva* (Napoli, Italy: Liguori): 329–356.
Deschamps, Jean-Claude; Moliner, Pascal, 2009: *A identidade em Psicologia Social* (Petrópolis: Vozes).
Fonseca, Isabel, 1996: *Enterrem-me em pé – a longa viagem dos ciganos* (São Paulo: Companhia das Letras).
Galinkin, Ana Lucía; Ismael, Eliana, 2011: "Gênero", in: Camino, Leoncio; Torres, Ana Raquel Rosas; Lima, Marcus Eugenio Oliveira; Pereira, Marcos Emanoel: *Psicologia social: temas e teorias* (Brasília: TechnoPolitik): 503–558.
Galinkin, Ana Lucía; Santos, Claudiene; Zauli-Fellows, Amanda, 2010: "Estudos de gênero na psicologia social", in: Galinkin, Ana Lúcia; Santos, Claudiene (Eds.): *Gênero e psicologia social: interfaces* (Brasília: TechnoPolitik): 17–30.
Jodelet, Denise, 1998, "A alteridade como produto e processo psicossocial", in: Arruda, Angela (Ed.): *Representando a alteridade* (Rio de Janeiro: Vozes): 47–67.
Jodelet, Denise, 2001: "Representações sociais: um domínio em expansão", in: Jodelet, Denise (Ed.): *As representações sociais* (Rio de Janeiro: EdUERJ): 17–44.
Jodelet, Denise, 2005: *Loucuras e representações sociais* (Petrópolis: Vozes).
Jovanović, Jelena; Kóczé, Angela; Balogh, Lidia, 2015: *Intersections of gender, ethnicity, and class: history and future of the Romani women's movement* (Budapest: Central European University).
Jovchelovitch, Sandra, 2011: *Os contextos do saber: representações, comunidade e cultura* (Petrópolis: Vozes).
Louro, Guacira Lopes, 2004: *Gênero, sexualidade e educação: uma perspectiva pós-estruturalista* (Petrópolis: Vozes).
Magano, Olga; Mendes, Maria Manuela, 2014: "Mulheres ciganas na sociedade portuguesa: tracejando percursos de vida singulares e plurais", in: *Revista Sures*, 3: 1–15.
Marková, Ivana, 2006: *Dialogicidade e representações sociais: as dinâmicas da mente* (Petrópolis, RJ: Vozes).
Mendes, Maria Manuela, 2000: "Um olhar sobre a identidade e a alteridade: nós, os ciganos e os outros, os não ciganos", Paper presented at the IV Congresso Português de Sociologia, Associação Portuguesa de Sociologia, Coimbra, Portugal.
Mendes, Maria Manuela, 2008: "Representações sociais face a práticas de discriminação: ciganos e imigrantes russos e ucranianos na AML", Paper presented at the V Congresso Português de Sociologia, Associação Portuguesa de Sociologia, Lisboa, Portugal.
Mendes, Maria Manuela, 2015: "Nos interstícios das sociedades plurais e desigualitárias: a situação social dos ciganos", in: Martins, Emília; Mendes, Francisco; Fernandes, Rosina; Fonseca, Susana (Eds.): *Modelos e projetos de inclusão social* (Viseu: Escola Superior de Educação de Viseu): 32–41.

Milhomen, Maria Santana F. dos Santos, 2011: "Enfoques de gênero no contexto indígena Xerente: algumas constatações. *Caderno Especial Feminino*, 24,1: 103–121.

Moscovici, Serge, 2012 [1961]: *A psicanálise sua imagem e seu público* (Petrópolis: Vozes).

Moscovici, Serge, 1978: *Representação social da psicanálise* (Rio de Janeiro: Zahar Editores).

Moscovici, Serge, 2003: *Representações sociais: investigações em psicologia social* (Petrópolis, RJ: Vozes).

Nicolaci-da-Costa, A.M., 1989: "Questões metodológicas sobre a análise de discurso", in: *Psicologia Reflexão e Crítica*, 4,1/2: 103–118.

Nóbrega, S. M., 2003: "Sobre a teoria das representações sociais", in: Moreira, Antônia Silca Paredes; Jesuíno, Jorge Correia (Eds.): *Representações sociais: teoria e prática* (João Pessoa: Editora Universitária/UFPB): 51–80.

Nogueira, Conceição, 2001: "Feminismo e discurso do gênero na psicologia social", in: *Psicologia Sociedade*, 13,1: 107–128.

Oliveira, Denize Cristina de, 2008: "Análise de conteúdo temático-categorial: uma proposta de sistematização", in: *Revista de Enfermagem*, 16,4: 569–576.

Pizzinato, Adolfo, 2007: "Identidade e gênero em famílias ciganas: negociações contemporâneas", in: Strey, Marlene Neves; Neto, Joao Alves da Silva; Horta, Rogerio Lessa (Eds.), *Família e gênero* (Porto Alegre: EDIPUCRS): 57–78.

Pizzinato, Adolfo, 2009: "Identidade narrativa: papéis familiares e de gênero na perspectiva de meninas ciganas", in: *Arquivos Brasileiros de Psicologia*, 61,1: 38–48.

Sá, Celso Pereira de, 1998: *A construção do objeto de pesquisa em representações sociais* (Rio de Janeiro: EdUERJ).

Santos, Maria de Fátima de Souza, 2005: "A teoria das representações sociais", in: Santos, Maria de Fátima de Souza; Almeida, Angela Maria de Oliveira (Eds.): *Diálogos com a Teoria das Representações Sociais* (Recife: Ed.UFPE): 15–38.

Scott, Joan, 1995: "Gênero: uma categoria útil de análise histórica", in: *Educação e Realidade*, 20,2: 71–99.

Sifuentes, Thirza Reis; Oliveira, Maria Cláudia Santos opes de, 2010: "O casamento e a construção do feminino na comunidade Xerente", in: Galinkin, Ana Lúcia; Santos, Claudiene (Eds.): *Gênero e psicologia social: interfaces* (Brasília: TechnoPolitik): 91–136.

Souza, Lidio de, 2004: "Processos de categorização e identidade: solidariedade, exclusão e violência", in: Souza, Lidio de; Trindade, Zeidi Araujo (Eds.): *Violência e exclusão: convivendo com paradoxos* (São Paulo: Casa do Psicólogo): 57–74.

Souza, Lidio de, 2008: "Alteridade, processos identitários e violência acadêmica", in: Rosa, E.M.; Souza, Lidio de; Avellar, L.Z. (Eds.): *Psicologia social – temas em debate* (Vitória: UFES-ABRAPSO): 169–198.

Tajfel, Henry, 1982: *Grupos humanos e categorias sociais: estudos em psicologia social I* (Lisboa: Livros Horizonte).

Tajfel, Henry, 1983: *Grupos humanos e categorias sociais: estudos em psicologia social II* (Lisboa: Livros Horizonte).

Toneli, Maria Juracy Filgueiras, 2012: "Sexualidade, gênero e gerações: continuando o debate", in: Jacó-Vilela, Ana María; Sato, Leny (Eds.): *Diálogos em psicologia social* (Rio de Janeiro: Centro Edelstein de Pesquisas Sociais): 147–167.

Trindade, Zeidi Araujo; Santos, Maria de Fátima de Souza; Almeida, Angela Maria de Oliveira, 2011: "Ancoragem: notas sobre consensos e dissensos", in: Almeida, Angela Maria de Oliveira, Santos, Maria de Fátima de Souza; Trindade, Zeidi Araujo (Eds.), *Teoria das representações sociais: 50 anos* (Brasília: Technopolitik): 101–121.

Vala, Jorge, 1997: "Representações sociais – para uma psicologia do pensamento social", in: Vala, Jorge; MonteiroMaria Benedicta (Eds.), *Psicologia social* (Lisboa: Fundação Calouste Gulbenkian): 353–384.

Ventura, Maria da Conceição Sousa Pereira, 2004: "A experiência da criança cigana no jardim de infância" (MSc dissertation, Universidade do Minho, Braga, Portugal).

## Chapter 11
# The Gendered Medicalized Body, Social Representations, and Symbolic Violence: Experiences of Brazilian Women with Artificial Contraceptive Methods

**Adriane Roso**

## Introductory Comment

**Susana Seidmann**

*Adriane Roso has developed a study[1] on contraception describing the practices and devices prevalent in historical periods prior to the Christian era. She argues that the changes that occurred in modernity not only encompassed innovations in practices related to biology, but also addressed behavioural and social transformations with regard to social representations of gender.*

*Within a historical perspective, she discusses the changes that are taking place in Brazil in relation to contraception. She highlights that prescription drugs are*

Adriane Roso is an associate professor at the Federal University of Santa Maria – UFSM (Graduate/Masters). She was a postdoctoral fellow at Harvard University (March 2018 to June 2019), focusing on postdoctoral studies in communication (UFSM). She undertook her PhD in psychology at the Pontifical Catholic University (PUC-RGS), and completed a Fulbright Fellowship at Columbia University. Her Master's degree was in social and personality psychology (PUCRS). She is a public health specialist (UFRGS) and a health management specialist (UFRGS). Blog at: www.psicologiasocialbrasileira.blogspot.com; Email: adriane.roso@ufsm.br.

[1]Susana Seidmann is Full Professor of Social Psychology at the University of Buenos Aires (UBA), Dean of the Faculty of Humanities, and Director of the Master's in Community Social Psychology at the University of Belgrano. She is also an associate researcher at the International Centre for Studies in Social Representations – Subjectivity and Education (CIERS-ed), Carlos Chagas Foundation, and coordinator of the research group of the International Centre for Studies in Social Representations – Subjectivity and Education in Buenos Aires. Email: susiseidmann@yahoo.com.ar.

© Springer Nature Switzerland AG 2021
C. Prado de Sousa and S. E. Serrano Oswald (eds.),
*Social Representations for the Anthropocene: Latin American Perspectives*,
The Anthropocene: Politik—Economics—Society—Science 32,
https://doi.org/10.1007/978-3-030-67778-7_11

*not just used to treat diseases but are a component which facilitates interference in private life projects. This component is one constituent of modernity, and it depends on the medical industry.*

*However, this process intersects with the gender perspective and the gender disparity of power – dominance – and its influence on identity construction through "constructions in interaction". She has developed a theoretical conceptualization employing two perspectives: Social Representations Theory (Moscovici) and Social Dominance Theory (Sidanius and Pratto). She formulates the emergence of different social representations through a dynamic social process, linked to the struggle for power and the hegemony of masculinity, which is rooted in Brazilian culture. The articulation of symbolic violence (Bourdieu) and social dominance (Sidanius and Pratto) polarizes the relationship between men and women, and leads to the construction of the legitimized myth of 'feminine beauty'. This myth sustains the relations of domination.*

*This articulation (symbolic violence and social dominance) is transferred to the relationship between medical doctors and female patients. Power relations generate realities through "constructions in interaction".*

*Applying snowball sampling, Adriane Roso carried out qualitative research with ten self-referenced middle-class Brazilian white women. Her primary research aim is to study social representations of femininity and masculinity based on women's experiences. The study also aims to look at the possibility of changing these social representations and the impact of the process of medicalization on women's thought and practices.*

*The author has developed an extensive academic and research pathway in health and gender studies from a feminist perspective. In this chapter, she integrates different theoretical frameworks, and articulates them with the research participants' discourses with clarity and creative depth.*

**Abstract** In this chapter, I want to present two theoretical approaches that can shed light on the phenomenon of the medicalization of women's bodies – Social Representations Theory (Moscovici 1961) and Social Dominance Theory (Sidanius/Pratto: 2012). Two key hypotheses sustain this reflection: (a) different kinds of social representations interfere in the decision-making process regarding contraceptive use; and (b) social dominance benefits from hegemonic social representations of masculinity and femininity when the legitimized beauty myth serves as a tool used by medical professionals to dominate women and limit their sexual and reproductive autonomy. The reflections on medicalization are underpinned by the experiences of Brazilian women regarding contraceptive methods.

**Keywords** Social Representations Theory · Social Dominance Theory · Women · Contraceptive methods

## 11.1 Introduction

Birth control and artificial contraceptive methods existed prior to the modern era. The historian Andrea Tone notes that the *Petrie Papyrus* (1850 BC), the oldest medical guide to contraception, "recommended vaginal suppositories made of crocodile dung, gum, or a mixture of honey and sodium carbonate". She also mentions that in Aristotles' writings women tended "to coat their cervixes with olive oil before intercourse" (Tone 2001: 13). However, artificial contraceptive methods gained another dimension with the invention of the Pill.

This invention is considered one of the greatest inventions of the modern era because it enhanced the opportunity for women to control their reproduction and experience their sexuality in a new way, even though the Pill has to be prescribed by medical doctors. Initially consumed in the US – after the Food and Drug Administration approved it in 1960 – it soon became available overseas. In the US, for example, data from 2006 to 2008 revealed that 11.2 million women and girls aged between fifteen and forty-four used oral contraceptives (Jones 2011 in Barnack-Tavlaris 2015).

In Brazil, the contraceptive pill (ENOVID™, produced by the laboratory Searle) began to be commercially available in 1962. By 1970, 6.8 million contraceptive pills had already been sold, and in 1980 this number increased to 40.9 million (Barbosa 1989 in Pedro 2003). The use of oral contraceptives declined between 1986 and 1996 (from 25.2 million to 20.7 million), while there was an increase in the number of sterilizations performed along with Caesarean sections (Potter 1999). Over the last decade, the use of hormonal contraceptives has been increasing. In 2015, Brazil had contraceptive prevalence levels of 70 per cent or more (United Nations 2015).[2]

The most recently reported data related to contraceptive use in Brazil is found in the study conducted by Minowa et al. (2015). They described contraceptive use patterns reported by Brazilian women in 2012, extracting data from the 2013 Brazil National Health and Wellness Survey (NHWS). Of the 12,000 total respondents, 4,560 were women aged 18–49, who were asked questions related to contraceptive methods. Results indicated that 63% of the surveyed women in this age group had used contraception during the past six months. This included married (39%) and single (33%) women. Young women aged 18–34 years (68%) were the most representative age group. Condoms and the Pill each represented 44% of the methods used and injection 9%. Use of vaginal rings, patches, and implants represented less than 1% each. Pills for non-birth control use were reported by 20%. The conclusion was that usage was lower in Brazil than in developed countries.

---

[2]Artifical contraceptive methods (ACM), such as third- and fourth-generation oral contraceptive pills, medical abortions, intrauterine devices, and patches, are considered drugs or devices that must be approved and monitored by the ANVISA/Ministry of Health – The Brazilian Health Surveillance Agency. ANVISA warns that contraceptives should only be sold under medical prescription and that women should have a thorough medical examination before using them.

One stimulating aspect of the invention of the Pill is that "For the first time, millions of otherwise healthy women took a 'medicine' unrelated to the prevention or treatment of a disease" (Tone 2001: 204), though some health care providers and policy-makers comprehend the "excess of fertility" as a disease in (under) developed countries (see Morsy 1995: 165; Tone 2001: 207). When "nonmedical problems become defined and treated as medical problems, usually in terms of illness and disorders" (Conrad 2007: 4), we have the development of a process called "medicalization". That means that the entity defined in the medical language framework "is not ipso facto a medical problem; rather, it needs to become defined as one" (Conrad 2007: 6). This process has been supported by the medical discourse, which, in turn, has been funded by the "medicalization industry", such as the pharmaceutical and the cosmetic industries (see Barnack-Tavlaris 2015) and educational foundations and scientific societies (see Ferguson 2002). Thus, medicalization describes a dynamic process, often facilitated and interplayed by technology and different social actors (Conrad 2007: 4), and is therefore a complex action. Social researchers have expanded their analysis, incorporating a relational, transnational and culturally sensitive perspective, where the medicalized body is constructed in allegiance to other processes – pharmaceuticization, biomedicalization, psychologization (e.g. Bell/Figert 2012; Biehl 2013; Biehl/Petryna 2013; Clarke 2010; Vos 2012).

Even though the pill can be seen as a beneficial tool to women's liberation, this new high technology[3] has placed women's health at risk (Saetnan 2000) and it impacts on women's well-being as well, because it teaches us to doubt our bodies' wisdom and abilities and to think of ourselves as unable to withstand common experiences (e.g., birthing, menopause, bereavement) without medical supervision or management (McHugh/Chrisler 2015: 7). It is a "potential means for disempowerment" (Stoppard/McMullen 2003: 56), since the decision-making process regarding contraception becomes more and more dependent on doctors' recommendations (Lopez 1998). As Tone recounts:

> in exchange for more trustworthy contraception and spontaneous sex, women gave up some of the social control they had previously exercised over their own pregnancy prevention. Before 1960, because of the popularity of the over-the-counter methods, the proportion of women who sought physicians' advice about birth control totaled no more than 20 percent. By the mid-1960s, the prescription-only pill had made visits to the doctor routine for millions of American women and had made physicians the chief custodians of new contraceptive technologies, confirming and heightening their authority as experts. (Tone 2001: 240)

This dependency on doctors means, in most contexts, dependency on men, since medicine is still dominated by men. Since women seek professional health care assistance more often than men, women's bodies "have been disproportionately medicalized" (Conrad 2007: 10), especially with regard to reproduction

---

[3]Artificial contraceptive methods are understood here as high technology that "may play an important role in stabilizing or destabilizing particular conventions of gender, creating new ones or reinforcing or transforming the existent performances of gender" (Oudshoorn 2003: 211).

control. The use of contraceptive methods is still considered "a women's responsibility" (Oudshoorn 2003: 211).

Why and how has it become women's responsibility? Do we take it for granted? How have women colluded in this assertion? Who makes this assertion 'true'? In fact, medicalization is "a truly gender concept" (Conrad 2007: 24),[4] and to answer these questions, we must pay attention to "the ways in which the social construction of sexual difference specifically impacts on social subjects, their identities, experience, life conditions, and power relations. It means questioning the universalizing male-female dyad and looking at specific subjects, groups and cultures" (Serrano Oswald 2013: 1). It requests understanding gender as not having an intrinsic quality. Gender is not "something that we are, but something we do. (...) [it is] a performative process" (Oudshoorn 2003: 210).

The contextual flow of gender relations contributes to the medicalization of women's and men's bodies. Neither women's or men's health demands nor professional health practice are placed in an a-historical environment; on the contrary, gendering is the making of historical "constructions in interaction" (Connell 1995: 35), which involves the interplay between personality and social relations (Connell 1995). The medicalization process is dependent on how, who, why, where, when and under which circumstances men and women interrelate, cooperate, disagree, etc. with other people, institutions, objects and (high) technologies.

In this process of medicalization, women are not passive actors. With the interdependence of the structural factors, they voice their opinions and desires, they resist and have agency over sexual and reproductive decisions. Some women are very aware of the side-effects of the Pill and decide to consume it nevertheless: "I do not care if you [the doctor] promise me cancer in five years, I am staying on the Pill. (...). For the first time in eighteen years of married life I can put my feet up for an hour and read a magazine" (Tone 2001: 245). Others, as we will show in the study I developed, are not well informed (by physicians or gynaecologists/obstetricians), or are not informed at all, about how the Pill will interact with women's bodies in the short and long term. Sometimes, the Pill is presented as the only and/ or the best contraception method available. However, some women instead decide not to take the Pill, either because of its side-effects or for other reasons.

In this chapter, we want to present two theoretical approaches that can shed some light on the phenomenon of the medicalization of women's bodies – Social Representations Theory (Moscovici 1961) and Social Dominance Theory

---

[4]In the book *The medicalization of society: on the transformation of human conditions into treatable disorders*, the sociologist Peter Conrad calls attention to the increasing medicalization of men's bodies, even though he recognizes "that more women's life experiences are medicalized than men's" (Conrad 2007: 24). However, the author does not make clear his definition of gender, as other medicalization theorists have (e.g. Bell/Figert 2012; Clarke 2010). In fact, the first wave of medicalization research was characterized by a genderless perspective (Rosenfeld/Faircloth 2006: 1), where gender meant female sex and femininity. Once the field of medicalization studies developed and attracted the interest of many social scientists, gender relations gradually became a focus of attention of some of them (e.g., Riska 2013, 2003).

(Sidanius/Pratto 2012). Two key hypotheses sustain this reflection: (a) different kinds of social representations interfere in the decision-making process regarding contraceptive use; and (b) social dominance benefits from hegemonic social representations of masculinity and femininity, when the legitimized beauty myth serves as a tool used by medical professionals to dominate women and limit their sexual and reproductive autonomy. My reflections were guided by the following questions: How do social representations of femininity and masculinity interact in women's decisions regarding contraceptive methods? Who are the agents involved in the decision-making process and how do they affect this process? What are the dynamics of the social representation systems that enable women to make a fair sexual and reproductive decision? How do symbolic violence and social dominance affect women's agency with regard to artificial contraceptive methods?

To underpin the reflections on medicalization, I cite extracts (vignettes) from the Brazilian women's accounts of their experiences with the use of contraceptive methods. The idea is that the vignettes might help to better illustrate the concepts and constructs of the theories I am working with, and extend their boundaries. The vignettes were deliberately chosen, based on feminist reflexivity (Hesse-Biber 2006), since I (the researcher) am considered a product of the social structures and institutions of my society.

After presenting some methodological notes, I divide the chapter into two parts. In the first part I differentiate four types of social representation – hegemonic, polemic, emancipated and alternative social representations, with the intention of showing how they produce different representations of women and different ways of thinking about contraceptive methods. In the second section I use two critical theoretical constructs, symbolic violence and social dominance, to demonstrate how they work in the process of medicalization. Within these constructs I highlight the legitimizing beauty myth and its consequences for the field of sexuality and reproduction.

## 11.2    Methodological Notes – The Collective Notebook of Women's Memories

The vignettes presented in this chapter were extracted from the main qualitative research "Precarious lives in the cyber world. Studies on violence, power, and intersectionalities of the hierarchical systems", approved by the Institutional Review Board of the Federal University of Santa Maria (Protocol 79231217.4.0000.5346). Starting in 2017,[5] a notebook called *Collective Notebook*

---

[5]In 2018 I decided to work with an online version of the notebook, because I began the post-doctoral research study "Medicalization of Women's Bodies and Sexual and Reproductive Rights" (a branch of the primary research cited before), in the United States, at Harvard University. The internet makes it possible to continue the research.

*of Women's Memories*, was handed to women so that they could write about their experience of using a contraceptive method for the first time. The first report was written by me with the intention of stimulating them to talk about sexuality and reproduction. Instead of "writ[ing] ourselves into the analysis" (Gilgun/ McLeod 1999: 185), working within a feminist epistemology, I included my personal story at the beginning of the research. The following questions were written in the instructions to guide the participants' writings, though I asked them to talk as freely as possible on the subject: When was it? How was it? Who participated in the decision? Who contributed to the use? What are the effects on your body? Why did you make that decision? What did you feel? What happened next? Once the participant had finished writing about her experience, she could either give the notebook back to me or pass it on to a friend (snowball sampling method). Most of them passed it on to a friend.

The part of the study presented here includes ten white (auto referred) middle-class Brazilian women, aged 25 to 50 years. Most of them live with their partner (60%), and 30% of them have at least one child, but no more than two children, which means they fit in "the common [Latin American countries] perception of the ideal family" (Lopez 1998: 252). They all major in psychology. Two of them have a PhD in psychology. All are acquainted with gender and feminist studies. They can be described as empowered women searching for better qualifications in the field of psychology.

The reports allowed the identification of some characteristics of the participants. The participants' age at which the first sexual intercourse with penetration took place varies from fourteen to twenty-two years old (one participant did not report). Before having intercourse, they were in the relationship for at least four months. Most of the participants got the Pill prescription from a gynaecologist, but some reported that they got the Pill from their sisters or friends. Most participants stated that they went to the gynaecologist for the first time accompanied by their mothers. Four participants mentioned that they went to the gynaecologist for the first time looking for acne treatment instead of contraception.

Regarding the sex of the gynaecologist consulted, the majority saw male, but two saw female gynaecologists. Only two participants said they used the Universal Health System (Public Health System). The majority of the participants reported that the first contraception method used was the Pill. Only three reported using the male condom during their first intercourse. Besides hormonal contraceptives (pill or injection) and condoms, other methods reported were the rhythm method, withdrawal, and IUD. Two participants reported using the morning-after pill. The side-effects of the hormonal method mentioned were a migraine and weight gain. Contraceptive brands prescribed by gynaecologists and cited in the reports were Yasmin®, Elani Ciclo®, Neovlar®, Level®, and Mirena®.

Because gender hierarchies, intergroup relations and institutional mechanisms are products of "a particular historical, political, geographical context" (Sidanius et al. 2008: 32), it is important to mention that the participants in my research are *Gaúchas*. That means they were born in the state of Rio Grande do Sul, Brazil, and at the time of this research they were all living in a medium-sized city of this state.

*Gaúchas* (females of *Gaúcho*) are women associated with the state of Rio Grande do Sul, in the extreme south of the country – a multicultural State settled mostly by Italian and German immigrants. The word "initially meant a vagabond and cattle thief, later on, designated a ranch peon and warrior, and refers, in a more gentrified mode, to any inhabitant of Rio Grande do Sul" (Oliven 2000: 128–129). In the Gaúcho culture, the adult male is responsible for furbishing the boy to become a Gaúcho man and for transmitting cultural values. There is an over evaluation of masculinity in the traditionalism culture when men and women attributes are fixed and hierarchized. The dominant pole (masculine) is the norm and the reference pattern, and the other pole (feminine) becomes invisible (Pacheco 2003). In general, we can say the *Gaúcho* culture resists modern values and preserves hegemonic social representations of masculinity and femininity. *Gaúchos* tend to reaffirm heterosexuality normativity and to be proud of their 'macho' behaviour. Moreover, the *Prenda* (female *gaúcho*) tend to be proud of their beauty, the way they dress and how they take care of their *gaúcho*.

## 11.3 From Hegemonic to Alternative Social Representations: Thinking of Contraceptive Methods

Social representation as a theoretical construct was developed by the Romanian-French social psychologist Serge Moscovici (1961). His thesis was the cornerstone for the development of the Social Representations Theory (SRT). Because it is a theory especially interested in communication and ideology, it provides us with a robust framework that not only helps to identify the representations intricated in how we experience sexual and reproductive issues, but also helps us to understand how social representations become "communicative action" (Habermas 1985: 137),[6] orientating inter/intra groups, actions and thoughts in a particular historical context.

Social representations are the continuously evolved contents of everyday thinking and the stock of ideas that give coherence to our beliefs (Moscovici 1988). They are not simple ideas or thought about something or someone, and they are *the* practice per se, moved by feelings, affections, emotions, and drives (like hate and love). That is the reason why Serge Moscovici argued that, as a system of communication, social representations orientate people in their material and social worlds, enabling them to interpret their world and master it (Moscovici 2008).

---

[6]Based on Habermas, communicative action intends to facilitate people's mutual understanding, coordinate efforts and socialization. It serves to transmit and renew cultural knowledge, to create and maintain social integration and solidarity and to form personal identities (Habermas 1985).

Through representing, humans search for meanings, and they construct, maintain and transform their reality (Marková 2008: 473).

It is beyond the scope of this chapter to explain SRT in depth – other authors have already done that very well (e.g. Howarth 2006; Jodelet 1991; Jovchelovitch 2007; Marková 2008) – but I want to highlight that there are at least four ways in which social representations, people or groups can be correlated. Even though it seems that in contemporary societies we can find more evidence of hegemonic representations (Pop 2012), other types of social representations are constructed to make sense of the world: emancipated, polemic and alternative social representations. This differentiation is essential to our parallel aim when I want to understand the interconnections of women's experiences with certain social representations of femininity and masculinity and how these same representations might change according to how society and women's thoughts change, and because of that they also affect the process of medicalization per se.

As Moscovici pointed out, hegemonic social representations seem to be uniform and coercive, homogenous and stable, but unlike collective representations, social "representation can be shared by all members of a highly structured group – a party, city or nation – without having been produced by the group" (1988: 221). However, hegemonic social representations are not static, like a photograph or a painting. Hegemony does not mean that the action of representing is uniform, massified, totalizing. Some prominences stand out in the representational field, and, although they have great force in the orientation of social discourses and practices, they do not determine them.

This kind of representation ends up "acquiring a truth status, and therefore blocking negotiation and dialogue" (Howarth et al. 2014: 28). Thus, it opens space for group-based social inequalities, social dominance and "symbolic violence" (I will return to this point later). Representations of masculinities fall into this category. Though hegemonic patterns of masculinity are not fixed, differing from one (sub)culture to another, "In contemporary mass society, nevertheless, a great deal of common ground is created by mass media, large-scale institutions, and economic structures. Therefore, the social representations of masculinity and femininity since the nineteenth century have undergone few transformations, despite changes in the social situation of women" (Poeschl et al. 2018). A familiar pattern of masculinity exists that is "hegemonic in society as a whole" (Schofield et al. 2000: 252).

Hegemonic masculinity emphasizes compulsory heterosexuality (Connell 1995; Rich 2003) and "men's mastery and control of sexuality rather than reproduction" (Oudshoorn 2003: 227), resulting in the exclusion of use and responsibility for the use of contraceptives by men (Oudshoorn 2003). Men still have control over the social, economic and political life of the community (Sidanius/Pratto 2012). They must be the financial and sexual providers in the family context (Zanello/Gomes 2010); they have to be the dominant and active part of the relationship (Welzer-Lang 2004). They are represented as more "competent, rational, assertive, independent, aggressive, ambitious, dominant, self-confident, direct, adventurous, and persistent" (Poeschl et al. 2018: 35) and braver and stronger (Connell 1995) than women.

Some of these representations are present in the reports of the women who participated in my study[7]:

*She asked me (the mother) if I was sure about the decision. I said "yes". Then, she started to say that "men are all the same", that "all they wanted was sex and then they let us go" (Morgana).*

*He [the first boyfriend] was too jealous; he thought he should be in charge of me, telling me what to wear, what to say, what to do. He measured his masculinity by the power he had over me. I had to be clean and perfumed to have sex with him, but he! He could be sweaty and stinky. It does not matter. For men, all that matters is to have sex whenever they want to* (Lola).

*Men are from Mars and women from Venus, as John Grey said. They believe they are more rational than women. They do not think with the heart. If they do not comply with the male repertoire, they are at risk of not being macho enough. They must provide safety for us because the world is unfair. Not that we are not capable of defending ourselves, but they are physically stronger than us. That is a fact I would like to change.* Nature is favourable to men; they do not menstruate, they do not get pregnant, they do not breastfeed... (Melanie).

Social representations are not only a system of thought, they are not "in the mind" of the people, they are the human action/practice per se. In the following vignettes we can see how the representations of masculinities assume this "responsibility" – and here lies the unquietness and at the same time the beauty of a representation, to play with the words Howarth (2006) and Jodelet (2008) used to talk about the uniqueness of SRT. These representations of man (or masculinity) contribute to the configuration and performance of their sexuality. In the case of the participants of our study, and probably of women in general, the father figure is mentioned as distant, a kind of myth:

*With my dad, I never talked about sex.* I never wanted to talk about it with him, and I think I would be ashamed to talk to him about it (Morgana).

*I love my father. He is the mainstay of our family; he is the one who decides everything. But I would never talk with him about sex.* And neither would he. He is a Gaúcho man, and Gaúcho does not speak of these things with the princess-Prenda. And there is this thing that the father has to be jealous of his daughter, he needs to protect his daughter, as if he knows of the bullshit a man can do (Lola).

*I grew up in an environment where the daughter is a mini-replica of her mother. My father was authoritarian, a critic of women who expose too much their body,* a critic of women who did not "respect themselves". So, talking about sexuality at home was impossible. And my mother just repeated the refrain. I needed to stay virgin to get a good man (Melanie).

This father-mother-daughter materializes the social representation of women as "objects to be won, prizes to be shown off, and playthings to be abused" (Berberick 2010: 2), as having an irrational and nurture nature, and their sexuality is associated with motherhood/fertility, lack of sexual desire, frigidity (Costa et al. 2007; Handwerker 1998; Scully/Bart 2012). They are also represented as less likely to control themselves, more easily to cede to sex – which characterizes them

---

[7]The reports were written in Portuguese. Some words or phrases had to be changed to fit the English language culture. Names have been changed to protect the speakers' identity.

as dangerous to the family, to civilization and the moral order (Costa et al. 2007), if they do not follow what is expected by society. They must be docile, loving, devoted, reserved, and value themselves by the other's gaze (Zanello et al. 2015). In Western countries, they are "seen as warmer, kind, friendly, helpful, sensitive, concerned with the well-being of others and willing to express positive feelings more than men" (Poeschl et al. 2018). In this case, they are represented positively as the "caring woman."[8]

These hegemonic representations also constitute the realm of the medical professions. Medical doctors (most of them men) usually occupy a hierarchical position superior to that of nurses and midwives (most of them women) (Baxi 2013; Cosminsky 2016; Pires 1989; Oliveira et al. 2010; Wolf 2012). I will return to this issue when I present the discussion on relations of dominance. Let's now try to understand how these social representations might become emancipated.

Emancipated representations are versions of existing representations, and though they do not directly challenge hegemonic views, they attempt to supplement the hegemonic representations, adapting knowledge and communication by offering a more detailed analysis of the subject (Devenney 2005). In this kind of representation, there is a certain degree of autonomy concerning the interacting segments of society, since each group creates its version and shares it with the others. When the concepts of sciences and the experiences of laypeople are brought together with those of the population at large, emancipated representations are developed (Moscovici 1988). "Social change occurs when emancipated representations evolve into polemic representations that render the 'self-evident' existence of hegemonic representations impossible, at least for some members of the group, and call for innovation and change" (Ben-Asher 2003: 6.4). The field of health contains several emancipated representations, according to Gillespie (2008). For example, the invention of the female contraceptive pill opened up the scenario to emancipated representations – women now could control fertility and to a certain extent manage their own sexuality. This emancipated representation evolved to polemic representation, characterized by antinomic thoughts: "A woman shouldn't control fertility because she is supposed to be a mother" vs: "A woman can't be a good mother if she can't support a child"; "A woman can't express sexual agency because the man is supposed to be in command" vs: "A women is a citizen; therefore she is entitled to express her desires".

The emancipated representations are always potentiality, placing in dispute practices and discourses that are never homogenous, producing singular and alive subjects. The relational fields (subject-object-world) enable the liveliness, dynamicity, and interaction of systems of thoughts through dialogue and negotiation

---

[8]This expression was created to 'play' with the alternative social representation of the male identity, the "Caring Man" (Oudshoorn 2000: 129). This representation has been articulated and defended by scientists (sometimes sustained by the pharmaceutical industries) to increase the acceptance of male hormonal contraception. The construction of a new representation is the "making of a New Man" (Oudshoorn 2000: 135), a man who cares about women's bodies and is therefore responsible, trustworthy and caring.

with others (dialogicity), and are fundamentally the "tension between different kinds of mutually interdependent antinomies that keeps their dialogue going" (Marková 2013).

In my empirical research with southern Brazilian women I also observed antinomies playing an essential role in the decision-making process regarding contraceptive methods. The antinomies have multiple functions: (a) to express hegemonic social representations; (b) to reflect the tension between traditional and progressive thought; and (c) to organize action according to temporality.

*The woman always wants to be a mother vs: The woman needs to find the right time to get pregnant*:

> *I was terrified of the idea of getting pregnant; I had many plans, and I was sure that having a baby at that moment it would destroy all of them* (Violeta).
>
> *My mother was always worried about the possibility of me getting pregnant [when I was a teenager]* (Virginia).
>
> *I remember writing in my diary that I want to have a girl… I was about ten years old. But I also recalled how I got terrified when a sophomore girl at my school got pregnant. She had to quit school… There's no way I would get pregnant during school years! That was one reason it took me so long to have sex with my boyfriend* (Lola).

*The woman has to say 'yes' to unprotected sex vs: The women is the one responsible for contraception*:

> *When I didn't use it [the condom], I used to think it was important to use a condom, but the truth is, I ended up having unprotected sex. I am not aware if I got any sexually transmitted disease. I confess in this aspect I neglected my health sometimes. Now, I am dating another guy that, in the beginning, he would say it was hard for him to wear condom. I think we used a few times, even so!* (Virginia).
>
> *All my boyfriends took it for granted that I was on the Pill, and they never questioned. It was like I had to be naturally protected, but not them. It was my responsibility* (Morgana).
>
> *[After taking the morning-after pill] I reflected about how hard it is to keep up the decision of not using the medication to end a pregnancy when we can't count on men's responsibility regarding their own health, and my health, and when we bet on the [reliability] of the [male] preventative* (Morgana).

Polemic representations necessarily antecede unconventional ways of dealing with any provocative issue. With regard to contraception issues, the furthest we went was to question why an effective male contraceptive was not developed, and the answer to this question rests on the hegemonic social representation of masculinity:

> Even scientists active in contraceptive research view men's psychological fragility as a significant impediment to the development of a male contraceptive. Around 1968, a Merck researcher remarked that the most difficult obstacle, perhaps, to a 'male' approach is the 'emotional-psychological' factor. The delicate male psyche equates virility with fertility, and it is believed that extensive education would be required to get men used to the idea of a 'male' contraceptive. (Tone 2001: 252)

While some authors criticize or reject the division into categories, others accept it (Pop 2012). For now, I position my view among the second, believing that these types of social representations can help us to comprehend the dynamicity of representing an object. Human beings are so complex that it would be challenging to find in the system of thought any representation in its pure, essentialist state. As Howarth keenly noted: "we do have multiple representations of the same social objects" (2006: 6).

One of the mechanisms that contributes to the fluidity of the process of representation is the relationship of power. Though we cannot "find a social psychology of power at the heart of social representations theory" (Howarth 2006: 13), the complexity of intergroup relations in "liquid times" (Bauman 2006) requires the analysis of how power is related to the process of representing the world.

Power can be understood as the individual or group capacity to interfere in the decision-making process of someone or of a whole community (e.g. Bourdieu 2001; Sidanius/Pratto 2012). It involves an emphasis on the study of (symbolic) violence, dominance, and ideology. In this case, when studying gender relations and medicalization, researchers should be paying attention to the hegemonic social representations.

## 11.4   Symbolic Violence and Social Dominance

Some types of violence are visible on the body, as in the case of torture, hate crimes, and rape. They may cause distress not only to the person who suffered the aggression, but also anyone who becomes aware of it. These kinds of physical violence can immediately trigger repulsion, pity, impotence and other sorts of feelings, but there is also another category of violence – symbolic violence: "a gentle violence imperceptible and invisible even to its victims exerted for the most part through the purely symbolic channels of communication and cognition (…), recognition, or even feeling" (Bourdieu 2001: 1–2). Through the notion of symbolic violence, Bourdieu "attempts to unravel the mechanism that makes individuals see the representations or dominant social ideas [hegemonic representations] as 'natural'" (Vasconcelos 2002: 80). It "is developed by the institutions and agents that animate them and on which the exercise of authority is based", and "The term symbolic violence appears to be effective in explaining the adherence of the dominated" (Vasconcelos 2002: 81).

Symbolic violence is more 'discreet', and, concealed through ideological discourses. Symbolic violence can be constructed through different strategies. One strategy is the withholding of technology. An example is provided by Cosminsky in her study in Guatemala. Technology was withheld from midwives on the assumption that they did not have the basic knowledge to use the technology properly. "Withholding is also a way of maintaining power and control through the possession of authoritative knowledge" (Cosminsky 2016: 230). Another form of symbolic violence is through silencing, also called euphemized violence. They are forms of erasure, for example: when the grief resulting from China's one-child policy is silenced; or the dissociation of men from the private domain (Ginsburg/Rapp 1995).

In my research, I observed euphemized violence interacting with the deci-sion-making process and in the restriction of women's agency. Euphemized vio-lence is at the core of hegemonic social representations of femininity. Silence is productive. It is capable of maintaining and revigorating these representations. The silence of their mothers is the first element pointed out by the participants. The mother does not discuss sexuality with her daughter and vice versa, forming a hereditarian cycle of euphemized violence:

> *My first sexual relation happened prior to my gynaecologist appointment. (...) My mother pretended she didn't know about it, and my family allowed me to talk about sex only after we were dating for a reasonable period* (Frida).
> *I rehearsed a long time before talking to my mom about this [sex]. For about four or five months I was already dating my boyfriend. My mom always said she was open-minded and accessible to talk about anything, but when the time came, she didn't react so openly. She got mad. She didn't say too much. ... She just handed me the same pill she was taking (Nordete™)* (Violeta).
> *I never talked about sex with my mom. ... Today I have a teen. ... When I tried to talk about sexuality, dating, sex and menstruation she said she prefers to talk with her friends* (Lola).
> *Contraceptive methods in my house were a veiled subject.* Before joining Higher Education I lived with my grandparents and my mother, who never talked to me about it. My mother always repressed the subject of the *contraceptive pill or a visit to the gynae-cologist* (Virginia).
> *My family was conservative, and we hardly talked about dating and sex.* Never. My mother has always worked so hard. So, practically, who raised me was my older sister, who was a furious person (...) (Antonia).
> *I listened to their lecture* [after they found the pills I was taking], *and then they threw it away. Then I lost control of my cycle, and sometimes I took the morning-after pill* (Olivia).

These vignettes illustrate relations of domination, which means that someone is trying to dominate or control the way someone else lives. To unpack how these relationships work, Social Dominance Theory (SDT) offers interesting theoreti-cal tools. This theory assumes that societies "tend to be organized as group-based social hierarchies, with one social group at the top of the social system and one or a number of social groups in the middle and at the bottom of the social system" (Sidanius et al. 2008: 20). "[T]he dominant group is characterized by its posses-sion of a disproportionately large share of positive social value, or all those mate-rial and symbolic things for which people strive" (Sidanius/Pratto 2012: 31).

SDT defends the existence of a shared grammar of social power in all soci-eties. This means that when we study the consumption of artificial contracep-tive methods of south Brazilian women, we have to take into consideration their experiences, controlled by international drug corporations, and also influenced by transnational feminist movements' ideas. In this way, SDT attempts to uncover the many interlocking processes responsible for the creation, maintenance, and re-cre-ation of the group-based hierarchy at the multiple levels of analysis, including the interactive psychologies of genders (Sidanius et al. 2008: 20).

SDT understands that among the stratification systems, it is in the gender sys-tem that "males have disproportionate social and political power compared with females" (Sidanius/Pratto 2012: 33).[9] This system seems to be present in all societies, and it is regulated by a shared grammar of social power (Sidanius/Pratto 2012) where the man is usually more positively valued than the woman.

Worth mentioning, in the process of medicalizing women's body, physicians still have a dominant role and considerable social responsibility as a result of their knowledge and expertise. This status provides them with the means to influence policy-makers in social and health care and highlight inequities in the reproductive and sexual health care of women and to advocate for the improved status of women (The International Federation of Gynaecology and Obstetrics (FIGO 2004: 1097). However, this status may also put them in a position which enables them to take advantage of this status to behave unethically or criminally, and worse, the same status sometimes grants medical doctors legal privileges, as in the case of Shafeeq Sheikh, a former physician at Baylor College of Medicine, in the US, who was recently found guilty of raping a heavily sedated patient in hospital, but got no jail time. Instead, he was sentenced to ten years on probation. According to *The Washington Post*,[10] his defence attorney, Lisa Andrews, said, "The facts are not black and white. …The truth is usually a version of gray. … Here we have this Latina woman with her fake boobs that came on to that little nerdy middle-aged guy, and he lost his mind." This social representation of Latina women reinforces discrimination against this minority and illustrates how a social representation can be ideological.

The reports in my studies indicate that the dominance of medical power is still active, even though the studies show that there are other institutional agents involved in medicalization (Conrad 2007; Bell/Figert 2012; Clarke 2010; Ferguson 2002):

> *The concern of my gynaecologist, at the time*, was that I did not get pregnant. I had to sign a thousand papers committing myself not to getting pregnant. And my mother, too. (…) But the gynecologist really wanted me to do that treatment. And my mother sat by my side in silence. Well, an ugly daughter is a daughter who does not marry. Better take that medicine! (Lola).
>
> *The doctor gave me no directions, nor questioned me about my decision. … He did not talk about the side-effects of the pill, nor did he tell me about other options I might have* (Violeta).
>
> *I started taking the Pill – the one the doctor himself said it was "strong" – until the cysts dissolved. Then I could change the medication. I did not participate in the decision. … The beginning of the sexual life was also a reason for him to prescribe it. … I do not remember the condom ever being mentioned. … I was discouraged entirely the moment I made mention of my willingness to continue with the Pill. The gynaecologist, a woman, said that even with the IUD I would have to continue to use contraceptives … for hormonal balance.* (Morgana).

---

[9]Other systems are the age system and the arbitrary-set system (see Sidanius/Pratto 2012).

[10]Phillips, Kristine, 2018: "A jury convicted a doctor of raping a patient at a hospital – and sentenced him to probation", in: *The Washington Post,* 19 August 2018, at: https://www.washingtonpost.com/news/true-crime/wp/2018/08/19/a-jury-convicted-a-doctor-of-raping-a-patient-at-a-hospital-and-sentenced-him-to-probation/ (10 May 2020).

> The gynaecologist examined my breasts for a long time (his specialty is breast oncology) and then examined my uterus (in the belly, above the pubic hairs), saying that I would feel discomfort as he would insert his finger into my anus to examine my ovaries, since I was "virgin" (his words). I really felt very uncomfortable and I remember wondering if that procedure was indeed correct. I still do not know the answer. ... After the examination, we returned to his desk and he prescribed me the Pill. He gave me no other possibility and gave no explanation of side-effects. I vaguely remember asking about varicose veins, but he said that interference would be minimal. He told me that I should take the Pill for at least three months before intercourse, so that my body would get used to it and there would be no risk of pregnancy (Morgana).

> The doctor prescribed me an injectable Depo-provera® (...). He did not explain the side-effects, just said that it was for me to take it, and I should not get pregnant again. Well, since I was not in a favourable situation (pregnant teenager), I accepted and started to take it, and I still take this medication (Antonia).

The vignettes illustrate how medicalization can be a top-down process in which someone (the gynaecologist) dominates the other (the woman). It might not be intentional, but once the doctor doesn't open space for dialogicity, the relationship becomes a relationship of dominance, a relationship of violence, when one side through communication maintains its status of authority and the other is afraid to challenge the unequal relationship. It is worth mentioning that the doctors referred to in the case studies weren't all men. Although SDT shows that men are consistently higher in anti-egalitarianism than women (Sidanius/Pratto 2012), it seems that once a woman enters a profession dominated by men, she has to adapt to that culture to survive, and her tendency to socially dominate seems to flourish.

## 11.4.1  Beauty as a Legitimizing Myth: Medical Power and "A Beleza da Mulher Brasileira"

> Look, the most beautiful thing (Olha que coisa mais linda)
>     Full of grace (Mais cheia de graça)
>     It is her, girl (É ela, menina)
>     That comes and goes (Que vem e que passa)
>     In a sweet swing (Num doce balanço)
>     On her way to the sea (A caminho do mar)
>
> Girl with a golden tan (Moça do corpo dourado)
>     [Got] from the sun of Ipanema (Do sol de Ipanema)
>     Her swing is more than a poem (O seu balançado é mais que um poema)
>     It is the most beautiful thing I have ever seen passing by (É a coisa mais linda que eu já vi passar)
>
> Song "Girl from Ipanema" (Garota de Ipanema) – Vinicius de Moraes\Tom Jobim (1962)

Since the time of the Ancient Greeks beauty has been regarded as a fine attribute of both males and females. "As the manifestation of bodily excellence, it betokens women's readiness for marriage and men for its male equivalent – the battlefield. It is therefore sought after in men and women alike. Yet beauty has special meaning for women" (Blondell 2013: 33). But nowadays, the myth of beauty is coloured by unique characteristics, as we will see.

The beauty myth is a subject well discussed in the social sciences (e.g. Del Priori 2000; Roso 2001; Wolf 1991), but here I want to look at this myth through the lenses of SDT, or better, taking into consideration the concept of legitimizing myths[11] (LMs) (Sidanius/Pratto 2012: 39). LMs are defined as "values, attitudes, beliefs, causal attributions, and ideologies that provide moral and intellectual justification for social practices" (Sidanius/Pratto 2012: 104), a definition very close to that of social representation. But the LMs definition is clearer than the notion of social representation regarding power relations. LMs are beliefs and practices that might "increase, maintain, or decrease levels of social inequality among social groups" (Sidanius/Pratto 2012: 104). Therefore, we can comprehend LMs as the engines that move the field of power relations.

For SDT, there are two classes of LMs: hierarchy-enhancing and hierarchy-attenuating. The first one serves to some people to "maintain and justify their superior group position (Hindriks et al. 2014: 538), for example, the criminalization of abortion in Brazil can be considered an hierarchy-enhancing LMs social policy, because generally black women seeking terminations are more at risk of dying than white women, since the latter are more likely to be able to afford clandestine clinics to terminate unwanted pregnancy. Pro-criminalization arguments emphasize the right of the foetus to live and the universal mothering nature of women (a hegemonic social representation), usually defended by Christians, which are the majority in Brazil. Hierarchy-attenuating LMs "reduce social hierarchy by delegitimizing inequality or practices that sustain it, or by suggesting values that contradict hierarchy" (Sidanius/Pratto 2012: 104), as in the case of *Católicas pelo Direito de Decidir* [Catholics for the Right to Decide] mentioned before. The differentiation between hierarchy-enhancing and hierarchy-attenuating LMs helps us to understand the tensions and conflicts within the social representation process and how the legitimizing myth of beauty interacts with the field of sexual and reproductive health.

In general, it is recognized that in Western societies a woman's body is constructed as a sexual object and constitutes the core of their identity. This myth is closely related to hegemonic social representations of women already pointed out in the above theoretical frameworks. But how different is the modern beauty myth – the myth of our times – from that of Ancient times?

Especially since Late capitalism, beauty has become a powerful tool for economically dominant groups (e.g. the cosmetic and pharmaceutical industries) to profit from women's beauty. By reinforcing hegemonic representations of women – they are born to married, to praise men, to conquer men through their beauty etc. – through printed and digital media, those groups are making high profits from selling 'magic' beauty potions (skin cleansers, anti-wrinkle lotions, hairspray etc.).

---

[11]According to Sidanius and Pratto (2012), the theory of legitimizing myths owes much to Moscovici's notion of 'social representations'.

Brazil has not escaped from the power of the beauty myth. And since the beginning of the twentieth century the forces of the media and scientific discourses have been reinforcing this myth (Poli Neto/Caponi 2007; Soares/Ferreira 2014). Consumers and medical interests are already allied, and consumers can become the dynamic force for market creation. In other cases, the public is a potential consumer that has to be transformed into a market. This involves persuading consumers about the need or utility of a product offered or creating consumer demand (Conrad/Leiter 2004). The role of patients in medicalization is neither neutral nor diminutive. The transmutation of the patient into the consumer in the medical market leads him/her to conform with the transformation of health goods and services into commodities (Albulquerque 2018).

The physician-consumer-beauty relationship was an essential aspect in the reports of the participants in my research. Participants who consulted a gynaecologist before having sex sought the service to deal with issues related to acne or polycystic ovaries (OP). Here we see what Albuquerque (2018) observed: The medicalization of certain conditions may help the patient to deal socially with her problem. Once the physical attribute becomes a medical concern instead of a cultural problem, it can contribute to the deconstruction of stigmas and prejudices. That is, when medicalization brings a visible benefit to the patient (reduction of acne or elimination of cysts), the cost of the side-effects is less relevant at the time. It appears women's search for the pattern of Brazilian beauty, and having a clear skin, contributes to this process.

But as we can read in the reports below, though the adolescents and young women might have benefited from the medication, they stated the gynaeocology consultation did not provide a space for them to discuss the side-effects, norr were they presented with other alternatives to treat the problem. The cultural search for a clear skin is not a factor considered by medical doctors; all they highlighted in their reports is the commitment the patient should make to not getting pregnant during the pill intake, which does not seem to pose problems for the mothers who accompany their daughters at that time:

> *At the time*, my friends would take a contraceptive pill to prevent acne. Due to the fact that I had *problems with acne since that age and the Pill was then considered a remedy for this, I wanted to use it as well. … At the age of nineteen, I began to have a dermatological treatment to solve the issues with acne, which included a powerful medicine that would create risks of abortion if I got pregnant. … Every twelve months I had to sign a form committing myself to not getting pregnant. That's what I did. From sixteen to twenty-three years old I took the pills* (Frida).
>
> *I had contact with the contraceptive pill even before I had any sexual intercourse. When I was fifteen or sixteen years old my periods stopped. I spent about nine months without menstruating until I decided to have my first contact with a gynaecologist. I discovered that I was suffering from ovarian micro-cysts. … As a teenager I sat and waited for an answer to my 'problem' in the gynaecological office. The contraceptive prescription came automatically, with the explanation that the cysts would be 'dissolved' into the hormonal form* (Morgana).
>
> *When I was fifteen, many blackheads and pimples appeared on my skin – something 'normal' for a teen with a body popping hormones. Since it was bothering me, my mother*

*took me to a gynaecologist. It was then that I started taking contraceptive pills because, according to the doctor, it would be the best method to control acne* (Maria).

*I used to read* Capricho [a magazine targeting young women]*, and saw all those pretty clean faces. I wanted so bad a skin like that. I had so many pimples and blackheads on my skin that made me feel ugly. I took that awful medicine* (Melanie).

Taking hormonal contraceptives to treat disease or a condition is a common practice. Findings from the National Survey of Family Growth in the US (2006–2008) "revealed that 11.2 million women and girls age 15 to 444 use oral contraceptives; about 1.5 million of them reported that they use pills for noncontraceptive reasons" (Jones 2011 in Barnack-Tavlaris 2015). Whether these medications are tested for that specific purpose, and whether the results of the tests in patients with acne are shared with the adolescents are questions that remain open. With regard to that, the participants in my study don't report awareness of any such tests. Even if they obtaom effective immediate results, what are the long-term effects of the medication? This kind of care is a "twisted medicine", where the medicine is produced for one reason but used for another, and in this way, it has has the potential to become a form of symbolic violence.

This "twisted medicine" works because it provides a cultural reward – the possibility of conforming to the hegemonic social representation: the Brazilian woman has to be beautiful. Beauty is the norm. It is hard to recognize this idea as violence, to refuse or resist it, because beauty becomes a legitimizing myth in the Brazilian context.

Brazilian beauty has been appraised since the colonization of Brazil, and the representation of the Brazilian women as beautiful runs across its borders. This myth was constructed and is tied to fundamental values of our society. The song I highlighted at the beginning of this section is known internationally and it serves to reinforce this representation. Brazilian beauty, among other things, is closely related to the skin appearance, a clean skin, neither black nor brown by nature, but suntanned, and subsequently without any other mark, protuberance, etc. Propaganda, cinema, soap operas, magazines, and medical discourses also reinforce this myth. And if a woman doesn't feel this way, she will make an effort to conform to this hegemonic representation, even if it means placing her health and life at risk.

It is important to highlight that the more firmly tied a social belief is to the basic values ofa society, the more difficult it is to change the LM and the more powerfully the LM will drive social policy (Sidanius/Pratto 2012: 105). Is that the case with the beauty myth? It seems that in a powerful way the legitimized beauty myth helps to drive social policy in the field of women's health. Cosmetic treatments are not available in the Unified Health System, but people can manage to get them through the system by requesting pills for contraception, but consuming them, in fact, to treat acne. The myth of beauty is legitimized by the *"jeitinho brasileiro"* (Brazilian way), a creative way to survive dominance in certain situations. The same seems to happen with plastic surgery. According to Ferreira (2011), the Brazilian Plastic Surgery Association states there has been a substantial increase in

the number of plastic surgery procedures in Brazil – the country in the world with
the second-highest number of plastic surgeries, behind only the USA. Every year,
approximately 350,000 aesthetic surgical interventions are performed in Brazil.
The questions that remain open are: in a country where the majority of its population is on or below the poverty line, how can the health system (private or public)
afford "Beauty Medicine"? (Poli Neto/Caponi 2007: 1). Are we, women and society, willing to pay the price of a LM?

## 11.5   Conclusions

In this chapter, I have presented theoretical elements, exemplified by women's
reports, about the phenomenon of medicalization with regard to the consumption of artificial contraceptive methods. I have identified that the experiences of
the participants in the study, even though singular, are linked to hegemonic social
representations of masculinities and femininities, and 'tied' together by relations
of domination, both at the family level and in relation to health professionals,
generating symbolic violence. In fact, as Pierre Bourdieu (2001) said, symbolic
violence is a dimension of all domination. In this sense, I can conclude that
Social Representations Theory can benefit from joining the knowledge of Social
Dominance Theory and vice versa when seeking to understand the relations of
domination that permeate the complex relationship between medicalization and
health care.

The importance of this reflection consisted in presenting some information that
might contribute to the field of women's health and rights. I did not intend to gen-
eralize the constructed information, but I want to highlight how certain hegemonic
social representations make it difficult to obtain sexual and reproductive rights and
how such representations reproduce stereotypes and discrimination. The idea is
not, as one participant reported, to 'demonize' artificial contraceptive devices, but
to align the decision to consume (or not consume) them to the sexual and reproductive rights of women.

For future studies I suggest diversifying the empirical corpus to include participants from different backgrounds, so that we also shed some light on how the
different stratification systems (age, gender and arbitrary-set) interact with the process of medicalization and are affected by different social representations, like rac-
ism and ageism. Beauty as a legitimizing myth is another issue that deserves to be
investigated more deeply in order to understand contextually the representational
process. Conceivably, through a study of traditional culture (Rio Grande do Sul) in
its relation to progressive ideas (feminist and others) we may better understand the
dynamicity of social representations.

To conclude this chapter, I want to share with you that while I was writing
my own report in the Collective Notebook of Women Experiences, I remembered the stories my High School friends told me about their suffering for having
to do everything hidden, in silence. Being a teenager is a challenging adventure,

not only because of hormonal changes but because of the hegemonic social representations we grow up with. While writing my report, I remembered the videos teachers showed us about abortion... teenagers aborting and throwing their babies into garbage cans... I remembered teachers harassing classmates... I remembered friends being raped when going home after school. A lot of memories were triggered. All memories must be kept in silence because women must be "pretty, maiden-like and a housewife".[12] And because I am propelled by polemic representations and in search of new equitable representations, I used this Notebook as a tool for research and for empowering women at the same time. I believe that we, women, need to share our experiences, to recognize that we are not alone in this world where hegemonic social representations still prevail. So, listening to the experiences of the participants helped me re-encounter my own life story, and I hope that this chapter may help you to activate your memories about your sexual and reproductive encounters.

# References

Albulquerque, Aline A., 2018: "O impacto da medicalização sobre os direitos humanos dos pacientes", in: *Revista Iberoamericana de Bioética*, 6 (February): 1–13.

Arruda, Angela, 2014: *Angela Arruda e as representações sociais: estudos selecionados* (Rio de Janeiro/Curitiba: Fundação Carlos Chagas/Champagnat).

Barnack-Tavlaris, Jessica, 2015: "The medicalization of the menstrual cycle: menstruation as a disorder", in: McHugh, Maureen; Chrisler, Joan (Eds.): *The wrong prescription for women: how medicine and media create a "need" for treatments, drugs, and surgery* (Santa Barbara, CA, USA: Praeg).

Bauman, Zygmunt, 2006: *Liquid times: living in an age of uncertainty* (Cambridge/Oxford/ Boston/New York: Polity).

Baxi, Pratiksha, 2013: *Public secrets of law: rape trials in India* (Oxford: Oxford University Press).

Bell, Susan E.; Figert, Anne E., 2012: "Gender and the medicalization of healthcare", in: Kuhlmann, Ellen; Annandale, Ellen. (Eds.): *The Palgrave handbook of gender and healthcare* (Basingstoke: Palgrave Macmillan): 127–142.

Bell, Susan; Figert, Anne, 2012: "Medicalization and pharmaceuticalization at the intersections: looking backward, sideways and forward", in: *Social Science & Medicine*, 75,5: 775–783.

Ben-Asher, Smadar, 2003: "Hegemonic, emancipated and polemic social representations: parental dialogue regarding Israeli Naval Commandos Training in polluted water", in: *Papers on Social Representations*, 12: 6.1–6.12.

Berberick, Stephanie Nicholl, 2010: "The objectification of women in mass media: female self-image in misogynist culture", in: *The New York Sociologist*, 5,1: 1–15.

Biehl, João, 2013: "The judicialization of biopolitics: claiming the right to pharmaceuticals in Brazilian courts", in: *American Ethnologist*, 40,3: 419–436.

---

[12]Expression used in a headline of a popular magazine in Brazil that caused discontentment in many people, because the underlying message reinforces the hegemonic social representation of women as subordinate to men. See the news at: https://veja.abril.com.br/brasil/ marcela-temer-bela-recatada-e-do-lar/.

Biehl, João; Petryna, Adriana, 2013: *When people come first: critical studies in global health* (Princeton, NJ: Princeton University Press).

Blondell, Ruby, 2013: *Helen of Troy: beauty, myth, devastation* (London: Oxford University Press).

Bourdieu, Pierre, 2001: *Masculine domination* (Stanford, CA: Stanford University Press).

Bresser-Pereira, L. Carlos, 2009: "Assault on the State and on the market: neoliberalism and economic theory", in: *Estudos Avançados*, 23,66: 7–23.

Clarke, Adele E., 2010: "From the rise of medicine to biomedicalization: US healthscapes and iconography, circa 1890-present", in: Clarke, Adele E.; Mamo, Laura; Fosket, Jennifer Ruth; Fishman, Jennifer R.; Shim, Janet K. (Eds.): *Biomedicalization. technoscientific transformations of health, illness, and US biomedicine* (Durham, NC, USA: Duke University Press): 104–146.

Connell, Raewyn W., 1995: *Masculinities* (Berkeley, LA: University of California Press).

Conrad, Peter, 2005: "The shifting engines of medicalization", in: *Journal of Health and Social Behavior*, 46,1 (March): 3–14.

Conrad, Peter, 2007: *The medicalization of society: on the transformation of human conditions into treatable disorders* (Baltimore, MD: The Johns Hopkins University Press).

Conrad, Peter; Leiter, Valerie, 2004: "Medicalization, markets and consumers", in: *Journal of Health and Social Behavior*, 45 (Special Issue): 158–176.

Cosminsky, Sheila, 2016: *Midwives and mothers: the medicalization of childbirth on a Guatemalan plantation* (Austitn, TX: Universiy of Texas Press).

Costa, Tania; Stotz, Eduardo Navarro; Grynszpan, Danielle; Souza, Maria do Carmo Borges de, 2007: "Naturalization and medicalization of the female body: social control through reproduction", in: *Interface*, 3: 1–16.

Del Priori, Mary, 2000: *Corpo a corpo com a mulher: pequena historia das transformações do corpo feminino no Brasil* (São Paulo: SENAC).

Devenney, Michael J., 2005: *The social representations of disability: fears, fantasies and facts* (Cambridge: Cambridge University Press).

Ferguson, Susan J., 2002: "Deformities and diseased: the medicalization of women's breasts", in: Kasper, Anne; Ferguson, Susan J. (Eds.): *Breast cancer: society shapes an epidemic* (New York: Palgrave): 51–86.

Ferreira, Francisco Romão, 2011: "Cirurgias estéticas, discurso médico e saúde", in: *Ciência & Saúde Coletiva*, 16,5: 2373–2382.

Gideon, Jasmine, 2006: "Accessing economic and social rights under neoliberalism: gender and rights in Chile", in: *Third World Quarterly*, 27,7: 1269–1283.

Gilgun, Jane F.; McLeod, Laura, 1999: "Gendering violence", in: *Studies in Symbolic Interactionism*, 22: 167–193.

Gill, Rosalind, 2007: "Postfeminist media culture: elements of a sensibility", in: *European Journal of Cultural Studies*, 10,2: 147–166.

Gillespie, Alex, 2008: "Social representations, alternative representations and semantic barriers", in: *Journal for the Theory of Social Behaviour*, 38,4: 375–391.

Ginsburg, Faye D.; Rapp, Rayna, 1995: "Introduction: conceiving the new World order", in: Ginsburg, Faye D.; Rapp, Rayna (Eds.): *Conceiving the new order: the global politics of reproduction* (Berkeley, LA, CA: University of California Press): 1–17.

Gomes, Patrícia Delage; Zimmermmann, Juliana Barroso; Oliveira, Lizandra Maris Borges de; Leal, Kátia Aureana; Gomes, Natália Delage; Goulart, Soraia Moura; Rezende, Dilermando Fazzito, 2011: "Hormonal contraception: a comparison between patients of the private and public health network", in: *Ciência & Saúde Coletiva*, 16,5: 2453–2460.

Gomes, Vanessa Pereira; Silva, Marcus Tolentino; Galvão, Taís Freire, 2017: "Prevalence of medicine use among Brazilian adults: a systematic review", in: *Ciência & Saúde Coletiva*, 22,8: 2615–2626.

Guareschi, Pedrinho; Roso, Adriane, 2014: "Teoria das Representações Sociais: sua história e seu potencial transformador", in: Chamon, Edna Maria Querido de Oliveira; Guareschi, Pedrinho Arcides; Campos, Pedro Humberto Faria (Eds.), *Textos e debates em representação social* (Porto Alegre: ABRAPSO): 17–41.

Habermas, Jürgen, 1985: *The theory of communicative action – Lifeworld and system: a critique of functionalist reason* (Boston: Beacon Press).

Handwerker, Lisa, 1998: "The consequences of modernity for childless women in China: medicalization and resistance", in: Lock, Margaret; Kaufert, Patricia Alice; Harwood, Alan (Eds.): *Pragmatic women and body politics* (New York: Cambridge University Press): 178–205.

Hesse-Biber, Sharlene Nagy, 2006: "The practice of feminist in-deph interviewing", in: *Feminist research practice: a primer* (Thousand Oaks, CA: Sage): 111–148.

Hindriks, Paul; Verkuyten, Maykel; Coenders, Marcel, 2014: "Dimensions of social dominance orientation: the roles of Legitimizing Myths and national identification", in: *European Journal of Personality*, 28: 538–549.

Höijer, Brigitta, 2011: "Social Representations Theory: a new theory for media research", in: *Nordicom Review*, 32,2: 3–16.

Howarth, Caroline, 2006: "A social representation is not a quiet thing: exploring the critical potential of social representations theory", in: *British Journal of Social Psychology*, 1,45: 65–86.

Howarth, Caroline; Andreouli, Eleni; Kesi, Shose, 2014: "Social representations and the politics of participation", in: Nesbitt-Larking, Paul; Kinnvall, Catarina; Capelos, Tereza; Dekker, Henk (Eds.): *The Palgrave handbook of global political psychology* (Palgrave Macmillan): 19–38.

International Federation of Gynecology and Obstetrics (FIGO), 2004: "FIGO professional and ethical responsibilities concerning sexual and reproductive rights: international joint policy statement", in: *Journal of Obstetrics and Gynaecology Canada*, 26,12 (December): 1097–1099.

Jameson, Fredric, 1982: *Postmodernism and consumer society* (New York, NY: Whitney Museum of American Art).

Jodelet, Denise, 1991: *Madness and social representations* (Hemel Hempstead: Harvester Wheatsheaf).

Jodelet, Denise, 2008: "Social representations: the beautiful invention", in: *Journal for the Theory of Social Behaviour*, 38,4: 411–430.

Jovchelovitch, Sandra, 2007: *Knowledge in context: representations, community and culture* (Hove, East Sussex: Routledge).

Jovchelovitch, Sandra, 2008: *Contextos do saber: representações, comunidade e cultura.* (Petrópolis: Vozes).

Jovchelovitch, Sandra; Priego-Hernández, Jacqueline, 2013: *Sociabilidades subterrâneas: identidade, cultura e resistência* (Brasília, DF, Brasil: UNESCO).

Lopez, Iris, 1998: "An ethnography of the medicalization of Puerto Rican women's reproduction", in: Lock, Margaret; Kaufert, Patricia Alice (Eds.): *Pragmatic women and body politics* (New York: Cambridge University Press): 240–259.

Manerikar, Vijaya; Manerikara, Sumeet, 2014: "A note on exploratory research", in: a*WEshkar,* XVII (March): 95–96.

Marková, Ivana, 2008: "The epistemological significance of the Theory of Social Representations", in: *Journal for the Theory of Social Behaviour*, 38,4: 461–487.

Marková, Ivana, 2013: "On dialogue and dialogicality", Interview with Ivana Marková by T. Fondelli, Open Dialogical Practices, United Kingdom (13 January).

Martins, Laura. B. Motta; Costa-Paiva, Lúcia; Osis, Maria José; Sousa, Maria Helena de; Pinto Neto, Aarão M.; Tadini, Valdir, 2006: "Conhecimento sobre métodos anticoncepcionais por estudantes adolescentes", in: *Revista de Saúde Pública*, 40,1: 57–64.

McHugh, Maureen C.; Chrisler, Joan C., 2015: *The wrong prescription for women: how medicine and media create a "need" for treatments, drugs, and surgery* (Santa Barbara, CA: Praeger).

Minowa, Eimy; Julian, Guilherme Silva; Pomerantz, David; Sternbach, Nikoletta; Feijo, L. A.; Annunziata, Kathy, 2015: "PIH4 – Contraception patterns in Brazil: 2012 National Survey data", in: *Value in Health*, 18,7 (November): A832–A833.

Morsy, Soraya Altorki, 1995: "Deadly reproduction among Egyptian women: maternal mortality and medicalization of population control", in: Ginsburg, Faye; Rapp, Rayna: *Conceiving the new world order: the global politics of reproduction* (Berkeley: University of California Press): 162–176.

Moscovici, Serge, 1961: *La psychanalyse: son image et son public* (Paris, France: Presses Universitaires de France).

Moscovici, Serge, 1976: *Social influence and social change* (San Diego, CA: Academic Press).

Moscovici, Serge, 1988: "Notes towards a description of social representations", in: *European Journal of Social Psychology*, 18: 211–250.

Moscovici, Serge, 2008: *Psychoanalysis: its image and its public* (Cambridge/Oxford: Polity Press).

Oliveira, Ana Maria de; Lemes, André Moreira; Ávila, Bruna Teixeira; Machado, Carolina Rocha; Ordones, Elisa; Miranda, Fernanda Souza; Loyola e Silva, Fernanda; Goetz, Hermann Soares; Miranda, José Oscar Ferreira de; Barbosa, Juliane Moreira; Leão, Lahis Ribeiro, 2010: "Relação entre enfermeiros e médicos no Hospital das Clínicas da Universidade Federal de Goiás: a perspectiva do profissional de enfermagem", in: *Revista Latinoamericana de Bioética*, 10,2: 58–67.

Oliveira, Beatriz Rosana Gonçalves de; Collet, Neusa, 2012: "Relações de poderes (inter)profissionais e (inter)institucionais", in: *Revista Brasileira de Enfermagem*, 53,2 (April/June): 295–300.

Oliven, Ruben George, 2000: "The largest popular culture movement in the Westeern world: intellectuals and Gaúcho traditionalism in Brazil", in: *American Ethnologist*, 27,1: 128–146.

Oudshoorn, Nelly, 2000: "Imagined men: representations of masculinities in discourses on male contraceptive technology", in: Kirejczyk, Marta Stefania Maria; Saetnan, Ann Rudinow; Oudshoorn, Nelly (Eds.): *Bodies of technology: women's involvement with reproductive medicine* (Columbus, Ohio: Ohio State University Press): 123–146.

Oudshoorn, Nelly, 2003: "Clinical trials as a cultural niche in which to configure gender identities of users: the case of male contraceptive development", in: Oudshoorn, Nelly; Pinch, Trevor (Eds.): *How users matter: the co-construction of users and technologies* (Cambridge, MA: The MIT Press): 209–299.

Pacheco, Luis Orestes, 2003: *Como o tradicionalismo gaúcho ensina sobre masculinidade* (Porto Alegre, Rio Grande do Sul: Universidade Federal do Rio Grande do Sul).

Paulon, Simone Mainieri, 2005: "A análise de implicação como cerramenta na resquisa-intervenção", in: *Psicologia & Sociedade*, 17,3 (September-December): 18–25.

Pedro, Joana Maria, 2003: "A experiência com contraceptivos no Brasil: uma questão de geração", in: *Revista Brasileira de História*, 23,45: 239–260.

Pires, Denise, 1989: *Hegemonia médica na saúde e a enfermagem* (São Paulo: Cortez).

Poeschl, Gabrielle; Silva, Aurora; Clemence, Alain, 2018: "Representações da masculinidade e da feminilidade e retratos de homens e de mulheres na literatura portuguesa", in: *Psicologia*, 18,1 (January): 31–46.

Poli Neto, Paulo; Caponi, Sandra, 2007: "The medicalization of beauty", in: *Revista Interface*, 11,23 (September/December): 569–584.

Pop, Alina, 2012: "On the notion of polemic social representations – theoretical developments and empirical contributions", in: *Societal and Political Psychology International Review*, 3,2: 109–124.

Potter, Joseph E., 1999: "The persistence of outmoded contraceptive regimes", in: *Population and Development Review*, 25,4: 703–739.

Prado, Daniela Siqueira; Santos, Danielle Loyola, 2011: "Contracepção em usuárias dos setores público e privado de saúde", in: *Revista Brasileira de Ginecologia e Obstetrícia*, 33,7 (July): 143–149.

Rich, Adrienne Cecile, 2003: "Compulsory heterosexuality and lesbian existence (1980)", in: *Journal of Women's History*, 15,3 (Autumn): 11–48.

Riska, Elianne, 2003: "Gender perspectives on health and medicine", in: Segal, Marcia T.; Demos, Vasilike; Kronenfeld, Jennie Jacobs (Eds.): *Gender perspectives on health and medicine-advances in gender research* (Emerald Group Publishing Limited): 59–87.

Riska, Elianne, 2013: "Aging men: resisting and endorsing medicalization", in: Kampf, Antje; Marshall, Barbara L; Petersen, Alan (Eds.): *Aging men, masculinities and modern medicine* (New York, NY: Routledge): 71–85.

Rodriguez, Marta de Assis Machado; Maciel, Débora Aves, 2017: "The battle over abortion rights in Brazil's state arenas, 1995–2006", in: *Health and Human Rights Journal*, 19,1 (June): 119–131.

Rosenfeld, Dana; Faircloth, Christopher, 2006: "Introduction: medicalized masculinities: the missing link?", in: Rosenfeld, Dana; Faircloth, Christopher (Eds.): *Medicalized masculinities* (Philadelphia, PA: Temple University Press): 1–20.

Roso, Adriane, 2001: "Espelho, espelho meu. Beleza, moda e trabalho: uma tríade histórica", in: Roso, Adriane; Werba, Graziela C.; Mattos, Flora Bojunga; Strey, Marlene Neves (Eds.): *Gênero por escrito: saúde, trabalho e identidade* (Porto Alegre, RS, Brasil: EDIPUCRS).

Roso, Adriane, 2010: "Mulheres Latinas e transmissão do vertical do HIV: visão dos profissionais da saúde que atendem mulheres soropositivas nos Estados Unidos", in: *Interamerican Journal of Psychology*, 44: 321–341.

Roso, Adriane; Guareschi, Pedrinho, 2007: "Megagrupos midiáticos e poder: construção de subjetividades narcisistas", in: *Política & Trabalho. Revista de Ciências Socias*, 26 (April): 37–54.

Roso, Adriane; Santos, Verônica Bem dos, 2017: "Saúde e relações de gênero: notas de um diário de campo sobre vivência de rua", in: *Avances en Psicología Latinoamericana*, 35: 283–299.

Roso, Adriane; Gass, Rosineia L.; Orsato, Daniela; Alves, Thiago; Moraes, Maurício M., 2011: "Minorias étnicas e representações sociais: notas sobre a entrada do psicólogo social em uma comunidade Quilombola", in: *Psico*, 42,3 (July/September): 346–353.

Saetnan, Ann Rudinow, 2000: "Women's involvement with reproductive medicine: introducing shared concepts", in: Saetnan, Ann Rudinow; Oudshoorn, Nelly; Kirejczyk, Marta (Eds.), *Bodies of technology: women's involvement with reproductive medicine* (Columbus, OH: Ohio State University Press): 1–30.

Schofield, Toni; Connell, Raewyn; Walker, Linley; Wood, Julian F.; Butland, Dianne L., 2000: "Understanding men's health and illness: a gender-relations approach to policy, research, and practice", in: *Journal of American College Health*, 48 (May): 247–256.

Schwandt, Hilary M.; Skinner, Joanna; Saad, Abdulmumin; Cobb, Lisa: 2016, "Doctors are in the best position to know": the perceived medicalization of contraceptive method choice in Ibadan and Kaduna, Nigeria", in: *Patient Education Counseling*, 99,8 (August): 1400–1405.

Scully, Diana; Bart, Pauline, 2003: "A funny thing happened on the way to the orifice: women in gynecology textbooks", in: *Feminism & Psychology*, 13,1: 11–16.

Serrano Oswald, Serena Eréndira, 2013: "El potencial de la teoría de la representaciónes sociales (TRS) para la investigación con perspectiva de equidad de género", in: *Acta Colombiana de Psicología*, 16,2: 63–70.

Sidanius, Jim; Pratto, Felicia, 2012: "Social dominance theory", in: Lange, Paul van; Higgins, E. Tory; Kruglanski, Arie W. (Eds.): *Handbook of theories of social psychology* (New York, NY: Sage): 418–439.

Sidanius, Jim; Veniegas, Rosemary C., 2000: "Gender and race discrimination: the interactive nature of disadvantage", in: Oskamp, Stuart; Spacapan, Shirlynn (Eds.): *The Claremont symposium on applied social psychology* (Mahwah, NJ: Lawrance): 47–69.

Sidanius, Jim; Levin, Shana; van Laar, Colette; Sears, David O., 2008: *The diversity challenge: social identity and intergroup relations on the college campus* (New York: Russell Sage Foundation).

Soares, Ana Carolina Eiras Coelho; Ferreira, Neide Célia, 2014: "As propagandas da revista feminina (1914–1936): a invenção do mito da beleza", in: *Oficina do Historiador*, 7,1 (January–June): 106–120.

Stoppard, Janet M.; Gammell, Deanna J., 2003: "Depressed women's treatment experiences: exploring themes of medicalization and empowerment", in: Stoppard; Janet M.; McMullen, Linda M. (Eds.): *Situating sadness: women and depression in social context* (New York: New York University Press): 39–61.

Stoppard, Janet M.; McMullen, Linda M., 2003: "Introduction", in: Stoppard; Janet M.; McMullen, Linda M. (Eds.): *Situating sadness: women and depression in social context* (New York: New York University Press): 1–16.

Tone, Andrea, 2001: *Devices & desires: a history of contraceptives in America* (New York, NY: Hill and Wang).

United Nations (UN), 2015: *Trends in contraceptive use worldwide 2015* (New York: United Nations, Department of Economic and Social Affairs Population Division).

Vala, Jorge; Garcia-Marques, Leonel; Gouveia-Pereira, Maria; Lopes, Diniz, 1998: "Validation of polemical representations: introducing the intergroup differentiation of heterogeneity", in: *Social Science Information*, 37,3: 469–492.

Vasconcelos, Maria Drosila, 2002: "Pierre Bourdieu: a herança sociológica", in: *Educação e Sociedade*, 23,78 (April): 77–87.

Vos, Jan de, 2012: *Psychologisation in times of globalisation* (New York: Routledge).

Welzer-Lang, Daniel, 2004: "Os homens e o masculino numa perspectiva de relações sociais de sexo", in: Schpun, Monica Raisa (Ed.): *Masculinidades* (São Paulo; Santa Cruz do Sul: Boitempo Editorial; Edunisc): 107–128.

Wolf, Naomi, 1991: *The beauty myth: how images of beauty are used against women* (New York: W. Morrow).

Wolf, Naomi, 2012: *Vagina: a new biography* (New York, NY: Ecco).

Zanello, Valeska; Gomes, Tatiana, 2010: "Xingamentos masculinos: a falência da virilidade e da produtividade", in: *Caderno Espaço Feminino*, 23,1/2: 265–280.

Zanello, Valeska; Fiuza, Gabriela; Costa, Humberto Soares, 2015: "Saúde mental e gênero: facetas gendradas do sofrimento psíquico", in: *Fractal: Revista de Psicologia*, 27,3 (September–December): 238–246.

Zola, Irving Kenneth, 1972: "Medicine as an institution of social control", in: *The Sociaological Review*, 20,4: 487–504.

Zorzanelli, Rafaela Teixera; Ortega, Francisco; Bezerra Júnior, Benilton, 2014: "Um panorama sobre as variações em torno do conceito de medicalização entre 1950–2010", in: *Ciência & Saúde Coletiva*, 19,6 (June): 1859–1868.

## Other Literature

Phillips, Kristine, 2018: "A jury convicted a doctor of raping a patient at a hospital – and sentenced him to probation", in: *The Washington Post*, 19 August 2018, at: https://www.washingtonpost.com/news/true-crime/wp/2018/08/19/a-jury-convicted-a-doctor-of-raping-a-patient-at-a-hospital-and-sentenced-him-to-probation/ (10 May 2020).

# Chapter 12
# Social Representations Theory in the Field of Nursing: Professional Autonomy, Vulnerability and Spirituality/Religiosity as Representational Objects

Antonio Marcos Tosoli Gomes

## Introductory Comment

Denize Cristina de Oliveira

*Abstract*

*The development of a researcher[1] happens through theoretical and empirical deepening throughout his or her career. It must contain a strong and solid methodological basis of the reality found. When this group of things is combined with the capacity to be attentive to the current challenges of the social, professional and political field in which it is inserted or by which it is touched, it allows the researcher to make a unique and important contribution to society in general and to his group in particular. In this sense, Social Representations Theory has become an epistemological basis for the development of the research group Sociocognitive and Psychosocial Processes of Health Care and Nursing of Population Groups and its researchers. In this chapter the trajectory of a particular researcher will be*

---

Full Professor of the graduate programme in nursing at the Faculty of Nursing, State University of Rio de Janeiro. Email: mtosoli@gmail.com.

---

[1]Denize Cristina de Oliveira is full professor at the Rio de Janeiro State University (UERJ). She graduated in nursing and has a PhD in public health from the University of São Paulo (USP). Her postdoctoral research in social psychology was undertaken at the École des Hautes Etudes en Sciences Sociales (EHESS) in Paris, France. She is associate editor of diverse journals, certified by the National Council for Scientific and Technological Development (CNPq). Her research areas are public health, nursing, and social representations in topics such as: social images and public health practices, the symbolic incorporation of health systems, health promotion, children's health, and HIV-AIDS. Email: dcouerj@gmail.com.

© Springer Nature Switzerland AG 2021
C. Prado de Sousa and S. E. Serrano Oswald (eds.),
*Social Representations for the Anthropocene: Latin American Perspectives*,
The Anthropocene: Politik—Economics—Society—Science 32,
https://doi.org/10.1007/978-3-030-67778-7_12

*approached from three successive phases that have been configured as dimensions of work in the construction of a research line and a career. They are the study of nursing autonomy and identity, the concept of vulnerability as an opportunity to improve the organization of health care and nursing, and, as a conclusion, spirituality and religiousness in health as an issue, which professionals and users/clients come across in their daily life in different health units.*

### The Developing Context of a Field of Studies

*Taken as a field of intellectual production, the health field, particularly in those areas historically orientated by an interdisciplinary knowledge, such as nursing, physiotherapy, occupational therapy and collective health, has been structured as a function of permanent reflection on its object's interests. In this sense, conceived as social practices, these areas have sought to define the specificity of their research objects and their work processes in line with other health work processes in order to disclose their social and psychosocial aspects in building an interdisciplinary knowledge. This interdisciplinary construction presupposes a complex approach in which individual and collective subjectivity and an understanding of values and perceptions about health are indispensable. It is about incorporating not only categories of science, but also the various types of mythical, empirical, rational and logical thoughts, in a relational network that makes the subject emerge in constant dialogue with the object of knowledge. The present work has been developed by recognizing the need for communication between different areas of knowledge, thus understanding that order, disorder and organization are important and necessary phases of a process as a presupposition of a complex approach, as proposed by Michel Morin.*

*Therefore, this epistemological articulation demands an examination of how the societal and psychosocial contexts intertwine, that is, how the individual is situated in society and how socially shared values become part of the repertoire of the individual subject. On the other hand, this recursion contains the seeds for rethinking important aspects of the organization of the work process, the management, planning and construction of new knowledge and practices in the field of health. So, the construction of new knowledge in health implies the necessity of overcoming a clinical-biomedical and normative dimension and a need for the construction of an interdisciplinary dimension of knowledge.*

*Psychosocial thinking about health and its objects maintains an intimate association with the epidemiological changes observed in Brazil, the structuring and development of health and education programmes, the organization of health care services, the development of the actions of the Ministry of Health and the dynamic role of civil society organizations, resulting in different structures to support and care for people and social groups. From this perspective, understanding health problems depends on comprehending the way these phenomena are structured, in both the institutional dimension and the symbolic and psychosocial dimension of different social groups.*

*From this perspective, the theory of social representations constitutes a privileged conceptual framework for testing, at a theoretical and empirical level, the facet that deals with socially shared symbolic universes. It facilitates the study of the processes of formation, diffusion and updating of social thought on which the orientation of behaviours and practices in the health-disease space is based, and on which the practices of care and management of these actions should be based.*

*In this context, in the last twenty years the activities of the research group Sociocognitive and Psychosocial Processes of Health Care and Nursing of Population Groups have been developed through research programmes in national and international partnerships. This group was created in 1998 and is coordinated by this author; it is linked to the graduate programmes of nursing and social psychology at the State University of Rio de Janeiro, and has as its main target the field of studies related to social representations and health, developing studies on the following topics: the health and living conditions of working adolescents; representations and practices of professionals and users of the Brazilian public health system; health care practices aimed at people living with HIV/AIDS; and quality of life and temporal perspectives of people living with HIV/AIDS.*

*The creation of the research group in the 1990s was a strategy for introducing Social Representations Theory into the graduate programme of nursing at the State University of Rio de Janeiro in order to intensify the insertion of this theory in Brazil, particularly in the health field. This strategy was developed in partnership with the graduate programme of social psychology at the State University of Rio de Janeiro, through Professor Celso Pereira de Sá, with whom the partnership is maintained to the present day.*

*In the last two decades, this research group has been hosting postdoctoral studies, doctoral dissertations, master's dissertations, scientific initiation fellowships and the development of research projects focused on the analysis of health issues among health professionals and between specific social groups. In addition to the training activities, it promotes national and international exchanges in the development of multi-centre studies at international, national and state levels, as a strategy supported by the existence of significant differences in the development of the health-disease process between different Brazilian regions and cities. This means recognizing that psychosocial thinking about health maintains an intimate association with the dynamics of changes in the health-disease process observed in Brazil, as well as the need for constructing and developing different health care structures at the local and national levels.*

*The psychosocial approach to the health-disease process adopted by this group aims to use the concept of social representations as an instrument for analysing the components of the quality of life of people and groups, understanding that these objects have an objective view on indicators of daily life, but also a view on how these elements are symbolically constructed, and also that both are articulated for the effectiveness of practices that aim to improve the ways of living and facing diseases and health problems.*

*With regard to health issues and related behaviours, the construct of social representations facilitates the development of strategies for managing health problems in order to improve people's quality of life. Specifically, the challenge lies in studying the potential effects of psychological constructs on group trends in order to emphasize the positive and negative aspects of well-being in the current experience.*

*The understanding of the health-disease-care process encompasses many fields of human knowledge – biological, social, political, economic and medical, among others – in a constant interrelationship, revealing its contribution to the constitution of nursing and health knowledge. However, little is known about the role played by psychological constructs in determining health practices and in the increase of the vulnerability of groups to health problems, apart from informing the monitoring of health care and nursing of different groups.*

*Finally, the training of human resources and the space for teaching the scientific research process is an intrinsic part and mission of this research group, and a privileged space for postgraduate and undergraduate development, in both the social psychology programme and the nursing programme. This training activity includes the levels of scientific initiation, master's, doctorate and post-doctorate, at the theoretical, methodological and technical levels of the relationships between social representations and health. The master's and doctoral students are linked to the graduate programme in nursing and the graduate programme in social psychology at the State University of Rio de Janeiro and to the postgraduate programmes of the partner institutions.*

*The importance of this group at the scientific level lies in the production of knowledge focused on current and past problems placed in the field of health, as well as in the cruciality of developing multicentric studies at the national and state levels, as a contribution of the University to public health policies, especially considering the existence of significant differences in the development of the health-disease process between many Brazilian regions and cities. Additionally, the research group has shown itself to play a fundamental role in the process of building the future of Social Representations Theory in the field of health at national and local levels, and contributing to the continuity of this theory in Latin America through the developed partnerships.*

*As a result of the training activities developed over the last twenty years, as highlighted above, various researchers have been trained – masters, doctors and post-docs – and taken responsibility for the dissemination and development of Social Representations Theory in Brazil. Among these researchers, one should particularly be highlighted, namely Antonio Marcos Tosoli Gomes, who is holder of a PhD, professor of the Faculty of Nursing and of the postgraduate programme in nursing at the State University of Rio de Janeiro, a primary researcher at the National Council for Scientific and Technological Development (CNPq), a pro-scientist of UERJ, and a young scientist of Rio de Janeiro at the Foundation Carlos Chagas Filho Research Support of the State of Rio de Janeiro (FAPERJ), among many other credentials that could be highlighted. He received training in the context of the research group presented here, and today he plays the roles of member and collaborator of the group and currently represents the continuity of the work begun two decades ago. Marcos Tosoli, as he is known, stands out for his relevant contribution to the constitution of the interdisciplinary health field highlighted in this text, having as a tool the articulations between Social Representations Theory and the health-disease-care process.*

*Considering his outstanding role in the field of health and nursing, it is a pleasure to present his contributions to the development of this interdisciplinary field, which will be summarized after this text.*

**Abstract** 'Social representations' is a polysemic term with specific interpretations in the context of different knowledge fields, which tends to make the situation of first contact even more complex. In Brazil, the theory has been intensively used in areas such as health and education in order to aid understanding of their main objects of study, and also with a view to supporting interventions in different areas of practice. In health, nursing has gained prominence over the last few decades, given the amount of production in the area and the involvement of a large number of researchers in its study and application. From the explanation of this scenario follows the approach of three representational objects that mark, at different moments, the trajectory of a researcher in the context of the postgraduate programme in nursing at the State University of Rio de Janeiro in the city of Rio de Janeiro, Brazil.

**Keywords** Social Representations Theory · Health · Nursing · Professional autonomy · Vulnerability · Spirituality, religiosity

## 12.1 Introduction

The study of Social Representations Theory as proposed by Moscovici (2012) involves challenges that must be faced by neophyte researchers, given its own internal coherence and specific language. At the same time, 'social representations' is a polysemic term with specific interpretations in the context of different knowledge fields, which tends to make the situation of first contact even more complex.

After this initial moment, when balancing the concepts forming it and, at the same time, understanding the ways in which they are related, is unveiled before the researcher, a rich opportunity opens up to apprehend the ways in which subjects and their social groups construct naive theories to explain reality, select specific contents with their own logic in their development process, implement practices consistent with them, present attitudes in the face of different phenomena, crystallize images over time that condense important symbolic processes, and so on. One can see how all this is present in people's daily lives and is addressed in bar conversations, in bank queues, at social gatherings, in intergenerational relations, in recommendations of practices that must be adopted and avoided, in the construction of social identity, and in the evaluation of the group to which one belongs and of a distinct group of one's own.

In Brazil, Social Representations Theory has been intensively used in areas such as health and education in order to aid understanding of their main objects of study, and also with a view to supporting interventions in different areas of practice. In health, nursing has gained prominence over the last decades given the amount of production in the area and the involvement of a large number of researchers in its study and application. From the explanation of this scenario follows the approach of three representational objects that mark, in different

moments, the trajectory of a researcher in the context of the postgraduate programme in nursing at the State University of Rio de Janeiro in the city of Rio de
Janeiro, Brazil.

The first one relates to the professional autonomy of nursing for nurses themselves and was the first study developed in the career in the early 2000s. The second
relates to the concept of vulnerability for nurses and patients in a hospital unit with
the emphasis on people living with HIV/AIDS (PLHA) in 2011. Finally, the third
relates to spirituality and religiosity for PLHA treated in an outpatient clinic, which
was performed from 2015. In the present chapter, the idea is to show the reading of
the object f the theory by avoiding, whenever possible, the presentation of empirical
data. Thus, the desire is to show a certain evolution in the way of researching itself,
while at the same time making clear the multipotentiality of SRT.

## 12.2  Social Representations Theory and Its Use in the Comprehension of Different Objects in the Fields of Health and Nursing

As mentioned above, the following is an analysis of three objects reconstructed in
the light of Social Representations Theory.

### 12.2.1  The Nursing Professional's Autonomy as a Representation Construction

This discussion is related to the approach to professional autonomy in nursing, especially in the constitution of one's own role and the professional space.
This discussion was developed from empirical data throughout the researcher's
career (especially in the study completed in 2002, but replicated in other scenarios and with other subjects over time), and reconstructed and analysed in light of
the assumptions of Social Representations Theory, especially the propositions of
Moscovici (1978), Jodelet (2001) and Rouquette (2000).

Thus are proposed two dimensions of social representations of professional
autonomy in nursing, namely: the knowledge translated into several concepts of
professional autonomy in professionals' discourse, and the reported or observed
autonomous practices. This form of data organization is justified to the extent that
representations are structured social and mental contents that encompass the cognitive, evaluative, affective and symbolic dimensions (Wagner 2000) embodied in
attitudes, knowledge, images and practices (Jodelet 2001).

In this first moment are presented the conceptual categories constructed for
professional autonomy, which are materialized in the contexts of: (a) autonomy
as the proper knowledge of nursing; (b) autonomy expressed in the approach to

interacting with human beings; (c) autonomy as a good manual and technical performance; and (d) the professional identity and formation of one's own role.

The first concept of autonomy refers to the constitution and existence of specific professional knowledge. However, note that in all different studies conducted with nurses, nursing staff or nursing students, this knowledge assumes an imaginary sense of mosaic, since several types of knowledge are pointed out as constituents of nursing. Machado (1988) states that knowledge and power are determined mutually, as there is no power relation without constituting a knowledge field, while at the same time that knowledge constitutes new power relations. Foucault (1988) also considers the existence of a relationship between power and knowledge, not only to the extent that power requires a particular information or scientific discovery, but the exercise of power creates objects of knowledge or causes their emergence at the same time that it accumulates information and uses it. More explicitly, Foucault (1988: 142) states "it is not possible that power is exercised without knowledge, it is not possible that knowledge does not engender power".

Thus, understanding professional autonomy in nursing from the constitution of a specific knowledge of the profession has two meanings: the attempt to delimit an essential core of the profession, and the constitution, delimitation and specification of a space of power. The first would give meaning to professional practice, at least in its instrumental aspects, and the second would establish the space from which power relations experienced with the health team and society would be not only exercised, but also legitimized.

In their traditional study of nursing knowledge, Almeida and Rocha (1997) highlight several aspects related to the structuring and importance of this knowledge for the profession. Historically, the structure of nursing knowledge was established around the importance of adequate performance of tasks and procedures. However, evolution of the profession indicates that the knowledge structure is one of the elements of the profession's nucleus foundation that gives meaning to the execution of tasks and procedures in professional practice without them having primacy in the constitution of the profession's specific knowledge. In a second phase, nursing was organized around scientific principles in the attempt to prove the scientificity of its actions and knowledge. However, these principles were based on knowledge from other areas, especially biology, which may explain the emergence of the representation of the field of nursing knowledge as a mosaic of various areas of knowledge. This gives the impression that there is more an idea of juxtaposition of knowledge and information than their integration in the formation of specific knowledge that guides professional practice.

Nowadays, tension is experienced in nursing regarding the construction of its knowledge between the application of knowledge from other areas, such as biomedical and human sciences, and the development of its own knowledge embodied in nursing theories. This phase is different from the previous one because, in addition to using scientific principles as guidance for practice, it is also based on the appropriation of theoretical frameworks from other knowledge areas. A deeper analysis is needed here. The existence of tension between yielded knowledge and the own knowledge forms a practice with difficulties and disagreements that impact on the self and hetero valorization of the profession. However, these knowledge types can be complementary as long as the value of each is properly understood and adopted. Thus, the own knowledge has the purpose of organizing and systematizing the way in which nursing relates, is placed and established in society in general, and in the health team in particular. The given or borrowed knowledge is, through knowledge itself, scaled and appropriated as an instrumental and practical knowledge for experiencing the daily life of institutions and the health needs of clients.

Nursing theories and, consequently, the constitution of knowledge proper to the profession, propose the organization, systematization and understanding of care, which is considered the essence of the profession. In this sense, the tension existing in the constitution of nursing knowledge is established between biomedical sciences and human sciences. At the same time, it is important to avoid overlapping human sciences, especially psychology and sociology, and nursing knowledge. This would only change the focus of the problem from biological sciences to human sciences without delimiting what is particular and the essence of the profession. Hence the need to discover to what extent the human sciences contribute to nursing specific knowledge, and to what extent they are also instrumental knowledge.

At the same time, the concept of autonomy, when expressing the approach to the human being cared for, is centred particularly in the holistic view, that is, in understanding the human being attended and cared for beyond biological parameters. This approach has always been highlighted as a characteristic of nursing, given its philosophy of caring for human beings in all their dimensions. Regarding nursing practice from a holistic view, some observations deserve attention. First, for nurses, the holistic notion seems to relate more to recognizing the influence of social, economic, spiritual, and cultural factors in the health and disease process than to holistic concepts advocated by some authors or Eastern medicine traditions (Capra 1982, 1990).

Second, holistic customer care means understanding clients not only as a disease or pathophysiological disorder, but as subjects with a specific and multi-causal alteration (beyond the notion of single-cause of diseases) located in their being. It also means serving them by understanding all the human drama involved in the disease process, especially the emotional, family, and economic issues. Third, the holistic approach to nursing care is embodied in the understanding of health, no longer as an absence of disease, but as psychic, physical, social and mental well-being, as proposed by the World Health Organization (WHO).

The third concept associated with professional autonomy refers to good manual and technical performance in the exercise of the functions and execution of procedures within health institutions. At the same time, the marked separation between theory and technical performance, as well as its representation as a synonym of professional practice, indicates how one understands the particularities experienced in professional daily life and how one has acted in the face of power relations and resistance within the institutional framework. This concept is anchored in the relationship established between scientific practice and empirical practice in the field of nursing. That is, good technical and manual performance implies an explicit demonstration of overcoming professional practice based on intuition and experience, or the necessity of demonstrating the importance of practice in relation to the institution and the market (Rocha 1989). On the other hand, demonstrating mastery at activities and specialized technical procedures, especially if these are exclusive of nurses, embodies a power relation with the nursing team. This differentiates the nurse from the others by a technical imposition of competence and mastery over the other components, and gives visibility and a highlighted role to nurses before their team.

The fourth category of analysis is based on the important construction of the identity and proper role of nursing for the autonomous exercise of the profession. In this sense, the construction of professional identity is characterized as the genealogy of knowledge described by Foucault (1988: 172), which presents itself as an "enterprise to liberate historical knowledge from subjection, that is, to make them capable of opposition and fight against the coercion of a theoretical, unitary, formal and scientific discourse". That is, the genealogy of knowledge has real and fundamental importance in nursing as it exposes the know-how that characterizes its professional essence by making it capable of confrontations, oppositions and resistances in daily professional relations.

In this tension between what belongs to itself and what belongs to the other, the identity of the profession needs to be constructed not as a copy or consequence of other professions' identities, but as its own project of singular contribution to the health team and society. Rodrigues (1999) and Guitton et al. (2002) emphasize that in the social imaginary, nurses' identity is linked to the issue of feminine sex, and dubious and mythological images such as angels, saints, prostitutes, witches and curious women. Thus, society refers a pertinent consideration to nurses' identity as dubious and loaded with stigma and stereotypes. In addition, the aforementioned authors still refer to society's lack of delimitation between nurses and the nursing team. In some contexts, there is confusion about the identity and role of the nurse and the doctor, especially given the use of the consultation and the consulting room as technology and workspace, respectively. In this way, nurses' professional identity is primarily diluted by myths, in spite of the perception of the overcoming of this phase of construction of professional identity, but in reality, there is still a strong non-differentiation between nursing assistant/technician-nurse and, in some situations, doctor-nurse.

The construction of the professional identity moves through dimensions of the human, the professional and the citizen. The presence of this triple aspect in the construction of professional identity becomes important, given that human relationships are considered to be the basis of the profession, because of its procedural and still undefined aspects. In other words, the fact that this identity is under construction determines the need for characteristics beyond those indispensable for technical performance. Thus, in the routine of institutions, nurses absorb everything as their own, without a specificity of action or delimitation of their own role, which makes them invisible to the institution, health team and society. Furthermore, professional identity is characterized as an amalgamator of various professional practices of the institution, that is, cement that performs the proper juxtaposition not only of each professional, but also of their actions.

With regard to the representation of autonomy as autonomous practices with objectives, social representations are closely linked to practices. Rouquette (2000) emphasizes that social representations are a condition of practices and, in turn, practices are transformation agents of representations. That is, subjects act from how they represent the world, and this action in the world influences the structure and continuity of representation. Abric (2000) confirms the above by highlighting that representations allow the understanding and explanation of reality, and justify decision-making afterwards. The same author also states that representations guide behaviours and practices, and this guidance is presented in three basic forms, namely: (a) intervening directly in defining the purpose of the situation, and determining the types of established social relations and coping strategies used by individuals; (b) presenting a system of anticipations and expectations; and (c) prescribing compulsory practical behaviours and defining what is lawful, tolerable and unacceptable in a given social context (Abric 2000). Wagner (2000) complements this assessment by considering that the development of common sense, hence of the representation, rarely arises without practical need and the group members' collectively elaborate rules, justifications and reasons for establishing beliefs and behaviours in their daily practices.

In the context of the different studies conducted, two manifestations of autonomous practices stand out: the work space of health programming on the one hand, and nursing consultation and health education on the other. Thus, nurses' insertion in the health programming context is valid and important, and they experience the tension between objectives and proposals of work within the programming concentrated in its epidemiological axiom, as well as the forms and methods through which these objectives and proposals materialize in the practice of institutions, including individual consultation and standardization of conduct.

However, it is interesting and pertinent that given their profile and training, nurses offer a unique contribution to programmatically organized work. Nurses' understanding of human beings beyond biological aspects, their understanding of the influence of social and environmental factors in health and disease determination, their ability to amalgamate practice by coordinating and managing the work process of the health team within the unit, their generous behaviour towards patients by showing concern about their access to health services and the availability of a dialogical didactic-pedagogical relationship with the community, can constitute fundamental elements of a multi-professional and transdisciplinary practice.

The nursing consultation and the health education within it was another autonomous practice reported. That is, this professional technology materializes the idea of autonomous space (the consultation office) and independent technology (the consultation) in the context of health programming. The moment of the consultation creates a physical space in which the professional-client relationship happens under the unique orientation of the nurse, which makes him/her responsible for meeting the needs of the clientele in question. Furthermore, the nursing consultation is a technology of measurable results in two aspects, namely: its quantification at the central level and its subsequent financial charge to SUS (the Portuguese acronym for the Brazilian Unified Health System) or the employing institution; and secondly, by serial evaluation of the clientele, when the improvement or not of their health status is assessed.

In this sense, Nietsche and Backes (2000: 161) mention the nursing consultation "is an autonomous space in which professionals can develop their autonomy regarding the nursing action to be performed by nurses in the face of problems to be identified in the clientele, and an indicative of the legally and socially protected professional territory". Guitton et al. (2002) describe the social representation of the nursing consultation by its subjects also as an autonomous space guaranteed by both legal and social effectiveness aspects. Thus, the nursing consultation itself is a moment and a privileged space for reflection on the constitution of specific know-how, as previously discussed.

Health education is one of the central aspects of a nurse's role within the nursing consultation, and the implications of this practice for professional practice must be deepened. In addition, the power of the word stands out as autonomy in their routine. In this sense, health education as a strategy of interference in the health-disease process of individuals has two aspects of power: first, because in this process, power is given in a subtle way through persuasion rather than as an apparent imposition; second, the position and the figure of health professionals are used for defining criteria of truth, right and wrong and of who holds power through holding knowledge. Thus, power is exercised in the midst of a tension between the empathy of persuasion and the attempt to impose what is right, in which the correct is defined by the scientific knowledge or professionals' empirical experience.

The social representations of nurses' professional autonomy for professionals themselves and students in the field are considered an important issue to be deepened for the area. Social Representations Theory is a theoretical instrument that aids understanding of this phenomenon's complexity and allows the construction of a theoretical model that collaborates in the proposition of aid measures for the evolution and consolidation of this field of know-how. Therefore, professional autonomy is defined as the centring of the profession and the professional action in a know-how itself that is essential in this theoretical-practical field, while used knowledge and practices are simultaneously represented as instrumental for institutional and professional daily life.

## 12.2.2  The Concept of Vulnerability for Professionals and Users of the Health System from Social Representations Theory

Vulnerability is an important concept for analysis of the health field, especially regarding professionals' and users' insertion in this context, as mentioned by Ayres (1996, 1999), Gomes (2011) and Santos (2016). In studies developed with health professionals and health system users in recent years, the characterization of this concept passes through health and nursing care, and is not only a political and programmatic marker of health care, but also a marker for care itself in the hospital setting. In general, the vulnerability concept is understood as overcoming the concept of health risk by understanding the subjective, political, social, economic and programmatic context of subjects in order to understand their behaviour and illness process. Even considering the importance of the vulnerability concept in the evolution of the health field over the last few decades, the absence of the representational component in its formulation must be highlighted, and this is what the heart of this section seeks to deepen.

The initial argument defended here is that vulnerability is not the opposite of empowerment, i.e. widespread empowerment, but rather the autonomy of being and of this being's expression, especially through speech and discourse before the world in the process of constructing the word itself that provides meaning to living and the interaction established with others. The vulnerability that makes individuals dependent and inferior, linked to external determinations and unable to freely pronounce words that define them, is markedly a process that prevents the possibility of being and of being more existing in every human being, even in people who are near death or with a growing state of incapacity.

This demonstrates the importance of autonomy, because it makes nursing care unique, since no standard is adopted as a homogenizing factor, but adapted to each situation and to the state of each patient or nurse. The autonomy of those facing the process of dying may be the condition of maintaining minimal communication, the reception of feelings transmitted by the environment and the capacity for cognitive and affective understanding that keeps them lucid and in an intense process of communication with themselves and the environment, in spite of greater or less capacity of external body movement. Regarding nurses, autonomy may be related to performing the required technical procedures for survival, the establishment of human relations that meet ethical demands or, also, the need for taking actions that maintain the dignity of their work and their care vocation.

Therefore, as a result of the different studies and in an attempt to encompass what was unveiled, one can think of the following definition:

> Vulnerability is a dynamic and changeable state of fragilities and susceptibilities in different dimensions, the result of the intersection of several factors and situations, but especially of symbolic and representational constructions built by the different groups to which patients and professionals belong. In the case of patients, its increase generates an increasing degree of dependence on nursing care, and in the case of nurses, on care of other professionals and of patients themselves, and implies a certain position before the hospital unit and society in general, where it becomes difficult to concretize a form of being in consonance with one's own essentiality and the free and spontaneous expression of this same being.

> Given its strong symbolic and representational base built socially in daily conflicts, in everyday conversations, in information circulating in society and in images one has of oneself and others, vulnerability requires relational care from professionals and patients in which, through freedom and hope, autonomy is achieved in order to be more and to exercise the expression of this being. Finally, vulnerability has two dimensions, one objective

and measurable, and the other subjective, which, together, form the whole of this process and their separation is possible only theoretically or even with didactic objectives.

In this sense, the difficulty in studying vulnerability is related to the emphasis on only one of these facets (the objective or the subjective) instead of their interrelationship. Another problem is that the majority of studies address more general aspects and do not always approach the health area heuristically. Although this general approach can provide a broader view of coping with political, programmatic and collective challenges, usually it is not operable in phenomena more focused on a population group or health unit. In this direction, there was a proposal to apply a vulnerability concept in the context of nursing care developed in hospitals with a view to developing technologies to enable nurses and other health professionals to implement care with intimate regard for its social-humanistic essence and not just its biomedical dimension. It should simultaneously collaborate in defining the role of nurses in the context of the health team and in the conquest of larger autonomy spaces.

In order to overcome these challenges, these structures must be deepened with the purpose of making them more concrete and tangible for the development of other studies and their concrete verification. Thus, the objective and measurable structure of vulnerability can be understood in a division into three sets which, in turn, unfold in dimensions. The first is related to the basic characteristic of men and women, which is being human, and contains only one dimension, the ontological. The second set is related to the person who is sick and hospitalized or to nurses caring for them, and it unfolds in four dimensions, namely: the individual, the organic, the psychological, and the spiritual/religious. The third set is sociocultural-related and includes the occupational, economic, social and cultural dimensions.

The first dimension is the ontological and refers to the characteristic vulnerability of every human being as an inescapable consequence of existing and being alive. It is the recognition of human fragility present indiscriminately in all human beings in any context, society, or cultural and educational level, for example. It is an inherent aspect of humans' bio-psycho-socio-spiritual evolution, which, in parallel with its great complexity, also has significant fragility in the face of situations, events and relationships. In a more practical way, it is the human fragility of falling ill, suffering and dying, consubstantiated by permanent uncertainty ending in the indirect or immediate future (Torralba y Roselló 2009; Waldow/Borges 2008).

The second set is linked to the person facing the illness process and to nurses developing their activities with this person. It unfolds in four dimensions, namely: individual, organic, psychological and spiritual/religious. The individual dimension is related to actions and attitudes that expose subjects to the possibility of illness as a result of personal decisions or representational constructions in which the analysis of the situation arrives as a consequence of some degree of autonomy of these subjects. However, in general, it is determined by social and cultural constraints, representational constructions, or the presence of memories giving meaning to the present. Thus, as noted by Ayres (1996), the idea of absolute freedom or autonomy is eroded, along with the delusion that subjects can make decisions without being influenced by any interference.

For patients, the organic dimension of vulnerability is related to situations and morbid conditions that indicate the body's fragility due to illness and histological, anatomical, pathophysiological and systemic manifestations revealing this fragility. This dimension manifests the biomedical, pharmacological and pathophysiological dependence of nursing care so that individuals can maintain the continuity of their existence, and recover from molecular, cellular, tissue and systemic aggressions. And if scientific knowledge is not enough, the process of dying must be accompanied by ethical and humane principles that guarantee the necessary dignity for imminent and inevitable death. It is one of the most palpable dimensions of vulnerability throughout the interventional arsenal of health units and by the constant concern with the implementation of nursing care that identifies problems, it executes an objective plan of action and measures results.

For nurses, the organic dimension refers to the physiological, histological, anatomical and pathological conditions generating fragilities that may represent a threat to these professionals' physical health by the fusion of illness processes or the possibility of these as a result of their professional performance, the environment where they work or from contact with patients. This fragility may be related to previous pathological situations, to bodily changes behaving as access to pathogens, and to anatomical structures facilitating this process.

For users, the psychological dimension focuses on the process of mental, emotional and affective frailty suffered by the individual given their illness condition, the need for hospitalization because of the clinical manifestation, the possibility of carelessness as an inherent phenomenon in the nursing care process, and the health institution characteristics. It can be identified in cases of any degree of aggression and/or psychological damage to people, whether provoked, manifested, potentiated or even repressed at any stage of the health-illness-care process of oneself or others, ranging from the simple chance of illness to the impact of its diagnosis until the imminence of death. For illustrative purposes, fear, sadness, anguish, depression, despair and aggression are evident, as well as disorders of a psychic nature, such as psychosis and neurosis.

For professionals, this type of vulnerability is related to the stress caused by the ever present possibility, at least symbolically, of exposure to physical, chemical and biological agents of illness as a consequence of the work activity, which translates into fear, concern, despair and the report of neurotic actions in the routine of care. This psychological aspect assumes special significance in this sense, given nurses' experience of death and suffering, and where work can acquire the characteristics of confrontation of this same situation in its own corporeity.[2]

For nurses and patients, the spiritual/religious dimension refers to actions, activities, attitudes and/or rituals that generate greater fragility in the face of

---

[2]In this case, psychological vulnerability includes professionals' psychic suffering in continually facing organic degeneration and death, which in itself produces a marked degree of depletion and stress. In addition, the *shadow* of illness in the workplace increases their awareness of the real possibility of experiencing such a situation themselves.

illness, possible or imminent death and suffering, and to those causing greater sus-ceptibility to illness or aggravating symptoms. It can also be related to difficulties in the fundamentally human search for the meaning of life, and manifest in the search for the Divine in its different forms, religious or not, originated in the pro-cess of illness or death. This comprises an extremely important facet for human beings in general and, in particular for Brazilians, given the intimate experience of religion and the Divine as one of the joint axes of the population's daily life, espe-cially in critical situations. Unlike other cultures, in which there is a more rational apprehension of God, such as European, for example, the interpenetration of dif-ferent religions in Brazil and even the rich process of intercultural dialogue that composed the so-called syncretism experienced in the country, explain what Boff (2005) calls the *experience of God* and not a *speech or idea about God.*

However, the specificity for patients is usually due to the belief in divine inter-ventions which, if not materialized, allow a progressive worsening of the organic state that in turn, may allow the appearance of other morbid conditions and favour the occurrence of death. For practitioners, it is related to the notion of divine protection against sickness during the process of caring for patients, even if nec-essary precautions are not adopted. This attitude seems to be rooted in the sym-bolic legacy of important figures of Christendom in which this belief was present in an effective manner, such as St John Bosco, St John of God and St Roch of Montpellier.

The third set relates to sociocultural issues and has four dimensions, namely occupational, economic, social and cultural. For patients, the occupational dimen-sion is connected with the work activity or the impediment to doing it because of their state of illness as a consequence of limitations, constraints, stigmatization and exclusion. This dimension materializes in difficulties of inclusion in the world of work or even in the maintenance of their paid activities, which in most cases are fundamental to the gregarious, psychosocial, affective and identity aspects. It can also be related to the experience of this same situation for nurses, who, by working with certain social groups, such as people living with HIV/AIDS or other communicable diseases, acquire this status given their physical and/or symbolic proximity, which includes relations with other professionals, clients and their fam-ily members, and which, at various times, may involve animosities and difficulties.

For patients, the economic dimension encompasses the difficulties faced dur-ing the disease process, and the obstacles encountered in access to care or imple-mentation of the recommended therapy due to deficiencies of material resources, emphasizing here personal financial resources or those of relatives. This type of vulnerability is manifested, for example, in the deprivation of assets for transport, residence, food, and the purchase of medicines, condoms and personal and domes-tic hygiene materials, among other things. It is important given its association with the process of adherence to therapy, continuity of treatment, access to the neces-sary services and consumer goods for human existence, or the establishment of greater levels of comfort.

With regard to professionals, the economic dimension of vulnerability can cre-ate a situation where nurses feel the need to take several work shifts at different locations to reach a minimum income for survival. In addition, they face the impossibility of doing improvement courses, participating in congresses or acquiring the necessary technologies to gain a certain skill or extend their knowledge. Although there is no linear effect, this dimension may prevent a freer exercise of human creativity and be a considerable stressor for nurses.

The social dimension involves the relational and interpersonal aspects of the human condition that can develop in the area of friendship, love, respect and contemplation, but also in the real and concrete possibility of unfolding in violence and instrumentality. Social vulnerability materializes as insecurity, risk, exposure to outrage and non-protection. It is the possibility that human beings are the object of violence within society, including processes of exclusion (Torralba y Roselló 2009). In this sense, the aim is to make this concept more heuristic, since its broadness can leave it without the necessary specificity for practical application, just like the difficulty with its delimitation in the scope of scientific research.

In the case of patients, it means concrete situations of social exclusion, public humiliation or discrimination they have suffered due to their state of illness, including within health units, religious organizations and public-governmental institutions. It also implies isolation in order to cope with the diagnosis and implement the recommended treatment adherence as a fundamental condition for maintaining the process of hiding situations with a strong symbolic burden, such as HIV/AIDS. As for professionals, it is related to social and family embarrassment faced as a consequence of the care provided for assisted individuals both in hospital units and the basic health network.

The cultural dimension includes, for both subjects, the popular wisdom, beliefs, urban legends, stories, and specialist knowledge of certain groups about diseases, their treatment and the prognosis that can generate greater fragility in the context of their experience or therapeutic adherence, or even in the implementation of the care process, as well as in the adoption of protective actions in professionals' personal life. It arises from the intersubjective dynamics of a specific group or between different groups that give meaning to existence and the challenges faced, such as illness itself and death, as well as the urgent need to live with a chronic disease associated with stigma and symbolism. This dimension also manifests in the possibility of not having the cultural capital, knowledge and information to deal with the acquisition of infection, the illness process and the achievement of social and/or sanitary rights.

It is important to consider the cultural aspect in health professionals who, in principle, belong to the reified scientific and academic universe, as if they were devoid of symbolic constructions that did not coincide with the information presented by science as being true. In this case, it is interesting to study the formation of social representations which depend on the encounter between common sense and the technical-scientific knowledge that influences a certain way of thinking and the practices implemented by this social actor. Professionals are the point of intersection between the consensual and the reified universe by allowing particular symbolic constructions to be manifested as the belief in not needing personal protection equipment when patients are known well, the attribution of the syndrome exclusively to risk groups, and the blaming of patients for their current state without understanding the context that determined their current situation. An important discussion about health professionals and representational constructions can be read in the theoretical proposal on the subject by Oliveira (2001).

### 12.2.3 Social Representations Theory as a Grid for Reading the Spirituality and Religiousity Phenomenon in the Health Arena

In one of his last interviews available on the worldwide network Paulo Freire[3] approaches his own personal belief in the transcendental dimension and proposes the need to overcome a dichotomy separating transcendence from immanence. He also mentions that any word used about transcendence and the Divine can only be pronounced by being here in this world, that is, with flesh and bone feet planted on the ground, where each individual unfolds their own personal history. Freire touches an essential aspect for understanding the dynamics of the mystical and the Divine experience that is at the origin of different religions: the question whether it is the transcendent that presents itself to the immanent or the immanent that rises to the transcendent. In other words, in the understanding of social groups: Is it the Divine that shows itself to human beings? Or is it human beings who reach the sacred sphere?

This reflection is supported by theoretical propositions of theologians and historians of religion, who state that there are religions in which the origin is attributed to a divine epiphany that reveals itself to human beings in their essence, even if not in its fullness (MESLIM 2014). Christianity, for example, is one of these cases. On the other hand, other religions are characterized as a human effort, even if it is through a person considered enlightened like Buddha, in order to reach the Divine or the enlightenment of human beings in themselves.

A dialectic is observed at the origin of different religions, which constitutes its identity. At one pole, there is transcendence, the invisible, the intangible and the Divine, and at the other, the immanent, the visible, the tangible, and the world or the human. From this encounter or from the perception of such an encounter, are born religious and spiritual experiences that form a particular religion tradition or spiritual path. The definition of a dialectical mode of understanding this relation is due to the idea of a movement between the approaching poles without necessarily overlapping or mutually excluding, which means they tend to complement each other.

However, if there is desire to continue this epiphany experience over time, it brings within itself the need for its institutionalization, its permanence in the

---

[3]Interview with Paulo Freire, at: https://www.youtube.com/watch?v=lwIPLO2Yxmw (10 May 2020).

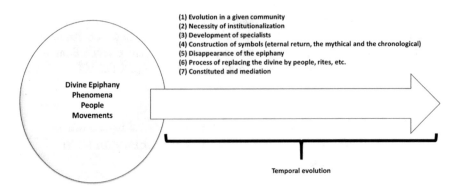

**Fig. 12.1** Temporal evolution of the Divine epiphany and spiritual phenomena in different communities *Source*: The author

frameworks of a given community and its social and cultural constructions. In this process a body of specialists must be constituted, along with a strong process of symbolization, the replacement of the Divine epiphany by rituals, people and specified actions over time, and the constitution of a strong mediation process of the own Divine, which, contradictory as it may seem, ends up disappearing within the process itself. This scheme can be better understood in the following Fig. 12.1.

With regard to the poles between spirituality and religiosity, the belief in Divine epiphany, its manifestation in human reality, tends to be more individualizing, whereas the experience of religion and religiosity is characterized as being communal and collective over time. The same consideration can be applied to the mystical experience, well documented even nowadays, but corresponding to a more individualizing pole of spiritual or religious experience. This consideration can be summarized in the following Fig. 12.2.

Christianity is an interesting case for analysis. In this set of beliefs characterizing the Christian faith, the almighty God incarnates himself in the figure of a specific man called Jesus of Nazareth and reveals himself as a consequence of this immanence of the transcendent. However, a concrete man with his bodily and human limitations was seen, or better put, represented as being the son of God or God himself. But nowadays, any intellectually serious Christian questions the access to the figure, belief and teachings of this Middle Eastern man, since the main accounts about his life, the Gospels, show, as cited by different authors (Boff 1986; Meslim 2014), a greater centrality in attending to the catechesis of the early Church than in the historical account itself. Just to exemplify, but without going into detail given the limited space of this chapter, the characteristics of the Jesus described in the synoptic Gospels (attributed to Matthew, Mark and Luke) are similar in each account, and based on his incarnation and human life as an apocalyptic prophet, which is different from the very original proposal of the Logos (Word) of God performed by the author of the Gospel of John, who even attributes to Jesus himself an equality with the unique and almighty God of the Jews (Lourenço

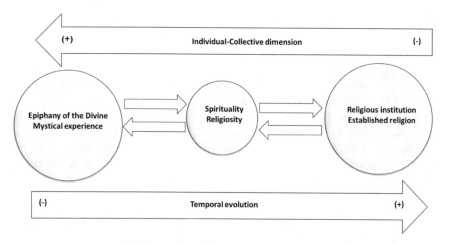

**Fig. 12.2**   The relationship between the mystical experience/epiphany of the Divine and the religious institution or established religion *Source*: The author

2017). These different visions have had philosophical consequences in the course of Christian history.

The point of saying all this is to demonstrate that even when a certain faith or belief admits a phenomenon of divine epiphany as essential for its essence, this epiphany is not apprehended in itself, in its unquestionable truth, but always and necessarily through the social, psychological, cultural and psychosocial analysis of the social groups that experience it and try to retranslate this experience to the others that follow. This is an important point for understanding the role of social psychology in general, and of Social Representations Theory in particular, in order to understand the mechanisms through which this process can occur, including the dialogue with critical theology as a way of mutual enrichment.

Hence the consideration that the Jesus to whom there is access is not historical in himself, but rather the one presented by the Gospels, as well as the face constituted throughout the history of the Church through the effort of diverse thinkers. Of these, the importance of Paul of Tarsus is noteworthy. He systematized a theoretical proposition in which not only the historical Jesus emerged, but the *Lord Jesus*, who, by his death and resurrection, is established as a universal standard of judgement, life, justification and salvation. The carpenter's son who walked through the dusty streets of ancient Israel has a transcendental function related to the fate of mankind in general and of every man in particular. He who came into conflict with the religious authorities and priests of his time now bears the title of high priest, as described in the Epistle of Hebrews, whose authorship is quite debatable.

It does not stop there. The early Christians, in a complex and refined process of building a specific social thought, have used what we would nowadays call anchoring in the context of Social Representations Theory. This makes the figure of the historical Jesus fit exactly the size of the Messiah of the Old Testament, both in aspects of his concrete life and his birth history, as well as in his function of the Lamb of God who takes away the sin of the world. Christians, one might venture to say, have proposed a complex theory in which human history is confluent since its origin until the existence of Christ, and extends from him by giving meaning to everything else (Boff 1986).

It is not by chance that one of the early theologians, St Justin, proposed the concept of *seeds of the Word*. The seeds of the Word or the manifestations of Christ, even if imperfectly and precariously, have always been present in all cultures, but were manifested in a decisive and fundamental way in the incarnation of the Word of God, as stated in the aforementioned Gospel of John. For example, the ancient goddesses, considered pagan by Christians, who were depicted with their children in their arms, were the prefiguration of Mary as Theotokos (Mother of God), and partially represented the truth that would reveal itself in Christ's incarnation. This process of anchoring was used by Paul of Tarsus as a strategy for the diffusion of Christianity. The most evident example is his dialogue with the Greek philosophers in Athens, where he anchored Jehovah, the God of the Jews and invisible by nature, with the unknown God that was in the Areopagus where this dialogue took place.[4]

At the same time, there is a marked process of objectification of the Divinity in Christianity from its origin to the present day: it is the figure of God as Father, who is considered by some theologians (Boff 1986) as the novelty proposed by Jesus in his relationship with the Divine. The expression used by him is Abba, i.e., the father of a small child, literally a *daddy* and not a judge, warrior or emperor, as he had been until then in different religions, including Jewish theology. This fact is a considerable break, in which the God whose name should not be said to Jews, assumes the shape of a man who has small children and cares for them with dedication and zeal. However, in the course of the first centuries of Christianity, this objectification coexisted with others, such as the Judge of the Apocalypse, for example.

With regard to the dialogue between Social Representations Theory and studies of religiosity and spirituality, there are some dimensions that will be outlined below:

(1) Cognitive dimension – There is need for an expression of faith and spiritual and mystical experience within a rational framework that can instantly be understood by the other members of the social group and, in most cases, by all other people. Even if this aspect does not make sense to other groups, dogmas and central beliefs are an effort to make the faith knowable from a certain perspective. The idea of the Virgin Mary is born of a logic in which a saint must necessarily be generated by a saint, while the conception of original sin is a rational way of understanding human finitude and spiritual decay as something

---

[4]For those who wish, the description of this scene may be read in the Acts of the Apostles, chapter 17, verses 18–23.

that stains all men through sex, and in Catholic theology Mary was properly preserved from such a stain. The important thing here is not the belief or not in these dogmas, but the rational process that supported their proposition.

(2) Attitudinal dimension – This is composed of quite rigid rules of conduct in some daily life situations, sometimes down to the smallest detail, depending on the religion. Examples of attitudes adopted in different environments include the initial considerations of demonizing the religion of others or the behaviour of a parliamentary bench from the foundations of their religion.

(3) Identity dimension – This refers to a social identity and sense of group belonging that is very strong and cohesive. Throughout history, one can see how much this dimension has generated the certainty that members of the group are the chosen ones elected to a new world or the only ones in possession of the truth, hence leading to cases of religious persecution and intolerance. At the same time, it is possible to highlight occasions when this identity has embodied deep social solidarity and group protection from existential and social dangers. Finally, there is the consideration of how maintaining this social identity has influenced the direction of groups in order to reinforce their identity and image, such as Christian religions developing charitable actions as a way of being identified as charitable.

(4) Normative dimension – This means the establishment of a normativity adopted by all members who follow a certain religious and spiritual belief and belong to a religious community. This adherence process is a right-and-wrong aspect that must also be followed as a form of identification with this group and its precepts. In the health area, for example, important bioethical issues were raised by Jehovah's Witnesses concerning blood transfusion, and by Catholics concerning the use of contraceptive methods. There is also the need for the adaptation of public contests on Saturdays for those who are Seventh-Day Adventists.

(5) Functional dimension – The functional dimension arises from the very relationship of worshippers with dogmas and their beliefs by generating an own way of acting in and on the world. Examples include some indigenous people who buried images of Christ and the Virgin Mary in their arable fields in different Latin American countries as a way of fertilizing the land and guaranteeing a good harvest, and the continued action of religious proselytism in an attempt to have as many conversions as possible.

(6) Dimension of sociability – This dimension addresses the coexistence and mutual help between worshippers of the same religion with the development of support and solidarity actions in difficult situations. Also highlighted is the construction of a specific communication and language process which, at times, may be relatively hermetic for other people.

(7) Affective dimension – This dimension is related to the experience of the affections and feelings of worshippers towards the Divine, towards those who belong to the same religious group and those who belong to other religions – for example, the love felt for the chosen God, and anger or compassion for those who do not follow the same religion.

(8) Dimension of construction and shared world-view – The religious vision becomes the sieve of the world-view shared by the group, and influences the way that situations are assessed and the practices that are used to reach this assessment. This dimension is characterised by the Manichaean view, in which events are judged according to the idea of good and evil, of god and devil that fight a fierce battle for all human beings and for each individual in particular. Another interesting issue in this topic is the different reports and beliefs in the end of the world that have consequences for the groups holding those beliefs.

Social Representations Theory provides heuristic and pertinent theoretical and methodological tools for deepening spirituality and religiosity studies, including the reality experienced in the Brazilian religious context. Among those that are noteworthy, the following are considered fundamental:

(1) Anchoring process – In addition to what has been discussed previously, it is important to highlight the anchoring processes that still occur as a way of making a certain religion encoded within symbolic reference frames of the population. This was the case when neo-Pentecostal churches began to use rituals and objects more closely related to popular Catholicism or *Umbanda* (Afro-Brazilian religion), such as coarse salt, blessed water, anointed roses, and cult-like nomenclatures of other religions, such as *descarrego* (a kind of purification in which bad spirits are evicted from a person), for example. However, at least in the first analysis, an important aspect that has not been properly apprehended yet stands out: that currently, there are two anchoring processes underway in the Brazilian religious field. The first has been mentioned above, regarding neo-Pentecostal churches through resources of the consensual universe of popular beliefs, similar to that proposed by Moscovici (1978). The second, however, seems to be different, since these same churches perform a process of anchoring phenomena and representational objects in the reified universe of biblical Judaism. Thus, Prosperity Theology, the breaking of hereditary curses and belief in a warrior God, only makes sense in this theoretical framework and not in the Christian theological heritage or in the theoretical systematization of the New Testament. This leads to the observation of a strong Judaizing process of the Evangelical Church in Brazil with the use of its symbols, the conformation of its temples and the adoption of its clothes.

(2) Cognitive polyphasia – In this instability of attributing meaning in the midst of the Brazilian religious scene, specifically within Christianity, whether focusing on the New Testament or on the Old, there is also a complex process of cognitive polyphasia taking root in a double way in the implementation of religious precepts. The analysis of religious intolerance situations nowadays in Brazil and in other parts of the world should take into account this ambiguity, in which the Christian law of love comes into conflict with the need to extirpate the unfaithful from the midst of people, as occurred in the Old Testament. Certainly there is a *decalage* between rational and affection in this process, which makes its analysis complex and, possibly, its implementation suffering. The law of not killing and the reality of homicide is an example that has always been present throughout Christian history.

Even the mystical and spiritual experience, as aforementioned, is located in one of the most individualizing poles, marked by social aspects, and crossed by psychosocial and cultural aspects that provide the borders and limits of this same experience. Noteworthy in this respect is the lack of reports from Protestants of mystical experiences with the Virgin Mary, or from Muslims or Jews with Jesus, or from Catholics with Buddha (Bingemer 2004). Another aspect to be considered is the adaptation of this experience to different contexts by giving familiarity to it, for example, the belief in the appearance of Our Lady of Fátima in Portugal with very European characteristics, or the belief in Our Lady of Guadalupe, who has significant indigenous traits.

The current reports of near-death experiences have the same guidelines, and the margins and limits of beliefs of people who have experienced them, in spite of reports of a basic pattern in different places where they were analysed (Fenwick 2013). There is also an important phenomenon to be analysed: cases of stigmatization, whereby certain people receive the stigmas of Christ's passion, as the Church reports about St Francis of Assisi in the thirteenth century, and St Pius of Pietrelcina in the twentieth. Regardless of the origins and even the veracity of these phenomena, it is a fact that the marks are not located where the wounds would actually have been provoked, but where they are officially represented in the sacred images. For example, Christ's crucifixion must have occurred on the wrist and not on the hand, but stigmata are all in the hands.

Different empirical studies have been developed by the Research Group on Spirituality and Religiosity in the context of nursing care and health, namely: discursive production and social representations with themes of spirituality and religiosity in the health area, most of them with subjects living with HIV/AIDS. The results will not be presented here, but if readers wish, they can search the various articles available online through databases. Only inferences and conclusions pertinent to the presented discussion will be mentioned.

First to be highlighted are the results of the first studies, in which, even preliminarily, spirituality and religiosity have the same representation and are therefore practically the same thing for the subjects studied. What the reified universe presents separately with conceptual specificities, the subjects' social thinking allows to be unified, as well as these subjects' transit between the two concepts. The faith present in the central elements of both representations seems to play an important role; in spirituality, it expresses its personal facet, while in religiosity this facet is social and communal. Further analytical efforts are needed to better understand these results, but faith appears to be a porous element which allows the passage of the same meaning between religiosity and spirituality.

A fact that draws attention is an opposition relation between the two content blocks in the set of the representational structure of religiosity. One is related to the dimension of the transcendent, love and faith, and the other is related to the social institution and community of faith. However, analyses indicate that the possible central elements are in the first block and overlap this representation with that of spirituality, as mentioned in the previous paragraph. The specificities within each representation should not be forgotten, and this is recognized, but the possible nuclear constitution of both allows stating that at least for now, they are the same representation.

Another interesting result is related to the quite profound idea among subjects, that both religiosity and spirituality have an intimate connection with the understanding of a human relationship with the Divine, that is, people establish a personal and affective relationship and are cared for by this transcendent instance. These representational constructions result from the definition of hegemonic religiosity in the West, which is the reconnection between human beings and God. For Christians, this connection is materialized through the Lamb sacrifice. Subjects' citations on this fact are recurrent in the results: there was an abyss between both, and the sacrifice allowed a bridge to be built between the two sides. Once again, this reinforces the symbolic constructions of Christ, like that of Pontiff, again placing him in the position of priest.

It was also found that the sociocognitive reconstruction of spirituality and religiosity concepts is not an abstraction, but quite concrete in the daily lives of people living with HIV/AIDS. It is related to the realization of daily practices without specificy of a relationship to spirituality or religiosity, and includes actions directed at people and social groups, oneself and the own Divine. In this same line, empirical data show that the representation of God is not organized from its cognitive and conceptual dimension, but, like spirituality and religiosity, is focused on a practical, concrete, contextualized and even bodily dimension.

In this case, God is an existential and social reality, almost tangible in the established relations and in personal and group challenges. Thus, daily life with the syndrome, from the moment of coping with the seropositivity diagnosis until the present time, is crossed by the presence of the Divine that manifests through different feelings (faith and hope), world-views specific to one's spirituality/religiosity (God has a plan and everything that happens is in his hands, or a world divided into good and evil) and language itself (the linguistic universe has a strong organization of this issue: sleep with God, may God be with you, God bless, God willing, and so on).

At this point, it is important to mention a recent result from a study conducted with religious leaders of a Pentecostal church about the social representations of HIV/AIDS. The group has a rather refined social thought about the relations established between God and his love, considered by them as infinite, the balance with divine justice and how men are placed in this process, whether distancing themselves or being embraced. Thus, love and justice are balanced in a specific relationship between God and men: infinite love embraces all men, but it does not include the idea of *anything goes*. It means love does not ignore justice as a form of distribution according to each one's merits. What determines which pole will be hegemonic in the relationship to each person is the sincere repentance of sins and the conformation of life according to standards deemed right by the social group to which one belongs. This conformation allows an even more precise regulation of divine action for physical healing, psycho-affective healing and the person's salvation in the post-mortem. Therefore, love and justice signify the divine movement of constant expansion and restraint towards the human being: love seeks the involvement of all, justice only of those who deserve it. However, deserving is not related to something important a person has done that deserves divine love, because this is considered by subjects to be an undeserved gift. It means the adoption of attitudes, behaviours and adherence to the faith professed by the community to which one belongs.

## 12.3   In Conclusion

By giving examples of its application to three distinct objects with at least three different social groups, the three topics presented in the previous section demonstrate the richness of Social Representations Theory and its explanatory and heuristic power in the field of health. When a researcher is approaching the same phenomena over a defined period, Social Representations Theory has the creative power and the capacity to provide insights and deepen understanding of, among other things, the way that social thought occurs in relation to temporality.

The research process from nurses' professional autonomy to studies of spirituality and religiosity in health and the approach to the concept of vulnerability do not necessarily imply a temporal succession, but, in most cases, a simultaneity that at times favours one object, and at other times another object. Common to all is the concern of the nursing care approach as a complex phenomenon that must be deepened in a diversified way. It involves: analysis of the proper role of the professional implementing it; the construction of a concept in which a dialogue is established between the weaknesses and susceptibilities characterizing patients and nurses in daily care; and demonstrations of moments when they approach and move away during this common path.

As for spirituality and religiosity in the area of care, it is necessary to understand the symbolic and cultural productions characteristic of care and the relationship established between professionals and patients in a daily routine in which objectivity and subjectivity are mixed with the transcendent and immanent, with a view to producing meaning in the face of suffering, sickness and death. As previously mentioned, certain theorists and empirical data explain that, for many Brazilian people, the Divine is not an abstract idea or concept, but rather a tangible and existential reality, and its presence is felt in the smallest details and every hour.

At this moment there is a possibility of deepening the spiritual and religious reality in the context of health in its multiple facets by understanding how the divinity and the relationship established with it are represented. In addition, understanding is expected of the complex syncretism process that has occurred since the first moments of what was conventionally called Brazil. Also noteworthy is the empowerment provided by *walking* with divinity when facing situations with which it is not possible to cope, and the creativity of solutions – not always effective, it must be recognized – found in people who have some faith. In any case, as Gilberto Gil affirms, *walking in faith I'll go cause faith doesn't usually fail.*

# References

Abric, Jean-Claude, 2000: "A abordagem estrutural das representações sociais", in: Moreira, Antonia; Oliveira, Denize (Eds.): *Estudos interdisciplinares em representação social* (Goiânia: AB Editora): 27–38.

Almeida, Maria Cecília Puntel; Rocha, Semiramis Melani Melo Rocha, 1997: "Considerações sobre a enfermagem enquanto trabalho", in: Almeida, Maria Cecília Puntel; Rocha, Semiramis Melani Melo (Eds.): *O trabalho de enfermagem* (São Paulo: Ed. Cortez): 15–26.

Ayres, José Ricardo de Carvalho Mesquita, 1996: *Vulnerabilidade e avaliação de ações preventivas: HIV/Aids, DST e abuso de drogas entre adolescentes* (São Paulo: Fundação McArthur).

Ayres, José Ricardo de Carvalho Mesquita, 1999: "Vulnerabilidade e prevenção em tempos de Aids", in: Barbosa, Regina Maria; Parker, Richard (Eds.): *Sexualidade pelo avesso: direitos, identidades e poder* (Rio de Janeiro: Relumé Dumará): 50–71.

Bingemer, Maria Clara Lucchetti, 2004: "A mística Cristã em reciprocidade e diálogo: a mística católica e o desafio inter-religioso", in: Teixeira, Faustino (Ed.): *No limiar do Mistério. Mística e Religião* (São Paulo: Paulinas): 35–73.

Boff, Leonardo, 1986: *Jesus Cristo libertador* (Petrópolis: Ed. Vozes).

Boff, Leonardo, 2005: "O cuidado essencial: o princípio de um novo ethos", in: *Inclusão Social*, 1,1 (October–March): 28–35.

Capra, Fritjof, 1982: *O ponto de mutação* (São Paulo: Ed. Cultrix).

Capra, Fritjof, 1982: *O tao da física: um paralelo entre a física moderna e o misticismo oriental* (São Paulo: Ed. Cultrix).

Fenwick, Peter, 2013: "As experiências de quase morte (EQM) podem contribuir para o debate sobre a consciência?", in: *Revista Psiquiatra Clínica*, 40,5: 203–207.

Foucault, Michel, 1988: *Microfísica do poder* (Rio de Janeiro: Edições Graal).

Gomes, Antonio Marcos Tosoli, 2011: "A vulnerabilidade como elemento organizador do cuidado de enfermagem no contexto do HIV/Aids: conceitos, processos e representações sociais" (PhD dissertation, State University of Rio de Janeiro, Nursing College).

Guitton, Beatriz; Figueiredo, Nébia; Porto, Isaura, 2002: *A passagem pelos espelhos: a construção da identidade profissional da enfermeiro* (Niterói: Ed. Intertexto).

Jodelet, Denise, 2001: "Representações sociais: um domínio em expansão", in: Jodelet, Denise (Ed.): *As representações sociais* (Rio de Janeiro: EdUERJ): 17–44.

Lourenço, Frederico, 2017: *Tradução da Bíblia: –Novo Testamento: os Quatro Evangelhos* (São Paulo: Companhia das Letras).

Machado, Roberto, 1988: "Introdução: por uma genealogia do poder", in: Foucault, Michel (Ed.): *Microfísica do poder* (Rio de Janeiro: Edições Graal): 7–23.

Meslim, Michel, 2014: *A experiência humana do Divino* (Petrópolis: Vozes).

Moscovici, Serge, 2012: *A psicanálise: sua imagem e seu público* (Petrópolis: Vozes).

Moscovici, Serge, 1978: *A representação social da psicanálise* (Rio de Janeiro: Zahar Editores).

Nietsche, Elisabeta Albertina; Backes, Vania Marli Schubert, 2000: "A autonomia como um dos componentes básicos para o processo emancipatório do profissional enfermeiro", in: *Texto e Contexto*, 9,3 (August–December): 153–174.

Oliveira, Denize Cristina, 2001: "A teoria das representações sociais como grade de leitura da saúde e da doença: a constituição de um campo interdisciplinar", in: Almeida, Ângela Maria de Oliveira; Santos, Maria de Fátima Santos; Trindade, Zeide de Araújo (Eds.): *Teoria das Representações Sociais: 50 anos* (Brasília: Ed.Technopolitik): 585–623.

Rocha, Juan Stuardo, 1989: "Prefácio", in: Almeida, Maria C. P.; Rocha, Juan Stuardo (Eds.): *O saber de enfermagem e sua dimensão prática* (São Paulo: Ed. Cortez): 9–15.

Rodrigues, Maria do Socorro Pereira, 1999: *Enfermagem: representação social das/os enfermeiras/os* (Pelotas: Editora e Gráfica Universitária/UFPel).

Rouquette, Michel-Louis, 2000: "Representações e práticas sociais: alguns elementos elementos teóricos", in: Moreira, Antonia; Oliveira, Denize (Eds.): *Estudos interdisciplinares de representações sociais* (Goiânia: Ed. AB): 39–46.

Santos, Érick Igor, 2016: "Autonomia profissional do enfermeiro e suas representações sociais elaboradas por estudantes de universidades públicas federais do Rio de Janeiro: contribuições ao ensino superior em enfermagem" (PhD dissertation, State University of Rio de Janeiro, Nursing College).

Torralba y Roselló, Francesc, 2009: *Antropologia do cuidar* (Petrópolis: Vozes).

Wagner, Wolfgang, 2000: "Sócio-gênese e características das representações sociais", in: Moreira, Antonia; Oliveira, Denize (Eds.): *Estudos interdisciplinares de representações sociais* (Goiânia: Ed. AB): 3–26.

Waldow, Vera Regina; Borges, Rosália Figueiró, 2008: "O processo de cuidar sob a perspectiva da vulnerabilidade", in: *Revista Latino-Americana Enfermagem*, 16,4 (August): 765–771.

# Chapter 13
# Children, Multiple Ordinations of Reality and Social Representations: Dialogues from Lévy-Bruhl

Daniela B.S. Freire Andrade

## Introductory Comment

Angela Arruda

*By relying on the way children make*[1] *moral judgments, revealed by Piaget's research, in the way they explain the relations between man and woman, revealed by Freud, as well as in the way that distant peoples think of the world around them, discussed by Lévy-Bruhl, Moscovici lays the foundations of his thinking: small, distant, non-specialists formulate world-views – produce worlds – as often as the 'qualified'. This is a framework for the de-hierarchization of knowledge: popular, everyday lay knowledge is worth as much as specialized (scientific or other) knowledge. They are diverse productions, with different purposes, for operation in different spaces, but of equal value. 'Unqualified' knowledge has as its qualification its vocation for communication, for moving in the world, for the resolution of the immediate, for the economy of affections.*

---

Daniela B. S. Freire Andrade is an associate researcher at the Federal University of Mato Grosso (UFMT), Cuiabá campus, and coordinator of the Research Group in Child Psychology. She is also a guest researcher at the International Centre for Studies in Social Representations and Subjectivity – Education (CIERS-ed) and vice-coordinator of the Brazilian Association for Research and Postgraduate Studies in Psychology (ANPEPP). Email: freire.d02@gmail.com.

---

[1]Angela Arruda is a pioneer in the study of social representations in Latin America, and a senior researcher at the Federal University of Rio de Janeiro and the University of Évora. She has extensive experience in the study of social psychology, with the emphasis on social representations and qualitative methodologies. She gained her BSc in psychology at the Federal University of Rio de Janeiro (UFRJ), her MSc in social psychology at the École des Hautes Études en Sciences Sociales (EHESS) in France, her PhD in social psychology at the University of São Paulo (USP), and conducted postdoctoral studies at the University Institute of Lisbon (ISCTE-IUL) in Portugal. Email: arrudaa@centroin.net.br.

© Springer Nature Switzerland AG 2021
C. Prado de Sousa and S. E. Serrano Oswald (eds.),
*Social Representations for the Anthropocene: Latin American Perspectives*,
The Anthropocene: Politik—Economics—Society—Science 32,
https://doi.org/10.1007/978-3-030-67778-7_13

*The fundamental originality of Daniela Andrade's text is its reaffirmation of these assumptions regarding children, who, despite being the object of fundamental attention in the engineering of Social Representations Theory, do not appear as often as adults in field research. It is possible that one of the reasons for this is the preference of for verbal methodologies (oral or written). Attempts to overcome this choice, in some cases, resort to images, drawing, which would work to leverage the explicitness by the word, indispensable when trying to know the meaning of the graphic production for those who execute it. In these cases, children's thinking is not always the main interest, as can be seen in De Rosa's (1987) work on mental illness, when she focuses on children.*

*The originality of the text then expands because it does not reside only in the fact of placing the child as a research subject, but in the interest in him as an agent of cognition, a producer of knowledge. For Daniela, it is not a matter of capturing only the meaning of an object, but also her thinking as a process, i.e. as an elaboration of thought and action, and as part of a development framework.*

*The search for underground actors in order to identify and understand their process of producing meaning, and their creation/action as part of the condition of being human, is an expression of openness to these Others who mobilize our universe as if they did not mobilize it, did not make it move. It is part of the vigour of the new generation of researchers working on Social Representations Theory. The generation to which Daniela Freire Andrade belongs and to which she contributes with quality attests to the vitality of this field.*

# Reference

De Rosa, A., 1987: "The social representation of mental illness in children and adults", in: Doise, W.; Moscovici, S. (Eds.): *Current issues in European social psychology*, vol. 2 (Cambridge: Cambridge University Press).

**Abstract** This chapter begins addressing a question posed by Jovchelovitch – Is the knowledge of a child a primitive form of the knowledge of an adult? It does so in order to propose a dialogue between Lévy-Bruhl and Moscovici on multiple orders of reality and on the concept of cognitive polyphasia. This approach is presented as an argument that approaches different rationalities in the analysis of children's logic and the coexistence of knowledge. The thesis of the discontinuity of the mind and the relations of participation contributes to reduce the boundary between scientific thinking and alternative rationalities. The text is based on the notion of scientific visibility of children's ability to think in social representational research contexts, privileging children as subjects of social representations.

**Keywords** Levy-Bruhl · Socialization · Children · Social representations

## 13.1 Introduction

The expression 'universes of socialization' proposed by De Lauwe (1991) reveals the importance of social representations without delineation of children's learning or coming close to what Brougére (2004) terms perspectives – possibilities available to children and their capabilities of perceiving. Social representations, beliefs and values related to childhood present and are linked to the contexts in which a child lives daily, explicitly, either from speech or in a natural way, through social practice and space organizations, which end up setting universes of socialization, filled with structures of opportunities for human learning and development.

The field of studies on social representations and education Gilly (2001) denotes an idea of didactic contract when referring to the mediation of the representations in the communication systems between adults and children in an educational context. However, studies in the field of social representations regarding the process of children's interpretations of social reality in connection with social representations initially appeared in the works of Gerard Duveen (1996).

In the words of De Lauwes (1991), the so-called universes of child socialization are tightly related to the children and the group to which they belong.

> The child's way of perceiving and thinking influences his or her conditions of life, their status, and the adult behavior toward them. In a given society, ideas and images relating to the child, however varied, are organized into collective representations,[2] which form a system at multiple levels. A language "about" a child is created as a language "for" the child, since ideal images and models are proposed to him. (De Lauwe 1991: 1)

The studies on the representational dimension that constitute the universes of child socialization have traditionally prioritized the perspective of adults – their social expectations on and for children. In the Brazilian educational context, such studies have mainly focused on targeting adults, family members, professionals, government officials and managers responsible for the elaboration of public policies as well as media producers, taking them as the main source of information. This finding is not different when the reflection turns to the context of paediatric health. Consequently, the paucity of studies that take the child as a social player of social representations reveal the fact that they know little or nothing about social reality and the child's perspective on reality is usually invisible.

In considering this brief overview, I understand the relevance of the debate about the status of the child in studies on social representation, given the breadth of research that mostly announces the child as an object of social representations of adults regardless of their condition as subject of social representations.

---

[2]Collective representations refer to dogmatic thought impermeable to the processes of change. Moscovici (1978), relying on this formulation and, at the same time, opposing it, proposed investigating social representations, characterizing them as common-sense knowledge, more diverse and permeable to the new and communicational processes typical of modern societies.

The incursion proposed in this text refers to the analysis of Lévy-Bruhl's contributions to the collective representations of primitive thought for Social Representations Theory (Moscovici 2003). It also takes into account the current studies by the anthropologist Tambiah (2013) on multiple reality in order to highlight the notion of peer culture (Corsaro 2009) as a space for sharing meanings within which the adaptation and creation unit (Vygotsky 2009) is inscribed in the cultural development of the child.

The proposed debate aligns with the discussions of Vygotsky (2009) on the unit adaptation and creation of cultural development announcing the existence of a process called creative re-elaboration.

> The child's play is not a mere remembrance of what he experienced, but a creative re-elaboration of impressions experienced. It is a combination of these impressions and, based on them, the construction of a new reality that responds to the aspirations and longings of the child. Just as in play, the child's drive to create is the active imagination. (Vygotsky 2009: 17)

On the other hand, Corsaro (2011), revisiting the classic notion of socialization as a process of the child's adaptation to reality, contrasts with the notions of interpretative reproduction and peer culture, proposing an alternative approach to the theme.

When using the term 'interpretative reproduction', Corsaro (2011) means the innovative aspects of children's participation in society, since they create and participate in their peer cultures (an expression used to signify children who produce and create their own collective worlds with relative autonomy) in a way that now captures the information of the adult world. According to their perspective, this equates to a shift towards the internalization of culture, whereby they are now actively contributing to cultural production and change.

The social, civic and scientific invisibility of the child, denounced by Sarmento (2007), is recalled to consider the extent to which children's knowledge is assumed to be an important source of social influence for researchers in social representations. For that, a return to the basis of the theory – to the thought of Lévy-Bruhl – is proposed, based on the reflections of the anthropologist Tambiah (2013) and the dialogue with social representations proposed by Moscovici (2003).

This is the scenario in which there is a debate about the relationship established between primitive thought and modern thought, orientating questions that are still shocking, as Jovchelovitch (2004: 12) states:

> would it be that the knowledge of a child is a primitive form of adult knowledge? Or would the knowledge of other cultures (primitive or "inferior" as the nineteenth and early twentieth century literature called them) be a rudimentary form of logic found in so-called civilized Western societies?

Jovchelovitch's questioning (2004), when analysed from the perspective of the knowledge produced by children and adults, will be illustrated through a brief analysis of research (Andrade 2006) into the social representations of different groups regarding the school space with the objective of highlighting the multiple arrangements and their effects on the social construction of reality.

## 13.2 Causality and Participation: About the Multiple Arrangements of Reality

In analysing Lévy-Bruhl's contributions and their unfolding in the form of propositions of the multiple settings of reality, Tambiah (2013) considers it important to emphasize that Lévy-Bruhl denounced philosophical inquiry and the applied sciences because they seemed to have raised Western thought to such a high level that it was referred to as a model for other systems of thought. From this point of reference, Victorian anthropologists stated that their purpose was to explain the origin of the misconceptions of 'simpler' peoples, believing that errors of 'primitive' thinking would naturally be corrected since the evolution of thought consisted of the 'savages' copying the European standards of observation and speech.

In opposition to this perspective, Lévy-Bruhl advocates abandoning the idea that primitive thought is irrational or misguided with regard to the application of the laws of thought. He proposes that primitive thought has its own organization, coherence and rationality, and bases this argument on relations of participation.

Participation, according to Lévy-Bruhl (in Tambiah 2013: 194) is an association between people and things in primitive thought, to the point of meaning identity and consubstantiality.

Primitive thought fuses distinct aspects from the point of view of Western thought into a single mystical unity. Tambiah (2013) exemplifies this phenomenon from the conception of society to the primitive thought constituted by the living and the dead who continue to live somewhere in the neighbourhood and play an active role in social life before they die a second time. As an example, he cites: "When a Bororo claims to be a macaw, which is exactly what he believes when he wants to express an inexplicable mystical identity between him and the bird" (Tambiah 2013: 195). Therefore, it is not a metaphor; the sense of participation implies a physical and mystical union.

When considering Lévy-Bruhl's notes, Gerken (2012) asserts that it is still possible to consider participation relations from the predominance of the affective elements in the construction of their own references in a dynamic characterized as a complex fusion between the emotions and the cognitive representation. This characteristic is anchored in the fact that primitive mentalities have mental habits different from those attributed to the modern mentality, among which stands out the least logical requirement.

Tambiah (2013) clarifies that for Lévy-Bruhl the primitive mind is indifferent to the causal relation; the relation between cause and effect is immediate, and the intermediates are not recognized. Hence, primitives would not prove causal connections in a scientific way, which should not be interpreted as a kind of deficiency, but as a style of rationality in which causality does not guide social doctrines and is not considered a parameter of its systems of knowledge.

> From the strictly logical point of view, one finds no essential difference between the primitive mentality and ours. In everything that concerns the ordinary daily experience, transactions of all sorts, political life, economics, use of numbering, etc., they behave in a way that implies the same use of their faculties as we do ours. (Lévy-Bruhl 1949 in Gerken 2012: 136)

The analysis of participation relations allowed Lévy-Bruhl to postulate that other cultures, civilizations or periods may present categories and alternative systems of thought which possess connections and 'logics', or internal coherences. Such systems of thought would be completely different from the dominant form of modern thought, whose cognitive theories and logical systems would be insufficient to explain them.

The second postulation of Lévy-Bruhl asserted that mystical and logical-rational mentalities coexist in humanity with variations in their weight and preponderance from primitive times to modern times. Yet in his third postulation, Lévy-Bruhl suggested that there was a mystical mentality present in all human minds, although it was apparent and more easily observable among primitives than at present.

> This mystical experience is influenced by a characteristic emotion: the feeling of the presence and the action of an invisible power, or of contact with a reality other than that which is given in real or everyday circumstances. (Tambiah 2013: 199)

Such experiences of participation would be progressively subject, in Western thought, to demands that consider them in logical terms but that from them poetry, art, metaphysics and scientific invention itself are forged.

In this way, Lévy-Bruhl observed that scientific thought and mystical thought could be better compared as normative ideational systems in the same society, especially in contemporary society. According to this reasoning, one can, in one context, behave in a mystical way and in another present a practical, empirical, and everyday frame of mind.

Tambiah (2013) suggests extending this reasoning by replacing the term 'mentality' with *multiple orientations of reality or multiple orders of reality*, exploring the social construction of meanings and systems of knowledge. He still speaks of modes of construction and experimentation of reality or ways of making-the-world, according to Goodman (1985, apud Tambiah 2013).

By emphasizing the *existence of multiple frameworks, multiple versions of the world* with independent interests and import, Tambiah's proposition (2013), relying on radical relativism under the rigorous limits of Goodman (1985), is irreducible to a single basis, so the unity between the different versions of the world must be through a general organization that would embrace all of them.

In taking science as a frame of reference, the author cites at least two different ways of *making the world* or two modes of referential function: denotation and exemplification.

Denotation refers to the scientific, literal, linguistic, or mathematical description of the world, and exemplification or reference is typical of artistic and non-representational forms which denote nothing but 'show much' and convey feelings (Tambiah 2013: 210).

Thus, a third statement by Goodman is highlighted by Tambiah (2013) and refers to the idea that the 'truth' of science is a frame of reference that incorporates a certain aesthetic of truth and certainty based on syntactic and semantic density, and, in recognition of patterns, on making the world as a composition, ordering, weighing.

In light of these contributions, Tambiah (2013) proposes the existence of two realities/orders with hybridization, preponderances and complementarities which depend on individuals, groups in a given culture, and cultures taken as collective entities: participation versus causality.

Causality is represented by the categories, rules and methodology of the positive sciences and logical-mathematical discursive rationality, which includes affective neutrality and a certain abstraction in relation to the events of the world. Participation is linked to holistic and systematic apprehensions of totalities, integrated with aesthetics and mystical consciousness. Participation emphasizes affective and sensory communication and the language of emotions; causality underlines the rationality of instrumental action and the language of cognition.

In addition, it is emphasized that one does not exist without the devices of the other.

> [...] people of all cultures and societies engage in distinct genres of discourse that relate and are driven by different contexts of communication and practice (as defined by Bourdieu). According to the occasion and context, we invoke, employ and manipulate a *corpus* of languages and concepts, culturally available and adapted to fit different systems of knowledge, styles of rationality and rhetoric, and modes of emotional experience. In this sense, we are flexible and plural and engage in many ways of making-the-world. (Tambiah 2013: 214)

It follows from this that the elements of participation are not absent in scientific discourses, and causality is not necessarily absent in the performances of participation; they intertwine in various ways, and there may be contexts in which one or the other predominates.

This approach suggests that children are more vulnerable to fear than social pressure or social influence.

## 13.3  Lévy-Bruhl's Ideas and Some Inspirations from Moscovici

Moscovici's work (2003) on Lévy-Brulh's contributions to the Social Representations Theory depicts the effort of the anthropologist who, in dialogue with Durkheim, analyses the constitutive collective representations of primitive thought.

The debate on the term 'collective representation' from the contributions of Lévy-Bruhl prompted Moscovici (2003) to highlight four aspects relevant to the formulation of his concept of social representations.

The first aspect refers to the idea that the semantic content of each idea and each belief depends on its connections with other ideas and beliefs. By taking them as isolated facts, as Evans-Pritchard warned (in Moscovici 2003), they may seem strange. However, when viewed as an integral part of a set of ideas and behaviours, your senses gain contours. Context and communication processes are underlined here, highlighting the existence of meanings and networks forged culturally and historically, which in turn act as the anchors of social representations.

The second aspect is the idea that all the symbols present and active in a society are orientated by both the logic of the intellect and the logic of the emotions; even though these symbols may be based on different logics, they are intellectual constructions of thought. For Moscovici (2003) this principle applies to any culture and not just to the primitive ones, and also explains that: "When you discriminate against a group, you express not only your prejudices about this category, but also the aversion or contempt to which they are indissolubly linked" (Moscovici 2003: 184).

In this analysis of the relevance of the study of common sense and the nature of social representations as knowledge constituted in the tension between the consensual universe and the reified universe, Moscovici (2003) announces, through the notion of cognitive polyphasia, the coexistence, in the same social group and in the same individual, of different types of rationality involved in the construction of social representations. In this case, people, including scientists, would resort to one or another argument according to particular circumstances and interests. He further adds that this peculiarity of social representations should be understood more as a rule than an exception.

The third aspect refers to the notion that a collective representation is at the same time of someone, and creates a world in itself as it expresses not only the meaning of things that coexist but fill the gaps, the who are invisible and absent from these things and thus, are rooted in the concrete life of the people (Moscovici 2003: 184). In this way, Moscovici (1978) presents social representations as guidelines which are elaborated through the sharing of meanings within communication processes in order to act in the interpretation of social reality, making the strange something familiar.

The fourth outstanding aspect asserts that, for Lévy-Bruhl, "all collective representations have the same coherence and value. Each has its originality and its own relevance, such that none of them has a privileged relationship to the others" (Moscovici 2000: 134). Therefore, collective representations cannot be taken as criteria of truth or rationality, otherwise there would be a risk of earlier collective representations being considered inferior, incomplete or irrational compared with scientific or modern ones.

According to this logic, when thinking about the concept of knowledge inspired by the contributions of Lévy-Bruhl, Moscovici (2003) concludes that it is impossible to propose an absolute criterion of rationality that is independent of the content of collective representations and their inscription in a specific society. It is this impossibility which underlines the legitimacy of studies of social representations.

This reasoning is anchored in the principle of the discontinuity of primitive thought and scientific thought proposed by the anthropologist in arguing that each type of thought has its own categories and rules of reasoning that correspond to different collective representations. Against this fact, the researcher has to "feel from within" – Husserl's expression rescued by Moscovici (2003) – to understand how the mentality builds up and how it organizes the society which it is to analyse.

Moscovici's (2003) remarks on Lévy-Bruhl's studies of collective representations of primitive thought reveal that they facilitated the emergence of social psychology, in particular the psychology of the so-called primitive cultures, asserting the existence of alternative rationalities to scientific thought. Such rationalities would base thinking on three principles: 1) mythical thought; 2) memory; 3) participation.

Non-scientific representations would be embedded in a context that sensitizes people to the existence of the supernatural or mystical. This principle makes individuals impervious to the data of immediate experience.

Memory in turn plays a central role so that the world of mediated and inner perceptions dominates the world of direct and outer perceptions. Gerken (2012) characterizes it as having a concrete character, meeting the real needs of culture reproduction. Yet people are not constrained to avoid contradictions in their arguments; relationships of participation now drive thought.

When questioning whether the relationships of participation manifest in the thought of modern man as well, Moscovici (1978) highlights one of the fundamental aspects of Social Representations Theory: the importance of common sense in the process of the social construction of reality. In this respect, popular knowledge and scientific knowledge come close and interpenetrate.

## 13.4   The Child as the Subject of Social Representations and New Realities of Reality

Gerken (2012), citing Jahoda (1999), points out that Lévy-Bruhl's objective has focused on studies on modernity, and in his work the anthropologist never compared the primitive adult with the Western child; such an approach came from the contributions of psychologists such as Piaget in his studies on the psychology of children's representations and Vygotsky in his psychology of psychological functions.

> The attempt to establish that each form of knowledge corresponds to a fundamental set of social relations was central to the thought that developed under the influence of phenomenology and hermeneutics. It was also central to the psychologies of Piaget and Vygotsky, who had a clear understanding of the social nature of logic and set out to demonstrate in detail the manner in which society shapes the development of logical structures in the child. The two showed that different social relations lead to different ways of knowing. (Jovchelovitch 2004: 10)

In particular, the Vygotskian perspective was anchored in the idea of the mediation of collective representations present in the arguments about the historical-cultural dimension of the human psyche based on the social nature of the processes of construction of the logical categories that allow the human being to understand the world (Gerken 2012: 136).

The Vygotskian thesis on the historical-cultural dimension of the human psyche allows us to think about the mediation of collective representations and social representations, described as cultural elements, in the constitution of the logical categories that allow the human being to understand the world.

According to Gerken (2012), Vygotsky relied on Janet's formulation of the fundamental law of psychology that, throughout development, the child begins to apply in himself the forms of behaviour that have been applied to him beforehand (Gerken 2012: 136). In other words, everything that is part of the human psyche was previously part of the social relations in which the subject is inserted. In Vygotsky's work, one can clearly identify such influence in his notes on learning and development in children of pre-school age (Vygotsky 1993) and more precisely in the formulation of the general law of cultural development.

> [...] every function in the child's cultural development appears on the scene twice, on two planes; first in the social plane and then in the psychological, in principle among men as an interpsychic category and then within the child as an intrapsychic category. (Vygotsky 2000: 150)

Thus, according to Luria's analysis (1992 in Gerken 2012) ideas as space, time, and number would be complex forms of memory and not intrinsic categories of mind. In other words, it can be stated that the child says he relies on a cultural substrate that is constituted by a certain level of sharing.

In the context of Social Representations Theory, Duveen's studies with children prioritized older children who were capable of articulating their thoughts through speech.

Based on these studies, it has been possible to determine that children interact with the social representations elaborated by adults, through social relations established in intergenerational dialogues, and, in accordance with the understanding of Vygotsky (2009) and Corsaro (2011), subsequently elaborate their own hypotheses about reality.

Duveen (1995) asserts that being born in an already structured world does not mean that the child is born with the skills to be an independent actor in the world, and that these competences are forged throughout the child's development. Consequently, such competences are the result of the construction of social knowledge. At first, the child figures as an object for representations that others support, and he or she gradually internalizes these representations and then identifies his/her position within a set structured world.

Duveen (1996) highlights the notion of the development of social knowledge by children as a movement of resistance and tension, which makes the internalization carried out by them understood not by their passive role but as a process of relative autonomy through which a reconstruction activity is undertaken.

This understanding inserts the child into the social scene as an actor whose processes of symbolic construction deserve attention in the sense of considering it capable of inscribing new orderings about reality.

The analysis presented in the empirical study entitled *The feminine place in the school: a study in social representations* (Andrade 2006) allows researchers to understand the inscription of the multiple ordinances of reality by considering the meanings of adults and children on the same object of social representation: the school space.

Andrade (2006), interested in analysing the social representations of students and their teachers of the spaces of the school, found a significant difference in the representational contents of these two groups with respect to the school bathroom. For the teachers interviewed, besides defining it from the perspective of physiological needs, the bathroom characterized a space to be controlled in order to avoid the incidence of sexual games, especially during breaks between classes.

In turn, the children's representations of the bathroom were more complex, making it possible to identify the coexistence of participation and causality relationships, and therefore the phenomenon of cognitive polyphase. In addition to the bathroom being viewed as a place where secrets are told, and where one plays without the boys destroying the games, the children explained that there was a territorial dispute between boys and girls with respect to the playground; it had been established that this would be the territory of the boys, even though this was against the will of many girls. In response to this imposition, the girls, within the context of their peer culture, announced that the entrance to the boys' bathroom was the place where, in fulfilment of a certain ritual, the spirit of Maria Cotton (an urban legend) would appear, thereby frightening many children, especially boys.

The observation about the flow of children in the bathroom and the school practices announced by the teachers revealed that the children, frightened, asked to go to the bathroom in pairs. A request which, for them, meant support and proof of friendship was, for the teachers, anchored in the notion of transgression, and a cause for concern and discipline in view of the teachers' belief that there was a risk of sexual games taking place in that space. Consequently, children at that school often urinated in their clothes because they were afraid to go to the bathroom on their own and were refused permission to take a friend for moral support.

By ignoring shared content in peer cultures and neglecting the fact that children, as well as adults, creatively re-elaborate socially shared content, teachers fail to consider that the way to cope at school escapes the adult-centric perspective.

In general, the representations of the school space of the different groups sometimes reveal the prevalence of causal relations regardless of the relations of participation, and in other instances the opposite. Yet they reveal how responses to issues depend on causal relationships, which are often built through participatory relationships, hence to ignore them would be to remain alienated from the symbolic processes that make up group life and school culture.

## 13.5  Final Considerations

This text reflects on the status of the child in studies on social representations, questioning the scientific visibility of children as subjects of social representations. Such reflections are based on the foundations of Moscovician thought, which, through Lévy-Bruhl's analysis of primitive thought and modern thought, set the main guidelines for the study of common sense. This chapter places these ideas within the parameters of the coexistence of rationality, which is characteristic of the logical-rational mentality typical of scientific thought, and mystical mentality, which is driven by the emotional dimension and noncommittal, and it notes the absence of any contradiction between such mentalities.

Causality and participation are taken as the mindset of reality or modes of making the world. This introduces the concept of different styles of rationality, rhetoric, and emotional experience, a perspective that makes the foundations of infantile thought equivalent to adult thought.

The thesis of the discontinuity of the mind as well as the law or relations of participation were presented as presuppositions that approximate studies in developmental psychology and Social Representations Theory, reinforcing the notion of development as a historical and cultural process.

The analysis of the closeness between the Vygotskian assumptions, the sociology of childhood (Corsaro 2011) and the ontogenetic approach of social representations (Duveen 1995, 1996) all indicate that the child is a social actor who, through *creative re-elaboration* (Vygotsky 2009) and *interpretative reproduction* (Corsaro 2011), has relative autonomy with respect to the internalization of social representations to give visibility to his or her different ways of constructing the world.

The notion of peer culture (Corsaro 2011) draws attention to the fact that the construction of social knowledge by children occurs within the framework of intergenerational relations when the child's appropriation of information from the adult world, according to his or her perspective of it, occurs through internalization of the culture. However, children also produce and create their own collective worlds with relative autonomy, which demonstrates their contribution to cultural production and change.

In this respect, the formulation of Andrade (2014) is legitimate:

Will children be able to influence processes of significance of reality in order to compete for the construction of social representations and create social influence?

When this question is analysed empirically, taking the school space as an object of social representation by children and adults, the orientation of the different structures of reality in the construction of the social representations in question can clearly be observed.

Different arguments let us see the existence of two ways of making the school, specifically, two ways of thinking about the school and its relations of power. In particular, the lack of adult clarity regarding the highly shared symbolic constructions in the peer culture is highlighted, which, once accessed, contributes to a more complex analysis of sociability in the school scene. Far beyond the issue of sexual games as adults, the social representations of children over the bathroom reveal the relations of power: the struggle for territory as well as the resistance of girls who rely on mythical thought when they reproduce a narrative in social memory through the myth of Maria Algodão, and at the same time produce a form of empowerment.

This brief analysis aims to inspire researchers in social representations about the dialogical potential between research on children and research with children. In other words, it the intention is to argue in favour of children's scientific visibility (Sarmento 2007) in the contexts of research in social representations.

# References

Andrade, Daniela Barros da Silva Freire, 2006: "O lugar feminino na escola: um estudo em representações sociais" (PhD Dissertation, Pontifícia Universidade Católica de São Paulo).

Andrade, Daniela Barros da Silva Freire, 2014: "A infância como objeto de representações e as crianças como sujeitos que elaboram novos sentidos sobre a realidade: sutilezas de um debate", in: Chamon, Edna M.Q.; Guareschi, Pedrinho; Campos, Pedro Humberto (Eds.): *Textos e debates em representação social* (Porto Alegre: ABRAPSO): 145–157.

Chombart De Lauwe, Marie-José, 1991: *Um outro mundo: a infância* (São Paulo: Perspectiva, Editora da Universidade de São Paulo).

Corsaro, William A., 2009: "Reprodução interpretativa e cultura de pares", in: Müller, Fernanda; Carvalho, Ana Maria Almeida (Eds.): *Teoria e prática na pesquisa com crianças: diálogos com Willian Corsaro* (São Paulo: Cortez): 31–50.

Corsaro, William A., 2011: *Sociologia da infância* (Porto Alegre: Artmed).

Duveen, Gerard, 1995: "Crianças enquanto atores sociais: as representações sociais em desenvolvimento", in: Guareschi, Pedrinho; Jovchelovitch, Sandra (Eds.): *Textos em representações sociais* (Petrópolis, RJ: Vozes): 261–293.

Duveen, Gerard, 1996: "The Development of Social Representations of Gender", in: *Japanese Journal of Experimental Social Psychology*, 35,3: 256–262.

Duveen, Gerard; Lloyd, Barbara, 2008: "Las representaciones sociales como una perspectiva de la psicología social", in: Castorina, José A. (Eds.): *Representaciones sociales: problemas teóricos y conocimientos infantiles* (Buenos Aires: Ed. Gedisa): 29–39.

Gilly, Michel, 2001: "As representações sociais no campo da educação", in: Jodelet, Denise (Eds.): *As representações sociais* (Rio de Janeiro: Ed. UERJ): 321–341.

Gerken, Carlos Henrique de Souza, 2012: "Razão e o Outro em Lévy-Bruhl: notas para um diálogo com a psicologia histórico-cultural de Vygotsky", in: *Pesquisas e Práticas Psicossociais*, 7,1 (January–June): 130–138.

Jovchelovitch, Sandra, 2004: "Psicologia social, saber, comunidade e cultura", in: *Psicologia & Sociedade*, 16,2 (May–August): 20–31.

Moscovici, Serge, 1978: *A representação social da psicanálise* (Rio de Janeiro: Zahar).

Moscovici, Serge, 2000: *Social representations: Explorations in social psychology* (Cambridge: Polity).

Moscovici, Serge, 2003: Representações sociais: investigações em psicologia social (Petrópolis, RJ: Vozes).

Müller, Fernanda, 2007: "Entrevista com Corsaro", in: *Educação & Sociedade*, 28,98 (January–April): 271–278.

Sarmento, Manuel Jacinto, 2007: "Visibilidade social e estudo da infância", in: Vasconcellos, Vera; Sarmento, Manuel Jacinto (Eds.): *Infância (in)visível* (Araraquara: Junqueira & Marin): 25–49.

Tambiah, Stanley, 2013: "Múltiplos ordenamentos de realidade: o debate iniciado por Lévy-Bruhl", in: *Cadernos de Campo*, 22: 193–220.

Vygotsky, Lev, 1993: *Lo sviluppo psichicho del bambino* (Roma: Ed Riuniti).

Vygotsky, Lev, 2000: *Obras Escogidas III* (Madrid: Visor Dis. SA).

Vygotsky, Lev, 2009: *Imaginação e criação na infância: ensaio psicológico: livro para professores* (São Paulo: Ática).

# Chapter 14
# The Contribution of Social Representations Theory to Science Education

Alcina Maria Testa Braz da Silva

## Introductory Comment

Edna Maria Querido de Oliveira  Chamon

*In times of[1] fake news, post-truth, and aversion to argumentative debate, science has been at the centre of attention, due to the polemics created about the conclusions it presents and its consequences. Thus, we have seen the growth of manifestations in favour of evidence-based science and a return to rationality, which should be supported by science itself.*

*In the Introduction of his book* What is this thing called science?, *Alan Chalmers asserts that "Science is highly esteemed" (Chalmers 1999: xix), indicating an apparent "widely held belief that there is something special about science and its methods" (idem). However, much of the time, the appropriation of science seeks to legitimize positions – commercial, political and even scientific. As Bruno*

Alcina Maria Testa Braz da Silva is a professor and researcher in the Science, Technology and Education Postgraduate Programme (PPCTE) of the Federal Center for Technological Education Celso Suckow da Fonseca (CEFET), Rio de Janeiro. Email: alcina.silva@cefet-rj.br.

[1]Edna Maria Querido de Oliveira is a full professor at the University Estácio de Sá and the University of Taubaté (UNITAU). She collaborates at the University of Campinas (UNICAMP) in the Department of Architecture and Construction at the college of Civil Engineering, Architecture and Urbanism. Her Master's degrees in Education, and her PhD in Psychology, were all undertaken at the University of Toulouse II (Le Mirail), with a postdoctoral research stay in Education at the University of Campinas (UNICAMP). Her research into Social Representations Theory and education is widely cited in the field. Email: edna.chamon@gmail.com.

© Springer Nature Switzerland AG 2021
C. Prado de Sousa and S. E. Serrano Oswald (eds.),
*Social Representations for the Anthropocene: Latin American Perspectives*,
The Anthropocene: Politik—Economics—Society—Science 32,
https://doi.org/10.1007/978-3-030-67778-7_14

*Latour argued, there is much more to science than scientific propositions: there is a mixture of knowledge, interest, politics, and power (Latour 1993).*

*Thus, in approaching scientific literacy and school science teaching as a neutral and disinterested area, linked to reified knowledge, there is a risk of forgetting that science is a human production with historical and social features.*

*It is from this perspective that the discussions on the relationship between social representation and teaching in science are explored in the text by Alcina Silva. Embracing a social view of questions on science education, Alcina is interested in the influence of public policies and formal institutions (schools, scientific institutions), as well as different social groups, in forms of dissemination, participation and decisions related to science and technology, joining the proposals sponsored by Snow and Dibner (2016), but strongly based on Social Representations Theory. This perspective discusses the public policies for education in general and teacher training and students' learning in particular, exploring the relations between State structures and the various communities related to science and technology.*

*In researching social representations of education and their relations with scientific and technological literacy, Alcina questions the ways in which scientific and technological knowledge is constructed by teachers (and, consequently, by students). This construction has a direct impact on the teaching and learning processes, at both the level of contents and the level of the mediation processes, particularly in the strategies of use/construction of science teaching.*

*The two papers discussed in Alcina's text, related to the use of technology in science teaching and to the quality of science education, exemplify the potential of Social Representations Theory for analysing the relationships of a social group – teachers in these two cases – with science teaching. Public policies, organization of the curriculum, deficiencies in infrastructure, lack of teacher training, and inadequacies in didactic construction are some of the elements that a qualitative analysis – specifically Semantic Networks technique, supported by Social Representations Theory – allows to be discussed.*

*Alcina calls, in her conclusions, for an agenda to advance in this line of study. Based on Social Representations Theory, she points out the topics that need further study in science education. These topics include the relationships between human and non-human elements, borrowing the terms of Latour (1993), such as teachers, textbooks, information and communication technologies, practical knowledge and public policy. These studies ask for new analytical tools which make it possible to include the action of the teacher in the classroom and the practical knowledge of the students in the analysis.*

*Finally, Alcina's work proposes a reflection on the usefulness of Social Representations Theory as a basis for new insights on science studies and opens a wide and interdisciplinary road for discussions on science education.*

# References

Chalmers, Alan F., 1999: *What is this thing called science?* (Queensland: University of Queensland Press).
Latour, Bruno, 1993: *We have never been modern* (Cambridge, MA: Harvard University Press).
Snow, Catherine E.; Dibner, Kenne A. (Eds.), 2016: *Science Literacy: Concepts, Contexts, and Consequences* (Washington, DC: The National Academic Press).

**Abstract** In the context of the discussion to be developed in this chapter around the theme "The contribution of Social Representations Theory to science education", it is necessary to define the parameters of the approach. In the case of this chapter, I initially intend to contextualize in the current scenario of science education in Brazil the various assessment systems, and I am focusing on the PISA (Programme for International Student Assessment) evaluation in science. Then, the contribution of Social Representations Theory to science education will be discussed, in respect of knowledge, science and schools. Finally, some research results will be presented on the projects developed by the research group Education in Sciences and Social Representations (EDUCIRS), in the Science, Technology and Education Postgraduate Programme (PPCTE) of the Federal Center for Technological Education Celso Suckow da Fonseca (CEFET), Rio de Janeiro, and in the final considerations there will be a reflection on future developments in the interface of investigation between scientific education and social representations.

**Keywords** Scientific education · Scientific knowledge · School knowledge · Educational evaluation · Quality of education · Social Representations Theory

## 14.1   Scenario: Science Education in Brazil

In recent decades the Brazilian education scene was marked by attention displacement, previously focused only on educational evaluation, that is, on the evaluation of teaching and learning processes, towards the various forms of institutional evaluation, such as the evaluation of institutions themselves, systems, projects and public policies.

Therefore, the discussion about the quality of education has also moved beyond academic and school centres, having conquered new spaces in various segments of society. In this context, the teaching of science has been the subject of discussions and reflections in the scientific community and in education systems. It is a concern that in part is related to the monitoring of the analysis of the results coming from the different instances of education evaluation, covering different levels and systems of education. The data collection carried out by Alves/Franco

(2008) points to several of these instances at the national level – SAEB (acronym in Portuguese for System of Evaluation of Basic Education), SARESP (acronym in Portuguese for School Performance Evaluation System of the State of São Paulo), SIMAVE (acronym in Portuguese for System of Evaluation of the State of Minas Gerais), School Census of Basic Education, ENEM (acronym in Portuguese for High School National Examination), Census of Higher Education – ENADE (acronym in Portuguese for National Examination of Higher Education Students' Performance) – and international level – UNESCO/OREALC (Regional Office Education of UNESCO for Latin America and the Caribbean), PISA (Programme for International Student Assessment).

In this section, I focus on the analysis of the PISA results, according to the reports, made available by the National Institute of Studies and Educational Research Anísio Teixeira, Brazil (INEP, acronym in Portuguese). PISA consists of a comparative evaluation applied in a sample way to students in the fifteen-year-old age group, the age at which the completion basic schooling is assumed in most countries.[2]

PISA is funded and carried out by the Organization for Economic Cooperation and Development (OECD), which has national coordination in each participating country, and in Brazil is coordinated by INEP. The results of these studies are taken as a reference to verify the development achieved in the key competences defined in the European Union's 2020 Strategy and in the Educational Goals 2021 of the Organization of Ibero-American States (OIE).

## 14.1.1 Assessment Systems: The Results of PISA

PISA encompasses the fields of reading, mathematics and science, not just related to the curricular domain of each school, but also to the knowledge and skills that are considered necessary to the students in their future lives. The emphasis on the domain of procedures, understanding of concepts and the ability to respond to different situations within each field of knowledge are the main features of this assessment. Such evaluations are conducted in cycles every three years, and in each cycle a 'main' content area is focused on, to which is dedicated two-thirds of the time in the tests. The main evaluation areas were: reading in 2000, mathematics in 2003, science in 2006, reading in 2009, mathematics in 2012 and science in 2015. Brazil has been taking part in all programme evaluations since 2000. At the time of writing, the 2018 evaluation, in which reading was the main area, was held in the first semester, and the corresponding reports were due to be made available in 2019.

---

[2]Instituto Nacional de Estudos e Pesquisas Educacionais Anísio Teixeira (INEP), at: http://www.inep.gov.br/ (10 May 2020).

The 2000–2015 reports are available on the OECD[3] and INEP[4] websites. According to the data released by INEP, with respect to those years, the countries were distributed in three ranges, according to the performances in the tests. In the second year of the implementation of PISA, for example, a group of seventeen countries scored above the OECD total average, i.e. in the range of 509 to 550. The other group composed of four countries had results equivalent to the OECD total average in the range of 498 to 506 points. A third group of nineteen countries was below the OECD total average (356 to 495): Norway, Luxembourg, Poland, Hungary, Spain, Latvia, the United States, Russia, Portugal, Italy, Greece, Serbia, Turkey, Uruguay, Thailand, Mexico, Indonesia, Tunisia and Brazil.[5]

A comparative analysis of the years 2000 and 2003 makes it possible to draw a profile of the countries' development relative to their performance in the assessment. Brazil presented some progress in the second year of the evaluation. It was the fastest growing in two of the assessed areas of mathematics, improved in science, and maintained the performance of 2000 in reading. In science, it improved from 375 points in the year 2000 to 390 points in the 2003 evaluation. However, in 2003 Brazil was still among those that achieved results below the OECD total average (356 to 495).[6]

In 2006, the evaluation was applied in 57 countries (all OECD members and invited countries) with a focus on science, including assessment in reading and mathematics, as well as informative questionnaires on students, their families and institutional factors that could explain the differences in performance. The results of the application of PISA 2006 showed Brazil in position 52, remaining below the OECD total average.[7]

The results of the last evaluation released by the OECD when this chapter was being written showed that Brazil increased its average score in the science PISA test from 390 to 405 points in comparison to 2006, and evolved 33 points on average in reading, mathematics and science compared with its performance in 2000 and 2009. In 2000, the Brazilian average was 368 points, reaching 401 points in 2009. The country had a greater evolution in the average of the mathematics assessment, which changed from 334 points in 2000 to 386 points in 2009, increasing 52 points. The scores increased from 375 to 405 in science and from 396 to 412 in reading. According to government officials, considering these results positive, the country exceeded the goal established by the Education Development Plan (PDE) in the decade 2001–2010, which was to reach an average of 395 points in all three disciplines. Despite this, according to a critical analysis, even standing

[3]Organisation for Economic Cooperation and Development (OECD), at: https://www.oecd.org/ (10 May 2020).

[4]Instituto Nacional de Estudos e Pesquisas Educacionais Anísio Teixeira (INEP), at: http://www.inep.gov.br/ (10 May 2020).

[5]INEP, at: http://www.inep.gov.br/ (10 May 2020).

[6]INEP, at: http://www.inep.gov.br/ (10 May 2020).

[7]INEP, at: http://www.inep.gov.br/ (10 May 2020).

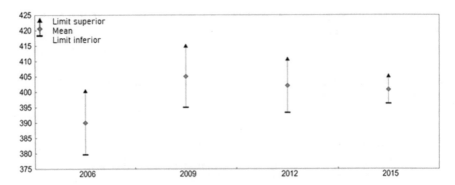

**Fig. 14.1** Evolution of the Brazilian students' proficiency in science, 2000–2015. *Source* The author

among the three countries that most developed in basic education, Brazil would still be far from strong in education. Among the 65 countries evaluated by PISA 2009, Brazil reached a total average of 401 points, occupying the 53rd position.[8]

In the decade 2011–2020, two PISA assessments had been carried out by the time this chapter was written: one in 2012 (main area mathematics), and the other in 2015 (science). There was an absence of empirical evidence to explain significant statistical differences between the performance of Brazilian students in science in PISA 2015 and the last three editions of the assessment, and "although the highest performing students showed an increase in average results when compared to PISA 2012, the lowest performers presented a reduction of 15 points in the score" (OECD 2016: 268). Figure 14.1 presents the evolution of the Brazilian student's average proficiency in science in the period of 2006–2015.

Many of the criticisms regarding the "supposedly scientific rankings of schools" (Lopes/Lopez 2010: 106) in the case of the science area, make sense when considering, in the scenario of student performance analysis, the seemingly old, but in fact quite current, discussions that point to the lack of experimentation in class, its role in the teaching and learning of scientific knowledge, book teaching, the insertion of technological resources in educational spaces, the absence of establishing relationships with the everyday knowledge in teaching and learning situations, and the hermetic curriculum, consisting mostly of a straitjacket for the development of the teacher's creative and critical work.

Other points are highlighted in the literature of the area, which, added to the entire structural context – recurrently unfavourable – involving the schools and the education system itself, only tend to complicate the analysis: the model of the transmission of scientific knowledge as neutral and absolute truths; students' lack of interest in the classes of scientific disciplines; psychosocial obstacles in the reconfiguration of the previous ideas held by the students; didactic approaches

---

[8]OECD, at: https://www.oecd.org/ (10 May 2020).

that disregard the historical and philosophical contextualization in the construction of scientific knowledge; lack of problematization on the issues that configure the scientific-technological formation; fragilities in the formative processes of teachers, their conceptions and representations; lack of articulation between training, research, extension and innovation; difficulties in overcoming the challenges for the understanding and development of interdisciplinary proposals (Carvalho/Gil Perez 1993; Chassot 2000; Lopes 1999; Franco 1989; Maldaner 2000; Mortimer 2000; Schnetzler 2000; Gil-Pérez et al. 2001; Neto/Fracalanza 2003; Ferreira/Selles 2004; Samagaia/Peduzzi, 2004; Cachapuz et al. 2005; Franco/Sztajn 2008).

## *14.1.2   Reflecting on Science Education*

In the last two decades, when discussing the formation and practice of teaching and learning in sciences, one of the polemical topics is to what extent it is necessary to think about a science of school – school knowledge. This school science, which in a broader and more problematized sense defines the symbolic elements that would characterize a scientific-technological culture, involves a process of reconfiguration of the knowledge of other sociocultural contexts.

An important contribution to the understanding of the relationship between scientific knowledge and school knowledge is based on the formulations presented by Lopes (1999), which develop his arguments based on the interpretation that scientific knowledge and school knowledge have their own epistemological configurations and that the epistemological approach is not enough for the investigation of school knowledge.

The author, in her initial formulation on the correlation between scientific knowledge and school knowledge, analyses the hierarchies of knowledge and defends this contextual character of knowledge. From this perspective, she asserts that school science involves a process of the re-elaboration of knowledge from other social contexts, aiming to meet the social purposes of schooling, which implies that school knowledge is not necessarily exclusively produced at or for school. As Chassot (2003) argues, one must look for a school science that is significantly different from that of university science, the so-called academic knowledge.

In a later discussion, Lopes (2002) analyses the role of contextualization in the PCNEM (National Curricular Parameters of High School) with respect to the valuation of daily life, from the relation of school knowledge to the concrete issues of students' lives, highlighting the lack of a political sense of the concept of daily life. In a critical approach in the curricular field, the author states that contextualization appears to be associated with the productive process of school knowledge through didactic transposition. This argument is based on the conception that school discipline is an intrinsic production of the school. According to the author, reflecting on the production of scholarly knowledge involves rethinking the various meanings that the question of quality implies in teaching practice, as well as in training, and allowing new perspectives on the educational context itself.

Recent works by Lopes and collaborators (2004, 2007, 2010, 2019) guide this analysis to demonstrate that epistemological distinctions between knowledge exist as a result of knowledge-power relations. They include the interpretation of the curriculum as a cultural production and critique the culture of performativity in curricular policies. The authors mainly discuss the texts of Ball (1997, 2003, 2003) and Foucault (1989) to make this critique, defining performativity as "a technology, a culture and a mode of regulation that employs judgements, comparisons and ends up revealing itself as means of control, wear, and change" (Ball 2003: 216).

In their analysis of performativity, Lopes and Lopez (2010) establish consistent arguments to discuss the articulation of the effects of globalization on curriculum and evaluation policies, taking the case of the documents related to the ENEM (National High School Exam) as a reference. The authors conclude that the focus of this examination is the formation of the individual who is responsible for the social efficiency of the educational system and the social system, and, unlike other times, focused on the self-regulation of their performances. Such reflections are pertinent to the various forms of evaluation that proliferate in the education system. In the face of curricular and evaluation policies with a focus on performance, what knowledge and senses are being legitimized to the detriment of others in the school and in the educational system in its multiple instances? What educational purposes does this knowledge serve?

## 14.2   Contribution of Social Representations Theory

The discussions that emerge in the contemporary context focus on the controversy over the role of evaluations in the educational sphere and have generated questions and reflections to explain the reality in which the school and its social actors are located. Understanding the meanings that circulate in this context, from the perspective of these social actors, constitutes an important step towards looking at the school's social commitment to the process of transformation necessary for educational proposals that excel, in order to enable equality and social justice to gain voice and space.

The theoretical-methodological orientation of social representations, in the approach proposed by Moscovici (1961, 1981, 1986, 1988, 2003, 2012), has the potential to deepen this understanding. Observation and investigation of the phenomena that arise in the social horizon are based on collective sociocultural constructions called social representations.

The study of social representations is becoming a field of research that establishes interfaces with several areas of knowledge: anthropology, economics, semiotics, justice, health and education. Its interdisciplinary character can be attributed to a psycho-sociological context from which the concept emerges and with which the relationships between individual-environment social-object are constituted (Jodelet 2001). These relationships collectively contribute to building, defining, and interpreting our daily reality.

Sá (1998) emphasizes that social representations should be understood as forms of knowledge about daily life. Their function is to enable communication between individuals and guide behaviour. Expanding this definition, Abric associates the functioning of social representations with a system of interpretating reality that aims to guide the relations of individuals with their physical and social environment, allowing people "to understand the interaction between the individual and the social conditions" (Abric 2000: 28) in which the social individuals act, and thus enabling them to determine behaviours and practices. Understanding the social representations that are disseminated and circulated in schools in the contemporary educational context can thus facilitate an understanding of teaching practices as social practices in the process of change.

According to the authors of the psychosocial perspective of Social Representations Theory, such as Moscovici (2003), Jodelet (2001) and Gilly (2001), the diffusion of an unfamiliar phenomenon, either because it is a novelty or sets a "new fashion", driven by a need to assimilate the 'novelty' in the social groups involved in this process, generates mobilization, conflicts, consensus, convergence and divergence, conversations and negotiations. The processes of making familiar what is not familiar are called objectification and anchoring and are producers of meanings and generators of social representations, which consist of a "form of knowledge, socially elaborated and shared, with a practical objective that contributes to the construction of a common reality to a determined social group" (Jodelet 2001: 22).

These representations are collective, multifaceted and polymorphic constructs, integrating cognitive, affective, symbolic, and value elements generated by the social subject, and therefore its importance in social life covers the relationship between cognitive processes and social interactions. Mazzotti (2002) reinforces the arguments that constitute a specific type of knowledge and have a guiding function of behaviour and communication between individuals.

Based on the concepts proposed by Moscovici (1976, 1978) of "reified universe" and "consensual universe", the field of social representations is a fertile ground for discussing the process by which the concepts produced by the sciences are reified, making it possible to analyse the diffusion and circulation of these symbolic representations in the daily reality. The "reified universe" consists of techniques, artefacts, and theories which correspond to knowledge elaborated according to logical and formal procedures, while the "consensual universe" consists of the 'locus' where social representations of reified objects are produced.

Gilly (2001) argues that the school system has always suffered, to a greater or lesser degree, the marks of social groups occupying different positions in relation to it – discourse of politicians and administrators, institutional agents from different hierarchy levels, and different speeches. Thus, according to the author, social representations are configured as autonomous systems of social meanings that are the product of contradictory compromises under the double pressure of ideological factors and impositions related to the effective functioning of the school system.

Gilly presents us with the successive 'decontextualization' and 'recontextualization' phenomena of knowledge, discussed in the work of Perret-Clermont et al.

(1981), when analysing the social transmission of scholarly knowledge, through processes of selection and organization of information. According to Gilly, successive social practices of selecting teaching content, writing textbooks, and preparing lessons activate reconstructions of a new object that generate successive social representations of the initial scientific knowledge, finalized by the objectives of the social practices.

As a privileged place of the production and circulation of social representations of various educational phenomena, and as a space of interaction, conversations, the production of knowledge and senses, school serves to legitimize some knowledge and meanings to the detriment of others, keeping them hegemonic. Teachers are one of the social groups that work in this space. In a shared way, they construct explanations about how teaching should be to achieve learning, in actions often thought to be unidirectional. These explanations affect the choices of what should be taught, the amount of time spent on each subject, the most appropriate didactic and evaluation strategies, decisions about whether to use technological resources or didactic and technological innovations, and how to use them, and they are influenced by the criteria of the evaluation exams.

Based on his research into the role of school evaluation as a corpus for the identification of social representations of the sciences, Mazzotti (2014) proceeded to analyse the mechanisms used to evaluate scientific knowledge. He identified the conceptual framework of the representation of evaluation in the arguments of those who legitimize it and those who oppose it, pinpointing the figurative model that coordinates and condenses the meanings of both groups. According to the position defended by this analytical process, the social representations of the sciences, technology, ideologies and the arts can be identified and described. Mazzotti points out that doing this reduces the sciences to a catalogue. This becomes a barrier to learning. Instead of encouraging students to conduct their own analyses, schools usually present scientific discourses as dogmas. This approach prioritizes the mechanisms of memorization and undervalues the importance of scientific methodology in the evaluative exams currently favoured by the education system.

## 14.3   EDUCIRS Group: Research Results

The research group on Education in Sciences and Social Representations – EDUCIRS[9] consists of professors and researchers in the field of science and mathematics education. Affiliated with the Science, Technology and Education Postgraduate Programme (PPCTE – acronym in Portuguese for Programa de Pós-Graduação stricto sensu em Ciência, Tecnologia e Educação) of the Federal

---

[9]Education in Sciences and Social Representations (EDUCIRS), at: http://educirs.webnode.com (10 May 2020).

Centre for Technological Education Celso Suckow da Fonseca – CEFET/RJ (acronym in Portuguese for Centro Federal de Educação Tecnológica Celso Suckow da Fonseca, Rio de Janeiro), its members consist of doctoral, masters and scientific initiation students linked to the PPCTE-CEFET, undergraduate students in physics, and also students of the postgraduate programme in science teaching at the Federal Institute of Rio de Janeiro (IFRJ – acronym in Portuguese for Instituto Federal de Educação, Ciência e Tecnologia do Rio de Janeiro).

The research developed by the group focuses on the formation of teachers and the teaching and learning process in natural sciences (biology, physics and chemistry) and mathematics. The theoretical and methodological techniques of Social Representations Theory offer unique insights which aid the understanding of several educational phenomena under study. The core subjects of the investigation are Teaching Knowledge, Technology, Natural Science and its Teaching, Mathematics and its Teaching, Inclusion Processes, Interdisciplinary.

This paper presents some results of the research that the group has developed from this dialogue between the areas of science education and social psychology, using Social Representations Theory as a guideline. These results are represented by semantic networks, elaborated from the thematic categorical analysis, according to Bardin (2002), whose thematic categories are articulated by meaningful relationships, with the support of software called ATLAS.ti (Muhr 1991).

## 14.3.1   Scientific Culture at School: The Interface Between Teacher Training and Professional Practice

Figure 14.2 shows the representative semantic networks of the teachers' speeches about the role of technology in their professional praxis when teaching science.[10]

The research was qualitative, and the data collection instrument consisted of open questions in a semi-structured questionnaire to access the discursive production of teachers on the role of technology in school, society and the contemporary world.

The study was carried out with ninety-six teachers of science courses (physics, chemistry and biology) in twenty public high schools in Rio de Janeiro.

These results are related to the following research theme: "The development of technology has dramatically grown, and the schools have been challenged to handle this scenario. Dear teacher, please write about your professional praxis in this technological insertion context."

---

[10]Braz da Silva, Alcina Maria Testa; De Jesus, Vitor L.B.; Oliveira, A.L., 2016: "The use of technologies in the teaching and learning process of science subjects under the teachers' perspective", in: *Anais do XVII IOSTE – International Organization for Science and Technological Education* (Braga: Universidade do Minho).

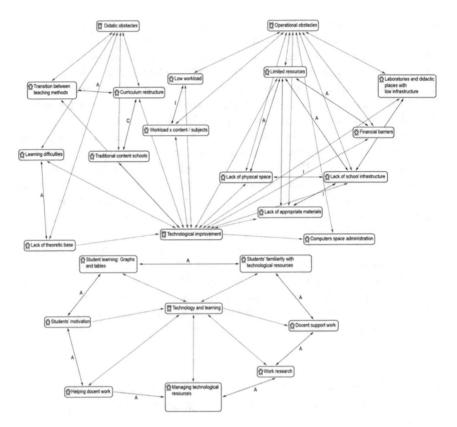

**Fig. 14.2** Semantic networks about technologies and science Labels: A = Association, I = Insertion, C = Contradiction. *Source* The author

Based on these built semantic networks one can point out that the role of technology is represented in the teachers' speeches as one of the obstacles they must face in their professional praxis in the process of teaching and learning the scientific knowledge. These obstacles appear to be associated with the schools' infrastructures, the curricular dimension, training paths and the gap between public policy and the actions developed in the schools.

## 14.3.2 Quality of Science Education from the Teachers' Perspective

The results presented in the semantic network of Fig. 14.3 indicate that the theme of quality was present in the discourses of the teachers of the scientific disciplines (physics, chemistry, biology) in the participating schools, judging from the

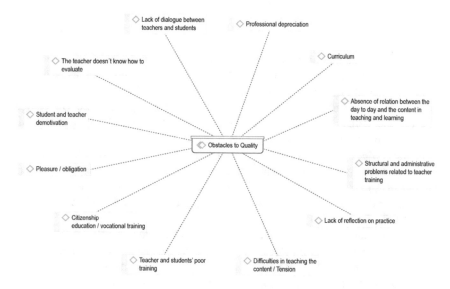

**Fig. 14.3**   Semantic network about quality – Rio de Janeiro, Brazil. *Source* The author

obstacles that they face, which appear to be associated with the elements of daily school life, the curricular dimension, the formative paths and public policies.[11]

The obstacles associated with a quality school daily life are defined by: the need to reflect on the practice of these teachers; the motivational factor, both for the teacher and the student, which is associated with the importance of learning with pleasure, of knowing how to seek information and establish moments of exchange and interlocution; the role of the teacher in the assessment process of learning situations; comprehension of the content and its relation to the daily life of the students.

The obstacles related to the curricular dimension are explained in the need to rethink the curriculum with a view to changing its structure, but without pointing the direction of change, while the obstacles associated with the formative pathways do not only indicate the need for better support but also question the quality of teacher training.

---

[11]Braz da Silva, Alcina Maria Testa; De Jesus, Vitor L.B.; Oliveira, A.L., 2016: "The use of technologies in the teaching and learning process of science subjects under the teachers' perspective", in: *Anais do XVII IOSTE – International Organization for Science and Technological Education* (Braga: Universidade do Minho).

## 14.4   Conclusions

I conclude this chapter by highlighting the importance of analysing the education scenario from the perspective of the social actors who are inserted and acting in it. In this way it should be possible to overcome the unilateral responsibility of individuals for the obstacles to both teaching and learning scientific knowledge, as a process collectively constructed in a context that involves social, political, economic and cultural constraints. This analysis should involve epistemological reflections, which, according to Maldaner (2000: 61) "can provide a broader view regarding knowledge, individuals interacting, curriculum, methodology, teaching and learning, in all processes of human development, which take place both in formal and informal educational processes, and in research and philosophical reflections".

Further expanding the range of action, it is necessary to place this analysis in an investigation context that involves the views of several social groups and includes the various defining elements of these conditions, thus making it possible to understand the multiple facets of pedagogical paths, teachers and their educational places, textbooks and their argumentative course, information and communication technologies and exclusion processes, the constitution of new knowledge, curricular conception and the field of conflicts, and, finally, social, political, economic and cultural scenarios, in which profiles, talents and identities are defined and, in turn, delineate new educational contexts.

I also add that work partnerships need to be formed and consolidated within the research groups, facilitating the analysis of theoretical and methodological references, their movements of convergence and divergence, and their limits as analytical tools. This scenario has the potential to foster the construction of theoretical frameworks in consonance with the practices, which suggests the necessity, in unfolding new projects, of both research and action designed to map the critical issues coming from the dialogical path between formation and practice.

## References

Abric, Jean-Claude, 2000: "A abordagem estrutural das representações sociais", in: Moreira, Antonia Silva Paredes; Oliveira, Denize Cristina de (Eds.): *Estudos interdisciplinares de representação social* (Goiânia: AB): 27–37.
Alves, Maria Tereza Gonzaga; Franco, Creso, 2008: "A pesquisa em eficácia escolar no Brasil: evidências sobre o efeito escola e fatores associados à eficácia escolar", in: Brooke, Nigel; Soares, José Francisco (Eds.): *Pesquisa em eficácia escolar: origem e trajetórias* (Belo Horizonte, MG: UFMG): 482–500.
Ball, Stephen John, 1997: "Policy sociology and critical social research: A personal review of recent education policy and policy research", in: *British Educational Research Journal*, 23,3: 257–274.
Ball, Stephen John, 2003: "The teacher's soul and the terrors of performativity", in: *Journal of Education Policy*, 18,2: 215–228.

Ball, Stephen John, 2004: "Performatividade, privatização e o pós-Estado do bem-estar", in: *Educação e Sociedade*, 25,89 (September–December): 1105–1126.

Bardin, Lawrence, 2002: *Análise de conteúdo* (Lisboa: Edições 70).

Braz da Silva, Alcina Maria Testa; De Jesus, Vitor L.B.; Oliveira, A. L., 2016: "The use of technologies in the teaching and learning process of science subjects under the teachers' perspective", in: *Anais do XVII IOSTE – International Organization for Science and Technological Education* (Braga: Universidade do Minho).

Cachapuz, Antonio; Gil-Pérez, Daiel; Carvalho, Anna Maria Pessoa, 2005: *A necessária renovação do ensino de Ciências* (São Paulo, SP: Editora Cortez).

Carvalho, Anna Maria Pessoa de; Gil-Perez, Daniel, 1993: *Formação de professores de Ciências* (São Paulo, SP: Cortez).

Chassot, Attico, 2000: *Alfabetização Científica: questões e desafios para a Educação* (Ijuí, RS: Editora UNIJUÍ).

Chassot, Attico, 2003: "Alfabetização científica: uma possibilidade para a inclusão social", in: *Revista Brasileira de Educação*, 22: 89–100.

Ferreira, Márcia Serra; Selles, Sandra Escovedo, 2004: "Análise de livros didáticos em ciências: entre as ciências de referência e as finalidades sociais da escolarização", in: *Educação em Foco*, 8,I–II: 63–78.

Foucault, Michel, 1989: *Microfísica do poder* (Rio de Janeiro: Graal).

Franco, Creso, 1989: "Os livros didáticos e a gravidade: uma queda pouco didática", in: *Revista Brasileira de Estudos Pedagógicos*, 70,165: 224–242.

Franco, Creso; Sztajn, Patricia, 2008: "Educação em ciências e matemática: identidade e implicações para políticas de formação continuada de professores", in: Moreira, Antonio Flavio Barbosa (Eds.): *Currículo: políticas e práticas* (Campinas: Papirus): 97–114.

Gil-Pérez, Daniel; Montoro, Isabel Fernández; Alís, Jaime Carrascosa; Cachapuz, Antonio; Praia, João, 2001: "Para uma imagem não deformada do trabalho científico: São Paulo", in: *Ciência & Educação*, 7,2: 125–153.

Gilly, Michel, 2001: "As representações sociais no campo da educação", in: Jodelet, Denise (Eds.): *As representações sociais* (Rio de Janeiro, RJ: EdUERJ): 321–341.

Jodelet, Denise, 2001: "Representações sociais: um domínio em expansão", in: Jodelet, Denise (Eds.): *As representações sociais* (Rio de Janeiro: EdUERJ): 17–44.

Lopes, Alice Casimiro, 1999: *Conhecimento escolar: ciência e cotidiano* (Rio de Janeiro, RJ: EDUERJ).

Lopes, Alice Casimiro, 2002: "Os parâmetros curriculares nacionais para o ensino médio e a submissão ao mundo produtivo: o caso do conceito de contextualização", in: *Educação & Sociedade*, 23,80: 389–404.

Lopes, Alice Casimiro, 2004: "Políticas de currículo: mediação por grupos disciplinares de ciências e matemática", in: Lopes, Alice Casimiro; Macedo, Elizabeth: *Currículo de Ciências em Debate* (Campinas: Papirus): 45–76.

Lopes, Alice Casimiro, 2005: "Política de currículo: recontextualização e hibridismo", in: *Currículo sem Fronteiras*, 5,2 (July–December): 50–64.

Lopes, Alice Casimiro, 2006: "Discursos nas políticas de currículo", in: *Currículo sem Fronteiras*, 6: 33–52.

Lopes, Alice Casimiro; Lopez, Silvia Braña, 2010: "A perfomatividade nas políticas de currículo: o caso do ENEM", in: *Educação em Revista*, 28,1: 89–110.

Maldaner, Otavio Aloisio, 2000: *A formação inicial e continuada de professores de ciências. Coleção Educação em Química* (Ijuí: Unijuí).

Mazzotti, Tarso Bonilha, 2002: "L'analyse des métaphores: une approche pour la recherche sur les représentations sociales", in: Garnier, Catherine; Doise, Willem (Eds.): *Les représentations sociales: balisage du domaine* (Montréal: Éditions Nouvelles): 207–226.

Mazzotti, Tarso Bonilha, 2014: "Ensino dos conceitos científicos ou de suas representações sociais?", in: Chamon, Edna M.Q.; Guareschi, Pedrinho; Campos, Pedro Humberto (Eds.): *Textos e debates em representação social* (Porto Alegre: ABRAPSO): 199–233.

Mortimer, Eduardo Fleury, 2000: *Linguagem e formação de conceitos no ensino de Ciências* (Belo Horizonte, MG: Editora UFMG).

Moscovici, Serge, 1976 [1961]: *La psychanalyse, son image et son public* (Paris: PUF).

Moscovici, Serge, 1981: "On social representation", in: Forgas, Joseph P. (Eds.): *Social cognition: Perspectives on everyday understanding* (London: Academic Press): 181–209.

Moscovici, Serge, 1986: "L'ère des représentations sociales", in: Doise, Willem; Palmonari, Augusto (Eds.): *L'étude des représentations sociales* (Neuchâtel-Paris: Delachaux et Niestlé): 34–80.

Moscovici, Serge, 1988: "Notes towards a description of social representations", in: *European Journal of Social Psychology*, 18: 211–250.

Moscovici, Serge, 2003: *Representações sociais: investigações em psicologia social* (Petrópolis: Vozes).

Moscovici, Serge, 2012: *A psicanálise, sua imagem e seu público* (Petrópolis: Vozes).

Muhr, Thomas, 1991: "ATLAS.ti: a prototype for the support of text interpretation", in: *Qualitative Sociology*, 14,4: 349–371.

Neto, Jorge Megid; Fracalanza, Hilário, 2003: "O livro didático de ciências: problemas e soluções", in: *Ciência e Educação*, 9,2: 147–157.

Organización para la Cooperación y el Desarrollo Económico (OCDE), 2016: *Brasil no PISA 2015: análises e reflexões sobre o desempenho dos estudantes brasileiros* (São Paulo: OCDE – Fundação Santillana).

Perret-Clermont, Anne-Nelly; Brun, Jean; Conne, François; Schubauer-Leoni, Maria-Luisa, 1981: "Décontextualisation et recontextualisation du savoir dans l'enseignement des mathématiques à de jeunes élèves", in: *Communication au colloque représentations sociales et champ éducatif* (Aix-en-Provence: UNINE).

Sá, Celso Pereira, 1998: *A construção do objeto de pesquisa em representações sociais* (Rio de Janeiro: EdUERJ).

Samagaia, Rafaela; Peduzzi, Luís Orlando de Quadro, 2004: "Uma experiência com o projeto Manhattan no ensino fundamental", in: *Ciência e Educação*, 10,2: 259–276.

Schnetzler, Roseli Pacheco, 2000: "O professor de ciências: problemas e tendências de sua formação", in: Schnetzler, Roseli Pacheco; Aragão, Rosália Maria Ribeiro (Eds.): *Ensino de ciências: fundamentos e abordagens* (Campinas, SP: UNIMEP): 13–41.

# Chapter 15
# Possible Dialogues Between Social Representations and Educational Policies: The Dilemma of Data Analysis

Romilda Teodora Ens

## Introductory Comment

### Elizabeth Fernandes de Macedo

*The text by Romilda Teodora Ens[1] summarizes a set of concerns that have been characteristic of her research trajectory, which speaks for itself. Romilda is a researcher who has devoted her work to challenging disciplinary barriers in the study of educational policies and teacher training. She went from the field of educational policies – teacher training – to undertaking field and empirical studies in order to understand the constitution of the theoretical and methodological schools that are instituted in the educational system. Thus, she participates actively in the Latin American Network of Epistemological Studies on Educational Policy*

Romilda Teodora Ens has a PhD in Education from PUC-SP and undertook post-doctorate research at the University of Porto-Portugal. She is currently a professor on the graduate programme in *stricto sensu* education at the Pontifical Catholic University of Paraná (PUCPR) and an associate researcher at the Carlos Chagas Foundation. Email: romilda.ens@gmail.com.

[1]Elizabeth Fernandes de Macedo is Full Professor of Curriculum Studies at the State University of Rio de Janeiro (UERJ), and was formerly Associate Provost for Graduate Programmes (2008–2015). She was a Visiting Scholar at the University of British Columbia (UBC), Canada, in 2007, and at Columbia University, USA, between 2013 and 2015. She received her PhD in education at State University of Campinas (1996). She is President of the International Association for the Advancement of Curriculum Studies. From 2010 to 2013, she edited the journal *Transnational Curriculum Inquiry*, sponsored by that Association. She currently serves as an Associate Editor of the *Journal of Curriculum Studies* and as a member of the Editorial Board of *Curriculum Inquiry*, and coordinates an inter-institutional group of researchers who are working collaboratively with state-level bureaucracies, schools and teachers to produce situated public policies in response to the compulsory national curriculum. Email: bethmacedo@pobox.com.

© Springer Nature Switzerland AG 2021
C. Prado de Sousa and S. E. Serrano Oswald (eds.),
*Social Representations for the Anthropocene: Latin American Perspectives*,
The Anthropocene: Politik—Economics—Society—Science 32,
https://doi.org/10.1007/978-3-030-67778-7_15

*(ReLePe), and this dialogue forms the foundation of her chapter. In it, there is a very successful effort to present methodological possibilities for the study of educational policies in relation to identity and subjectivity issues. Reappraising the literature on educational politics, Romilda makes her personal reflections in a wider framework than is discussed in Brazil.*

*Some central tensions in analyses of and for educational policies are revisited by Romilda throughout her text: conservation and change; production and implementation; structure and subjectivity. These are not new dilemmas in the field or in Romilda's research trajectory. In her Master's dissertation, as a newcomer, she devoted herself to establishing the relationship between the production and the implementation of educational policies, with a specific interest in implementation and the lived aspect of politics. Since then, her work in the field of teacher training has often examined teachers' professional practices.*

*This set of tensions in Romilda's research is what led her to approach the use of social representations in education. In recent decades, the researcher has been an active participant in the International Centre on the Study of Social Representations and Subjectivity – Education (Centro Internacional de Estudos em Representações Sociais e Subjetividade – Educação, CIERS-ed), hosted at the Carlo Chagas Foundation. There, she also participates in the UNESCO Chair on the Professional Development of Teachers, which greatly intersects with CIERS-ed, given her trajectory as a researcher. It was due to her inputs that the topic of educational policies gained increasing relevance in the studies undertaken at the Chair, and at CIERS-ed this can be seen in the publications stemming from the projects that are already finalized.*

*Social Representations Theory (SRT) has enabled Romilda to seek, rather than solutions, the theoretical foundations of the tensions referred to previously. Citing Moscovici, she highlights the extent to which SRT helps us to think about the relationship between "development and change, rather than [...] reactions to fixed environments" and the "struggle between tradition and innovation".*

*It is also in Social Representations Theory that Romilda seeks elements to study what Stephen Ball denominates context of practice, at the same time as the author explains its inextricable relation with all other contexts of politics. For more than three decades, Ball has been insisting on the negative political effects of studies that separate the production and the implementation of educational policies, which has been recognized by most researchers in the educational field. This recognition, by the way, does not imply that it has been possible to overcome the difficulty to produce research in which the circularity of directions in different contexts becomes clear. The contribution of Romilda's research, which shines through this text, is that "research proposals that study educational policies, based on the contexts proposed by Ball (1997), can broaden their understanding by adding the psychosocial aspects of the field of Social Representations". The bet of the author – following years of diligent work – is that whilst articulating the individual and the collective, Social Representations Theory makes viable a theoretical-methodological alternative which achieves something that we have been seeking to do for a long time in the field of educational policies. At the same time, reorientating*

*Social Representations Theory into a field in which it was not specifically conceived – with both the risks and the 'licence' that such dislocation implies – widens the dialogues that the theory can propitiate and bets on the interdisciplinary desire at its base. These are courageous bets, and the author makes them with diligence and skill, as can be seen in this text.*

## References

Moscovici, Serge, 2003: *Representações sociais: investigações em psicologia social* (Petrópolis: Vozes).
Ball, Stephen J., 2017: *The education debate: policy and politics in the twenty-first century* (Great Britain: University of Bristol Policy Press).

**Abstract** Based on Social Representations Theory (SRT) and educational policies, I aim to demonstrate the fertility of SRT for the analysis of policy texts and how these anchor the representations of teachers and undergraduate students on school and on being a teacher. I adopted a combined qualitative, theoretical-bibliographic and exploratory interpretative approach to support the present study, which yielded reflections that indicate ways to analyse texts about teaching policies through the lens of SRT. From this perspective, I present my attempt to explain the contribution of SRT to the analysis of these texts, since I presume the SR of the teachers on educational policies to be one of the macro-regulators of their craft which allows them to give meaning, signification and resignification to carry out these policies, because they are intrinsically related to ideas, representations, translations and simulacra situated according to the historical, political and economic moment in which the institution and its actors are inscribed.

**Keywords** Social Representations Theory · Education policy · Brazil

## 15.1  Introduction

Understanding the dilemmas that are present in the school space/time through the analyses of social representations and educational policies has been one of the concerns of those who are currently investigating ways to make the complex dynamics of the school in which it occurs comprehensible. There is "always a cross-fertilization of conjunctures, interests and intentions" (Moscovici 2003: 314) which weave and sustain the formation of social representations. In the field of policies we can see processes of rupture, expectations regarding educational reforms, and the definition of school practices that substantiate relationships in or outside the classroom. The tension between change and conservation, experienced by the school, its actors, and the policies in the spaces/times of the school constitute moments of reproduction or transformation (Almandoz/Vitar 2006), anchored in a network of representations of the past and present and on the future, which can be considered "… almost tangible entities" (Moscovici 1978: 41).

The transversality of the notion of Social Representation (SR) which emerges from an interdisciplinary perspective sparks the creation of a symbolic universe shared by individuals in their groups and with others in order to understand teaching policies, since Social Representations Theory (SRT), as a source for the analysis of psychosocial processes, makes it possible to perceive the main tendencies that guide social life, since its scope raises new possibilities for apprehending "what societies think of their ways of living, the meanings that they attach to their institutions and the images they share" (Moscovici 2003: 173).

In the policy field, the tension established between changing and preserving expresses a symbolic legacy, which (re)establishes glimpses of the past, present and future, since the teaching policies derive from "an interlaced complex with multiple influences" (Tello/Mainardes 2015: 44), because they are socially constructed by diverse actors and interests that express power struggles.

Based on SRT and this look at educational policies, I aim to demonstrate the fertility of SRT, proposed by Serge Moscovici (1961–2012), for the analysis of policy texts and how these anchor the representations of teachers and undergraduate students on school and on being a teacher. I used a combined qualitative, theoretical-bibliographic and exploratory interpretative approach to undertake the present study, which led to reflections that indicate ways to analyse teaching policy texts with the help of SRT. From this perspective, I present my attempt to explain the contribution of SRT to the analysis of these texts, since I presume the SR of the teachers on educational policies to be one of the macro-regulators of their craft which allows them to give meaning, signification and resignification to carry out these policies, because they are intrinsically related to ideas, representations, translations and simulacra situated according to the historical, political and economic moment in which the institution and its actors are inscribed.

The actors of the school scenarios in their times/spaces construct representations, and their interpretation from the theoretical contribution of SRT enables them to attend to the complexity of factors pertinent to their constitution and the respective contexts, as well as the role of the State in guaranteeing the right to education, aspects defined and constructed socially by various actors and interests, which in turn emphasize the importance of "recognizing that different policies, or more precisely the types of policies, position and produce teachers as different types of subjects of policies" (Ball et al. 2016: 131).

## 15.2 Policy

The concept of policy – of Greek origin, *pólis (politikós)* – means "everything that refers to the city and, consequently, what is urban, civil, public, and even sociable and social". What we are experiencing, however, is "State science, State doctrine, political science, etc. It is commonly used to indicate the activity or set of activities, which in some way has the *pólis* reference term, that is, the State" (Bobbio et al. 1998: 954).

Today, it can be said that policy is "a form of activity closely linked to the exercise of power and to the multiple consequences of this exercise, in which interests are transformed into objectives and objectives directed to the formulation and decision-making" (Vieira 2007: 141). To exercise this power, governments establish policies – economic, social, educational, housing, among others – through plans, projects, and programmes in the form of government strategies, directed to guide the processes and changes, which become "[...] guidelines or lines of action that define or guide practices, such as norms, laws and guidelines" (Ferreira 2008: 88). These actions or public policies are "... the result of the dynamics of the game of forces established within the framework of power relations, relations constituted by economic and political groups, social classes and other organizations of civil society" (Boneti 2011: 18). A correlation of forces from different sources – global, national and local – that is, many of them imposed by international organizations, civil society organizations and social movements, are felt in the dynamics of the productive structure and policies in a nation.

It is, in this sense that we take Education as a social public policy, which "is situated temporally, culturally and spatially within a certain form of State, suffering interference from it, although it is not thought only by its organisms, but by the society, its institutions and interest groups and influenced by international organizations" (Werle 2010: 57).

For the accomplishment of policy studies in the United States, around 1958, policy discipline emerged (Lasswell[2] 1984), as policy science, structured in three dimensions: polity, which emphasizes the legal, political administrative system of a country's political system; politics, which delineates political processes, permeated by political negotiations for the formulation of goals and agenda items, usually conflicting; and policy, through which the contents of the policy, materialized in government programmes, with goals and objectives to be fulfilled through actions (implementation/enacted), define its scope. By intertwining and influencing each other, these dimensions support the explanation of policy practice, shaping a rational view of policies (Martins 2013). This proposal is being followed by courses of Higher Education Institutions (HEI) that propose to research and analyse policies.

In order to understand and analyse policies, I selected as axes two dimensions – "analysis of policies" and "analysis for policies" – the first in order to understand policy and the policy-making process, the second to study and evaluate how we seek to identify the impact that policies have on the population, which by itself, given the descriptive and prescriptive aspects, marks the borderline between "analysis of policies and analysis for policies" (Ham/Hill 1993). The aim of these is to study forms

---

[2]Harold Dwight Lasswell (1902–1978) was a sociologist, political scientist and communication theorist at the University of Chicago, and is considered one of the founders of policy psychology. See: Lerner/Lasswell (1951) and Lasswell (1984).

of State management, the tensions between work and employment, crises of professional, personal and collective identities, trajectories of formation, insertion into the world of work, and ethical and moral values. From the social and economic point of view, these require new methodological designs focused on the analysis of everyday aspects in educational institutions (micro policies and/or micro-sociology). These analyses oscillate "between knowledge in the policies process and knowledge of the policies process, shaping three trends: the analysis of public policies based on theories of the State; explanations of how public policies operate and, finally, the evaluation of their impact" (Martins 2006: 383)

In short, policies can be analysed in different dimensions and different ways, but the focus of this examination must be defined and informed by the researcher. In Latin America, several studies focus on policies and/or relations between policies and public (governmental) policies that regulate education, such as, in Argentina, Paviglianiti (1996); and in Brazil, Azevedo/Aguiar (2001), Mainardes (2017a and b, 2018a and b), Mainardes et al. (2017), and Mainardes/Tello (2016). Other studies are more focused on international regulations and interferences that determine the actions of governments, such as Barroso (2005) and Reis (2013), who analyse the relationship between the concept of regulation and governance, besides pointing out how governments are submitted to the processes of globalization. As Gorostiaga (2017) indicates, based on Jans (2007), there are studies that analyse policies in the direction of top down, among them those of Feldfeber (2009), Oliveira (2005), and Miranda et al. (2006). In addition to these, I would like to highlight the work of Shiroma et al. (2005), which, among others, contributes greatly to the analysis of educational policies.

## 15.3   Teaching Policies and Social Representations

People are "at the same time *recipients* of political decisions and subjects as rights holders whose exercise not only imposes limitations on the political decision-maker's scope, but also directly conditions the basis of legitimization of political power" (Duarte 2013: 21–22, emphasis added).

Teaching policies can be understood as social praxis and the relevance of SRs, since they are anchored in multiple social processes by which we can explain the individual/group assignment of meaning, signification and resignification of the individual/group social reality, attributes which interfere in the interactions of the individual in his/her social, political and economic context, currently governed by neoliberal ideologies. These, in turn, define the proposition and execution of policies, such as those of teaching. In this scenario, SRs express the diversity between individuals, groups of individuals when they represent themselves, and how they represent the reality in which they coexist, representations that, when identified, provide the investigator with a fresh perspective on what supports interactions and their own subjects, especially when investigating communicational phenomena, namely those related to their actions.

School institutions and their spaces/time do not exist independently of their social reality, in which teaching policies, public or not, are engendered to regulate the practices of these professionals. Social representations may indicate how teachers in school space/time and undergraduate students experience "preparation for action" (Moscovici 2012: 46) with regard to how they conceive and perceive their professional reality and the policies which define it, since social representations are "[…] specifically defined phenomena that need to be analytically discovered … not only […] as guides to behaviour, but especially because they reshape and reconstitute the elements of the environment in which the behaviour must take place" (Marková 2006: 203).

In order to clarify what he means by social representation, Moscovici (2012: 27) explains: "In a word, social representation is a particular form of knowledge having the function of elaborating behaviours and communication between individuals", as well as their importance in affirming that "… they have a constitutive function of reality, the only reality that we experience and in which most of us move. Thus, social representation is each time the sign and the double of the socially valued object."

Examining policy texts requires the researcher to consider "multidisciplinarity and interdisciplinary complexity" (Tello 2013: 41), since, in order interpret the results, it is necessary to indicate both the positioning and the epistemological approach of the researcher.

The fields of educational policies and SRs are not only enriched by a discipline, they are both part of a multidisciplinary field, sometimes fragmented epistemologically and theoretically and, as Tello (2013) says, we can talk about "epistemologies" of educational policy. In turn, the SRs "imply a critical position and not an ironic position, which can lead to practical engagement", and therefore Moscovici (2003: 378), in dialogue with Markova, pondered, "if science became an ideology, what was ideology was replaced by knowledge or science", which, in his view, was not right. "Epistemology can also be a form of censorship, which in [other situations] may have cost the lives of many. In our times, it only touches on the issue of our intellectual ostracism." However, he does not believe that "a good epistemology, or a good ideology, leads to creativity", since, for Moscovici (2003: 379–380), science and philosophy are forms of art, whereby researchers and artists strive to create new notions, discover phenomena, etc. His ethos is that we should not destroy what we cannot replace. The author argues that criticism without theory is non-real, fictitious, misleading, considering that "men make history, but they do not know what stories they do".

We find that, in the vein of SRT, and according to Jovchelovitch (2008: 162), Moscovici (2000: 229), when seeking "inspiration for what social psychology could be: a science about development and change, rather that on reactions to fixed environments", showed that through social representations we can find "the struggle between tradition and innovation, between conformity and rebellion of active minorities". He also warned that when a new idea of scientific knowledge enters society, we perceive in cultural life what the Germans call "*kulturkampf*, cultural struggles [or fight for ideas], intellectual polemics and opposition between different modes of thought", which are responsible for "the process of transforming knowledge and the birth of a new social representation".

Situated contexts, configured by the multiple effects of specific social, economic, political and cultural processes, demand to be looked at from their local perspectives where historical, sociological and psychosocial aspects converge, making it possible to understand phenomena such as education and educational policies that are being presented in a structured way at a global, national and local level, so that scientific knowledge directs the orientation of educational practices and the definition of their policies. However, as indicated by Palamidessi et al. (2014: 51), "Educational research takes place in different institutional contexts and in terms of a complex relationship and many forms, not only in the field of academic knowledge of social and human sciences, but also in the field of technical-bureaucratic knowledge and pedagogical practices."[3]

By its very nature, research on social representations, education and educational policies requires multidisciplinary knowledge, as well as psychosocial tools, ranging from sociology to economics, in order to gain an understanding of the increasingly complex reality, the simpler and more restricted situations and phenomena, and the subjects or groups to be researched and examined. Such investigations need to have their multi/multidisciplinary character emphasized. Therefore, in order to contribute to the elaboration and implementation (enactment) of educational policies that meet the reality of the school, the student and the teacher, they cannot be uncritical, one-dimensional or decontextualized, but must be imbricated to social structures.

Studying educational proposals under the focus of social representations is analogous to what Moscovici (2003: 374–375) says about innovation theory, taking this as "a theory of influence [... and] a theory of communicative process that usually happens between supporters and opponents from different points of view" when we perceive understand communication as a social process and a social institution, as Moscovici (2003: 375) did in clarifying that the purpose of communication is not only to change people's minds, but, more than that, to constitute them. Moreover, if persuasion in communicative relations "is the part of the process that is related to the change of people", advertising is not at the same level, because it constitutes "something that an institution does continuously and changing the minds of people is part of it". For this reason, it is ritualistic in the exercise of its function to maintain or impose a *status quo*, such as those of an institution. For example, in the Brazilian high school reform proposal, official propaganda seeks to fortify representations that this policy will bring benefits to students, despite the resistance of various social segments, class entities and academics, among others, to such proposal.

---

[3]Translation from the original: "La investigación educativa se desarrolla en contextos institucionales diversos y en función de una compleja y multiforme vinculación, no sólo en el campo del saber académico de las Ciencias sociales y las Humanidades, sino también en el terreno de los saberes técnico-burocráticos y de las prácticas pedagógicas".

In their turn, processes of influence, says Moscovici (2003: 376) "operate, largely through persuasive communication", making use of science and ideology. As we can infer from the studies of Ball (1997), the effects and impacts of these policies on practices developed in the contexts of basic schools and/or universities can be better understood. In order to achieve this, we must realize that the enunciation of public policies is prescriptive in defining how organizations and society should structure themselves, even if they do not necessarily convey the representations of the subjects/groups for which they are intended, since, in the majority of cases they are often influenced by international organizations rather than by discerning, critical and situational evaluations of the proposition and implementation of policies, programmes, and actions, or by research.

Because they are based on these social and cognitive processes, policies anchor representations that function as macro-regulators, as great social constants (Campos 2003). The texts of public policies can thus be regarded "as products and producers of policies' orientations" (Shiroma et al. 2005: 427). In order to analyse the policy texts, I rely on Ball's (1997) proposal regarding the "policy cycle" constituted by a network of non-linear influences, which suggests analysing policy texts according to their social context (Table 15.1).

Neither the policy cycle nor social representations can be defined independently of their context. In other words, they must be in relation to something, as Moscovici (2003: 356) explains. "Such representations serve people on the one hand as paradigms in communication, and on the other as means of practical guidance." Hence, the author argues, if "the basis for knowledge [is not] a process of struggle and persuasion in the course of human history […this] raises another question: how is knowledge shared?"

Analysing the influence context which permeates the other contexts proposed by Ball is easier if we agree with Markova's summary, when dialoguing with Moscovici (2003: 367):

[…] processes of influence are based on two related ideas. In the first place, influence is exercised in two directions and reciprocal: 'from the majority towards the minority and from the minority towards the majority'. Secondly, following from the first, 'each part of a group must be considered to emit and receive influence simultaneously … each individual or subgroup … is at the same time being influenced or influencing, others at the same time, whenever there is influence'. [Emphasis added in the original]

Although Ball's studies (1997, 1998, 2017) are based on sociological analyses of policies, proposed through legislation, local or global, the reading of his texts impels interpretations, re-readings that can strengthen or not the senses of his proposition, since these are "reproduced and reworked over time through reports, speeches, 'moves', 'agendas' and so on" (Ball, 2017: 7). Because they are understood more as 'processes' and less as 'objects', policies are not translated directly into practice, but rather "contested, interpreted and enacted in a variety of arenas of practice". Therefore, according to Ball, in order to analyse educational policies from a sociological perspective and to be able to understand them, one must focus on concepts, ideas, arguments and investigations, since they can be experienced in different ways.

**Table 15.1** Contexts of analysis by the policy cycle. *Source*: Organized by the author based on Ball (1997: 27)

| Context | Characteristics |
| --- | --- |
| Influence context | Marked by the moment in which political speeches are produced at the level of political institutions |
| Policy text production context | This encompasses the production of different texts (normative and mediatic/not), most of which are indisputably contradictory and sometimes inconsistent, in an attempt to articulate diverse opinions and alliances |
| Practice context | A conjuncture in which practitioners intervene, adapting/ changing, resisting and sometimes rejecting what is imposed on them |
| Results or effects context | This involves situations relating to issues of justice, equality and individual freedom, in which policies are analysed in terms of their impact and in relation to existing inequalities. The effects of the policy can clearly be seen when changes and data based on observations of the practice are analysed. |
| Policy strategy context | This involves social and political activities that may contribute to the resolution of problems that may, for example, be caused by policies and practices of teacher formation and professionalization |

I concur with Ball, adding to the sociological positioning of educational policy a psychosocial perspective analysis that considers elements of subjectivity, assuming that the analyses proposed by Ball and collaborators (1997, 1998, 2017) help to demonstrate how representations of educational policies are constructed and anchored in social constants. In his discussion with Markova (2003: 374), Moscovici explains that in the formation of a representation there is a "transference of conjunctures, interests and intentions" that, according to Frankel (1987), do not translate into "a simple transference of but rather in a conjunctural transference of different global or local sets of interests that intersect with values, beliefs or patterns of behaviour, cultures of particular groups or individuals" (Ens et al. 2018: 104). Similarly, Jodelet (2007: 16) describes the production of representations in the field of education that result from a "situation of social symbiosis",[4] because they are embedded in a social and institutional context which defines the school, for example, as a space for the formation of the citizen, and of the teacher himself/herself.

Based on this design drawn up by educational policies, representations of individuals or groups portray regulations that, when put into practice, generate teacher accountability, such as, for example, the failure to achieve goals in institutional assessments, which points to rankings and becomes an arena for disputes between schools. Additionally, policies which comply with the guidelines of international organizations are focused only on aspects related to teacher formation. These

---

[4]Expression used by Schaff (1960).

regulations are anchored in the so-called "fourth industrial revolution",[5] also called Smart Factory, or Intelligent Factories, which, because it is anchored in the demands of political, economic, cultural and social aspects, brings this formation closer to that of the productive sector.

They are reforms which began in the 1990s and continue in the present century, since, according to Evangelista/Shiroma (2007: 537), they reach "all spheres of teaching: curriculum, textbook, initial and continuing formation, career, certification, locus of formation, use of information and communication technologies, evaluation and management". These policies reflect the precariousness and intensification of teaching work, with the extension of the responsibilities they attribute to the performance of the functions of 'being a teacher', professionals who, according to the authors, must now "attend to more students in the same class, sometimes with special needs; act as a psychologist, social worker and nurse; [...] educating for entrepreneurship, peace and diversity; [...] be involved in the elaboration of strategies to raise funds for the school" (Evangelista/Shiroma 2007: 537).

Given the exposition, I suggest that, based on the contexts proposed by Ball (1997), research proposals that study educational policies can broaden their understanding by adding the psychosocial aspects of the field of social representations, since it is in the space/time of the (re)articulations of the individual and the collective that the political texts, the epistemological tensions and the possibilities to realize them or not are situated, because one field is not reduced to another.

In this vein, Jodelet (2007) points out that SRT provides a viable path for the understanding of scientific knowledge, deepening awareness of subjectivity and representations (individual and/or groups) and indicating the most prominent conflicts and dilemmas, thereby helping to improve the political processes of teacher training. Both fields – social representations and policies – are subject to political changes, as Shiroma et al. point out (2005: 436), supported by Ball[6] (1994: 16), since it is "crucial to recognize that policies are products of compromises at various stages, in the micropolitics of the legislation formulation in the debate between parliamentarians, and the micropolitics of the interest groups' articulation". And because social representations "are complex phenomena always activated and at work in social life", when they are studied, they provide "informative, cognitive, ideological, normative, beliefs, values, attitudes, opinions, images etc.", as Jodelet (2001: 21) clarifies.

In search of possible dialogues between social representations and educational policies Icontend that people and fields of knowledge cannot be conceived in isolation, "but as active social actors, affected by different aspects of daily life, that develops itself in a social context of interaction" (Jodelet 2017: 118).

---

[5]The concept of the fourth industrial revolution, or industry 4.0, was created by the Germans in 2011. It is directed at new strategies that combine technology and means of production, focusing mainly on the reduction of costs. It is characterized by automation, cybernetic systems, the Internet of Things and cloud computing (Schwab 2016; Azevedo et al. 2001).

[6]Shiroma et al. (2005) discuss the conception of policies, based on the work of Stephen J. Ball (1994).

In pointing out the importance of these possible dialogues between Social Representations Theory and educational policies for the fields of Education and Politics, Valentim's arguments (2013: 163–164) indicate some of the reasons that show the importance of this dialogue for the analysis of data: "their paradigmatic characteristics [... are situated] in a unique position [...] for the understanding of human phenomena in contemporary societies", through "the central idea of a meta-system of social regulations that intervenes in the cognitive system", because it allows us to understand the "change of social representations and [...] of policies as well as [being] a useful concept for understanding the processes of transformation in societies and communities".

Despite the possibilities, this dialogue implies that it is necessary to continue investigating, even if it entails "transgressing accepted theoretical limits" (Zemelman 2006: 462), keeping an open mind, and making a commitment to avoid "any closure of meanings, so as not to incur easy extrapolations, which almost always prove to be fallacious". Historicity "obliges us to constantly trace the specificity assumed by phenomena", and "to conceive the moment as part of an unfinished process, open to its consequences, within which its development [allows us] to recognize different directions" (Zemelman 2006: 463).

I agree with Ruivo (2006: 598) that in this new research paradigm, "the application of scientific knowledge will no longer be simply technical, it will be edifying, since it does not prescind technical applications, but it is subject to ethical know-how" (Santos 1989).

# References

Almandoz Maria Rosa, Vitar Ana, 2006: "Caminhos da inovação: as políticas e as escolas", in: Vitar, Ana; Zibas, Dagmar; Ferretti, Celso; Tartuce, Gisela Lobo B.P. (Eds.): *Gestão de inovações no ensino médio* (Brasília, DF: Líber Livro Editora): 12–36.

Azevedo, Janete Maria Lins de; Aguiar, Márcia Angela, 2001: "A produção do conhecimento sobre política educacional no Brasil: um olhar a partir da ANPED", in: *Educação & Sociedade*, 22,77: 49–70.

Ball, Stephen J., 1994: *Education reform: A critical and post-structural approach* (Buckingham: Open University Press).

Ball, Stephen J., 1997: *Education reform: A critical and post-structural approach* (Oxford: Open University Press).

Ball, Stephen J., 2017: *The education debate: Policy and politics in the twenty-first century* (University of Bristol: Policy Press).

Ball Stephen J.; Maguire, Meg; Braun Annette, 2016: *Como as escolas fazem as políticas: atuação em escolas secundárias* (Ponta Grossa: Editora UEPG).

Barroso, João, 2005: "O estado, a educação e a regulação das políticas públicas", in: *Educação & Sociedade*, 26,92: 725–751.

Bobbio, Norberto; Matteucci, Nicola; Pasquino, Gianfranco, 1998: *Dicionário de política* (Brasília: Ed. Universidade de Brasília).

Boneti, Lindomar Wessler, 2011: *Políticas públicas por dentro* (Ijuí: Ed. Unijuí).

Boneti, Lindomar Wessler, 2012: "As políticas públicas no contexto do capitalismo globalizado: da razão moderna à insurgência de processos e agentes sociais novos", in: *PRACS: Revista Eletrônica de Humanidades do Curso de Ciências Sociais da Universidade Federal do Amapá*, 5 (December): 17–28.

Duarte, Maria Luisa, 2013: *União europeia e direitos fundamentais* (Lisboa: AAFDL).

Ens, Romilda Teodora; Oliveira, José Luis; Naguel, Jaqueline Salanek Oliveira; Couto, Edina Dayane Lara, 2018: "Representações sociais de estudantes de licenciatura: 'a corda bamba' do futuro", in: Sousa, Clarilza Prado de; Marcondes, Anamérica Prado; Jardim, Anna Carolina Salgado; Coêlho, Vanusa dos Reis (Eds.): *Qual futuro? Representações sociais de professores, jovens e crianças* (Campinas, SP: Pontes Editores): 67–96.

Evangelista, Olinda; Triches, Jocemara, 2014: "Professor: a profissão que pode mudar o país?", in: Evangelista, Olinda (Ed.): *O que revelam os slogans na política educacional* (Araraquara, SP: Junqueira & Marin Editores): 47–82.

Evangelista, Olinda; Shiroma, Eneida Oto, 2007: "Professor: protagonista e obstáculo da reforma", in: *Educação e Pesquisa*, 33,3 (September–December): 531–541.

Feldfeber, Myriam, 2009: "Nuevas y viejas formas de regulación de los sistemas educativos", in: Feldfeber, Myriam (Ed.): *Autonomía y gobierno de la educación: perspectivas, antinomias y tensiones* (Buenos Aires: Aique): 25–50.

Ferreira, Naura Syria Carapeto, 2008: "A gestão da educação e as políticas públicas de formação de profissionais da educação: desafios e compromissos", in: Ferreira, Naura Syria Carapeto (Ed.): *Gestão democrática da educação: atuais tendências, novos desafios* (São Paulo: Cortez Editora): 119–140.

Frankel, Boris, 1987: *Los utopistas postindustriales* (Buenos Aires: Ediciones Nueva Visión).

Gorostiaga, Jorge, 2017: "La formación de investigadores en el campo de la política educativa: problemas y desafios", in: Ens, Romilda Teodora; Ribas, Marciele Stiegler; Favoreto, Elizabeth Dantas de Amorim (Eds.): *Políticas educacionais, representações sociais e docência na interface com a escola* (Curitiba: CRV): 29–42.

Ham, *Christopher*; Hill, Michael, 1993: *The policy process in the modern capitalist state* (London: Harvester Wheatsheaf).

Jans, Martin Theo, 2007: "Politics beyond the state", in Deschouwer, Kris; Jans, M. Theo (Eds.): *Politics beyond the state: Actors and policies in complex institutional settings* (Bruselas: VUB University Press): 7–24.

Jodelet, Denise, 2001: "As representações sociais: um domínio em expansão", in: Jodelet, Denise (Ed.): *As representações sociais* (Rio de Janeiro: EdUERJ): 17–44.

Jodelet, Denise, 2007: "Contribuições das representações sociais para a análise das relações entre educação e trabalho", in: Pardal, Luís; Martins, Antonio; Sousa, Clarilza; Dujo, Angel Sel; Placco, Vera: *Educação e trabalho: representações, competências e trajetórias* (Aveiro: Editora Universidade de Aveiro): 11–26.

Jodelet, Denise, 2017: *Representações sociais e mundos de vida* (Curitiba/São Paulo: Ed. PUCPRess/Fundação Carlos Chagas).

Jovchelovitch Sandra, 2008: *Os contextos do saber: representações, comunidade e cultura* (Petrópolis, RJ: Vozes).

Lasswell, Harold, [1936] 1984: *Politics: Who wins, what, when, how* (Brasília: UNB).

Lerner, Daniel; Lasswell, Harold (Eds.), 1951: *The policy sciences* (Stanford: Stanford University Press).

Mainardes, Jefferson, 2017a: "A pesquisa sobre política educacional no Brasil: análise de aspectos teórico-epistemológicos", in *Educação em Revista*, 33: 1–25.

Mainardes, Jefferson, 2018a: "Reflexões sobre o objeto de estudo da política educacional", in: *Sorocaba*, 4,1 (January–April): 186–201.

Mainardes, Jefferson, 2018b: "A pesquisa no campo da política educacional: perspectivas teórico-epistemológicas e o lugar do pluralismo", in: *Revista Brasileira de Educação*, 23: 1–20.

Mainardes, Jefferson; Tello, César, 2016: "Research on the field of education policy: Exploring different levels of approach and abstraction", in: *Archivos Analíticos de Políticas Educativas*, 24,75: 1–14.

Mainardes, Jefferson; Stremel, Silvana; Rosa, Gregory; Luis, Rolim, 2017: "A pesquisa sobre a disciplina política educacional no Brasil: situação e perspectivas", in *Revista Brasileira de Política e Administração da Educação*, 33: 287–307.

Marková, Ivana, 2006: *Dialogicidade e representações sociais: as dinâmicas da mente* (Petrópolis: Vozes).

Marková, Ivana, 2017: *Mente dialógica: senso comum e ética* (São Paulo/Curitiba: Fundação Carlos Chagas-PUCPRess).

Martins, Angela Maria, 2006: "A agenda da educação nos tempos atuais: considerações sobre o cenário e as políticas de formação docente", in: *Revista Iberoamericana de Educación*, 39: 1–12.

Martins, Angela Maria, 2013: "O campo das políticas públicas de educação: uma revisão da literatura", in: *Estudos de Avaliação Educacional*, 24,56 (September–December): 276–299.

Miranda, Estela María; Lamfri, Nora Zoila Senén; González, Silvia N. de; Nicolini, Mariana Alejandra, 2006: "Construcción de la regulación política en educación en la década postreforma: procesos emergentes y efectos en los sistemas educativos provinciales", in: *Cuadernos de Educación*, 4: 47–58.

Moscovici, Serge, 1961: *La psycanalyse, son imagem et son public* (Paris: PUF).

Moscovici, Serge, 1978: *A representação social da psicanálise* (Rio de Janeiro: Zahar).

Moscovici, Serge, 2000: "Ideas and their development: A dialogue between Serge Moscovici and Ivana Marková", in: Moscovici, Serge (Ed.): *Social representations: Explorations in social psychology* (Cambridge: Polity Press): 208–286.

Moscovici, Serge, 2003: *Representações Sociais: investigações em psicologia social* (Petrópolis: Vozes).

Moscovici, Serge, 2012: *A psicanálise, sua imagem e seu público* (Petrópolis: Vozes).

Oliveira, Dalila Andrade, 2005: "Regulação das políticas educacionais na América Latina e suas consequências para os trabalhadores docentes", in: *Educação & Sociedade*, 26, 92: 753–775.

Palamidessi, Mariano; Gorostiaga, Jorge; Suasnábar, Claudio, 2014: "El desarrollo de la investigación educativa y sus vinculaciones con el gobierno de la educación en América Latina", in: *Perfiles Educativos*, 36,143: 49–66.

Reis, Isaura, 2013: "Governança e regulação da educação: perspetivas e conceitos", in: *Educação, Sociedade e Culturas*, 2,39: 101–118.

Ruivo, Maria da Conceição, 2006: "A ciência tal qual se faz ou tal qual se diz?", in: Santos, Boaventura de Sousa (Ed.): *Um conhecimento prudente para uma vida decente* (São Paulo: Cortez).

Shiroma, Eneida Oto; Campos, Roselane Fátima; Garcia, Rosalba Maria Cardoso, 2005: "Decifrar textos para compreender a política: subsídios teórico-metodológicos para análise de documentos", in: *Perspectiva*, 23,2 (July–December): 427–446.

Schwab, Klaus, 2016: *A quarta revolução industrial* (São Paulo: EDIPRO).

Tello, Cesar; Mainardes, Jefferson, 2015: "Políticas docentes na América Latina: entre o neoliberalismo e o pós-neoliberalismo", in: Ens, Romilda Teodora; Villas Bôas, Lucia; Behrens, Marilda Aparecida (Eds.): *Espaços educacionais: das políticas docentes à profissionalização* (Curitiba/São Paulo: PUCPRess/FCC): 31–62.

Tello, Cesar, 2013: "La epistemologia de la política educativa-notas históricas y epistemológicas sobre el campo", in: Tello, Cesar (Ed.): *Epistemologías de la política educativa: posicionamentos, perspectivas y enfoques* (Campinas: Mercado de Letras): 23–68.

Valentim, Joaquim Pires, 2013: "Que futuro para as representações sociais?", in: *Psicologia e Saber Social*, 2,2:158–166.

Vieira, Evaldo, 2007: *Os direitos e a política social* (São Paulo: Cortez).

Werle, Flávia Obino Corrêa, 2010: "Reforma, inovação e mudança: delineando questões na área de políticas educacionais", in: Martins, Angela Maria; Werle, Flávia Obino Corrêa (Eds.): *Políticas educacionais: elementos para reflexão* (Porto Alegre: Redes Editora): 49–62.

Zemelman, Hugo, 2006: "Sujeito e sentido: considerações sobre a vinculação do sujeito ao conhecimento que constrói", in: Santos, Boaventura de Sousa (Ed.): *Conhecimento prudente para uma vida decente: um discurso sobre as ciências revisitado* (São Paulo: Cortez): 457–468.

# Chapter 16
# Social Representations of Violence among Public School Students

Andréia Isabel Giacomozzi, Amanda Castro, Andrea Barbará da Silva Bousfield, Priscila Pereira Nunes and Marlon Xavier

## Introductory Comment

Brigido Vizeu Camargo

*This chapter illustrates the dynamism of[1] the new generations in the Social Psychology of Communication and Cognition Laboratory (LACCOS). One characteristic of this Lab, which specializes in the study of social representations, is teamwork. As authors, we have two professors from the Psychology Department, our coordinator Andrea Barbará da Silva Bousfield and Professor Andréia Isabel*

Andréia Isabel Giacomozzi is a professor, on the postgraduate programme in psychology in the Psychology Department of the Federal University of Santa Catarina (UFSC), Brazil. She is a member of the Social Psychology of Communication and Cognition Laboratory – LACCOS. Email: agiacomozzi@hotmail.com.

Amanda Castro is a professor at the University of the Far South of Santa Catarina (UNESC) in group training and in the specialization on psychodrama at the Viver Psicologia Psicodrama School. She is a undertaking her PhD in psychology via the graduate programme of the Federal University of Santa Catarina (UFSC). She is also a member of LACCOS. Email: amandacastrops@gmail.com.

[1]Brigido Vizeu Camargo is a pioneer in the study of social representations in Brazil, and a professor on the postgraduate programme in psychology in the Psychology Department at the Federal University of Santa Catarina (UFSC), where he founded the Laboratory of Communication and Cognition (LACCOS). He trained in psychology and undertook his doctoral and postdoctoral studies in France. His contributions to the Brazilian Society of Psychology (SBP) and the Ibero-American Federation of Associations of Psychology (FIAP) reflect his interest in bridging social psychology in Brazil, France and Italy. He is currently interested in analysing the diffusion of Social Representations Theory in Brazil, as well as the construction of ideas and ideologies as representational systems. Email: brigido.camargo@yahoo.com.br.

© Springer Nature Switzerland AG 2021
C. Prado de Sousa and S. E. Serrano Oswald (eds.),
*Social Representations for the Anthropocene: Latin American Perspectives*,
The Anthropocene: Politik—Economics—Society—Science 32,
https://doi.org/10.1007/978-3-030-67778-7_16

*Giacomozzi. They are two former LACCOS alumnae. Other LACCOS members and collaborators have also taken part in this research: Professor Marlon Xavier, PhD student and Professor Amanda Castro, and undergraduate student Priscila Pereira Nunes.*

*The presence of former alumnae and of a number of collaborators of our Social Psychology of Communication and Cognition Laboratory highlights the continuity and vitality of the research group's work on social representations. Yet, as can be observed from the theme of this chapter, it is not only about continuing the work that is being done, but also about changing and updating the understanding of social phenomena.*

*The theme of this research is part of one of LACCOS's new lines of research: the social representations of risks. Such a line is the end result of an extension of the old lines of research on the social representations of chronic diseases, resulting from the contribution two of the LACCOS professors who are authors of this chapter. The study of chronic diseases led the research group to engage with the problem of the social representations of risks, and the aforementioned authors increased the project's scope beyond the risks only connected with health issues. The new line of research also encompasses environmental risks (issues related to water and water pollution, natural disasters, etc.) and social risks, such as the manifold forms of violence.*

*This chapter deals with a significant social risk: violence. The study on the social representations of violence among teenagers, albeit of an exploratory nature, describes what young students from a public school in a state capital in the south of Brazil think about such social phenomenon.*

*As the authors stress, in order to promote a culture of non-violence it is necessary to understand the theories of everyday life regarding violence, above all among youngsters, who are the main victims and agents of violence, for, in a certain sense, such theories may be conducive (or not) to the dissemination of this phenomenon in the social space. Thus this study represents an initial approach towards the youngsters' social representations of violence.*

Andrea Barbará da Silva Bousfield is a professor on the postgraduate programme in psychology in the Psychology Department at UFSC, Brazil. She is also a member of LACCOS. Email: andreabs@gmail.com.

Priscila Pereira Nunes is an undergraduate student in the Psychology Department at UFSC, Brazil. She is also a psychology assistant at the Júnior Autojun Company, involved in the Scientific Initiation Scholarship Programme (PIBIC), and a member of LACCOS. Email: priscila.ppnunes @gmail.com.

Marlon Xavier is a substitute professor in the Department of Psychology, Federal University of Santa Catarina (UFSC), and an associate professor (on sabbatical) at the University Caxias do Sul. He is a member of LACCOS. Email: marlonx73@gmail.com.

*The study produced two important findings. Violence is mostly associated with its visible manifestations, and is seen as a monocausal phenomenon by the students. A large proportion of them seem to believe that an internal disposition, of character, is behind acts of aggression and destruction.*

*On the one hand, developing the studies on this phenomenon is needed so as to observe systematically the dynamics of the production of social representations of violence. On the other hand, it is important to verify whether there is – and if so, how it occurs – interference between the social representations and the youngsters' living experiences of concrete violent situations – not only physical violence, but also its less visible forms, such as those of a psychological, moral, and ethical nature.*

**Abstract** Violence is a social phenomenon, for it exists in specific contexts and is effected in relationships with others, due to socio-economic, political and cultural factors. This study, of a quantitative, qualitative and descriptive nature, aimed to investigate the social representations of violence among 349 students from Florianópolis, Brazil, through a self-administered questionnaire. The participants' average age was 16 years and 3 months old. It was found that violence is part of the students' everyday life. Their elaboration and dissemination of social representations of violence focused on three main aspects: physical violence; verbal violence; and internal causes (perpetrator characteristics) and external causes (social inequality, drugs, and games) of violence. It is important to know the social representations of violence and their associated factors in the context of adolescents' daily lives in order to contribute to preventive works and to promote a culture of peace in schools.

**Keywords** Social representations · Violence · Teenagers · School · Education

## 16.1  Introduction

Violence is a social phenomenon (Campos et al. 2004; Moser 1991; Velho 2000), for it exists in specific contexts and is effected in relationships with others, due to a network of socio-economic, political and cultural factors (Boulding 1981; Domenach 1981) that are articulated and become manifest in the life conditions of social groups. It is generally understood as an expression of conflict within a power dynamic (Tavares dos Santos 1999, 2004; Velho 2000).

In Brazil, external causes of death, which include accidents and violent events, are the leading causes in people aged 1–49 (Souza et al. 1997; Minayo/Souza 1999), and also account for years of potential life lost (Reichenhein/Werneck 1994). Violent death (homicide, suicide, and violent accidents) is now the leading cause of mortality in the youth group (aged 15–24), according to the Health Ministry's *Mortality Information System* (SIM 2011).

For the teenager who is living through a process of socialization, a violent context may promote what Elias (1994) called a social modelling of teenagers according to adult patterns – for it embeds the formation of conceptual and attitudinal parameters or frameworks based on knowledge and attitudes structured upon common sense (*sens commun*). This can generate social representations that will provide rationales and underlie practices in relation to social reality, which may reaffirm violent styles of socialization. Such representations, which are built in a spontaneous and practical manner by individuals based upon their attitudes and social actions in daily life, can alienate youngsters – as well as a large portion of the population – from the right to advance toward a higher stage of knowledge (Moscovici 2000); knowledge regarding both violence, as an object, and interpersonal relationships, as actions among subjects.

According to Jodelet (1989), a social representation always symbolizes something, i.e. it is a symbolic structuration of the object that it represents. For Moscovici (1961), it is through social representations that humans make their physical and social reality intelligible. According to Abric (2003), this reality is re-appropriated by the individual or group, reconstructed in their cognitive system, and integrated into their system of values, in accordance with their social history and context. Therefore, social representations are constituting aspects of reality: in elaborating a certain representation, individuals will try to create a reality that legitimizes predictions and explanations that stem from the representation (Moscovici/Hewstone 1984).

For Abric (2003), social representations are organized around central and peripheral elements; the first are more resistant to change. The central core of a representation is formed by normative elements (social patterns and ideologies) and functional elements (descriptive characteristics and social conducts). The closer the individual is to the object of representation, the more the central core becomes functional. The peripheral elements are organized around the core elements, and are more accessible and concrete; they play an essential role in the adaptation of the representation to contextual changes. As the central core is more stable, the peripheral elements constitute the mobile or flexible aspect of the representation; thus the transformation of a representation occurs first in relation to its peripheral elements (Abric 2003).

In studies on the social representation of violence in schools (Abramovay/Rua 2002; Campos/Guimarães 2003; Campos et al. 2004), violence is represented as a phenomenon embedded in life experiences in the context of both the school and the teenagers' daily lives, which include physical violence, use of weapons, influence of gangs and drugs, lack of security, and police brutality. In a study by Guimarães and Campos (2007) on social representations of violence among teenagers, violence was seen in situations marked by physical aggression, likely to be trivialized, and even expected in certain contexts by the teenagers, as a social norm.

In a study with adolescents using images (Ribolla/Fiamenghi 2007), social representations of violence were associated by children and youngsters with explicit scenes of destruction, crime, politics, and sports. The teenagers identified themselves with such scenes, pointing out the fear and insecurity provoked by the phenomenon of violence. Participants also mentioned that reducing violence requires co-operation and awareness-building. A research study with teenagers by Melo et al. (2011) found that their social representation of violence was connected to inflicting damage on someone; the physical form of violence was the most quoted one, and the use of drugs pointed as its main cause. The students reported more visible policing as a required strategy, and having a family was their most discussed project.

Considering this context and the issues that revolve around violence and youth, this study proposes to explore and analyse the social representations of violence – and their associated factors – among public school students in a southern Brazilian capital, with the aim of subsidising and grounding articulated actions by health and education professionals in the context of the governmental School Health Programme (*Programa de Saúde na Escola* – PSE) implemented in Florianópolis, especially with regard to its health promotion and prevention component, which deals with the issues of non-violence and Culture of Peace in schools.

## 16.2    Method, Participants, Instruments and Data Analysis

- *Method:* This study has a quantitative, qualitative and descriptive nature (Flick 2009).
- *Participants:* Participants were 349 public school students from a southern Brazilian state capital, who took part in the PSE and were in their last year of elementary school or in high school. Their age varied between 13 and 26; average age was 16 years and 3 months old (*SD* 1.3). 55.3% of the participants were girls. Regarding who they lived with, 52.2% lived with both their parents, 20.2% only with their mother, 13.5% with their mother and another legal guardian, 6.3% with their father only, 4% with their grandparents and 3.7% with other legal guardians.
- *Instruments:* A self-reported questionnaire was employed, with both closed-ended and open-ended items, beginning with questions on sociodemographic data such as sex, age, level of education, level of education of the parents, parents' occupation, and with whom the participant currently lives. The other questions addressed participation in fights, drug use, access to games and movies with violent content, and whether violent scenes had been witnessed. The questionnaire ended with an evocation question with the inductive theme "violence".
- *Data analysis:* Answers to closed-end questions were entered into a spreadsheet and analysed with the software *Statistical Package for the Social Sciences* (SPSS-17.0). Descriptive statistical analyses and analyses of association between variables were then employed. A comparison between subsamples

(sex, age, taking part in fights, etc.) was effected through contingency tables (non-parametric statistical tests) and by comparing averages.

- Answers to the evocation question were subjected to structural and similitude matrix analyses, employing the software IRaMuTeQ – *Interface de R pour Analyses Multidimensionnelles de Textes et de Questionnaires*[2] (Camargo/Justo 2013; Ratinaud/Marchand 2012). Similitude analysis is based on graph theory, and allows co-occurrences between words to be identified; its results give an indication of the words' connectivity and help to discern the structure of the representation.

- *Ethical aspects:* The research obtained assent from the Federal University of Santa Catarina (UFSC) Research with Humans Ethics Committee, protocol no. 2178/11, and all its ethical guidelines were respected, following norms contained in Resolution no. 196/1996, Brazilian National Council of Health.


## 16.3   Results

Regarding violence, 81.9% of the participants reported having witnessed violent scenes. With regard to the location of such scenes, 55.7% cited their neighbourhood, 23.5% different places, 9.8% their street of residence, and 7.1% school. Moreover, 32.9% of the participants said they had suffered some kind of discrimination at school; from this group, 38.9% mentioned prejudice due to body weight, 23.8% homophobia, 19.8% racial discrimination, 4.8% physical appearance, and 4% cited unspecified bullying. Crossing the variable sex with the types of discrimination suffered, it was found that girls are more likely to suffer body-weight discrimination (43.9% of the female participants who suffered discrimination, against 33.3% of the males), whereas boys cited homophobic discrimination more often (35% of males, 13.6% of females), as well as racial prejudice (23.3% of males, 16.7% of females). However, there was no significant statistical association between the variables.

With regard to the questions on leisure associated with violence, 42.2% of the participants said they watch violent movies and 29.3% often play violent videogames. Crossing these variables with sex, it was found that 50.3% of the boys and 35.6% of the girls watch violent movies, whereas 44.2% of the males played violent games, against only 17.3% of the females. There was a significant statistical association between these variables, indicating that boys are more likely to watch violent movies [$\chi2 = 6.96$; g l= 1; p = .000] and play violent games [$\chi2 = 28.46$; g l= 1; p < .005].

Regarding drug use, 17.4% of the participants mentioned having used at least once (19.4% boys, 15.8% girls). Moreover, 12.8% said they had taken part in

---

[2]"R-Interface for multidimensional analysis of texts and questionnaires".

**Table 16.1** Diagram of evocations for the inductive term "violence" ($n = 214$)

| AEO < 1.7 | | | | AEO >= 1.7 | | |
|---|---|---|---|---|---|---|
| | Element | $F$ | AEO | Element | $F$ | AEO |
| | Physical aggression | 207 | 1.4 | Verbal aggression | 156 | 1.8 |
| | Harmful act | 63 | 1.4 | Psychological aggression | 39 | 2.1 |
| $f >= 24.3$ | Aggression | 34 | 1.4 | Prejudice | 30 | 2.1 |
| | Disrespect | 21 | 1.3 | Sexual abuse | 17 | 1.9 |
| | Hurting | 6 | 1.5 | Feeling inferior | 13 | 1.7 |
| | Unnecessary reasons | 4 | 1.5 | Being forced (to) | 9 | 2 |
| | Discrimination | 3 | 1.3 | Stealing | 8 | 1.9 |
| $f < 24,3$ | Impose ideas | 3 | 1.3 | Shouldn't exist | 6 | 1.7 |
| | Lack of communication | 3 | 1 | Social inequality | 5 | 1.8 |
| | Lots of things | 3 | 1.3 | Suffering | 5 | 2.4 |
| | Lack of character | 2 | 1 | Playing | 5 | 1.8 |
| | Against the law | 2 | 1 | Drugs | 4 | 2 |
| | Anger | 2 | 1 | Cowardice | 3 | 2 |
| | | | | Bandit | 3 | 2 |

*Source* The authors

violent physical fights (17% boys, 9.3% girls); but there was no significant statistical association between the variables.

With regard to the evocation question, the inductive term "violence" induced 663 evocations of 27 different words. Average evocation frequency was 14.30 and average evocation order (AEO) was 1.7. Based on the analyses employed, it was found that the words that were promptly evoked and with higher frequency were related to objective aspects connected with violent social practices, such as third-party conduct that harms other people, such as: "physical aggression", "act of discrimination", and "aggression", as seen in Table 16.1.

Table 16.1 shows in its upper left quadrant the elements that possibly organise the social representation of "violence". Such elements are the most likely to constitute the central core of the representation, because they are the most frequently cited and the first to be evoked by the subjects (Sá 1996). In this quadrant, the highlighted elements are *physical aggression, harmful act*, and *aggression*.

The elements in the upper right and lower left quadrants occupy an intermediary position, indicating that though they do not constitute the central core, they correspond to its near periphery, being the most accessible part of the representation (Abric 2003). The elements in the upper right quadrant present violence in its non-physical manifestation: *verbal aggression, psychological aggression* and *prejudice. Verbal aggression* is the one with the highest frequency; the other two elements can also be expressed as verbal aggression. Thus it is possible that *verbal aggression* can possibly be seen as a relevant element for the constitution of the social representation of violence.

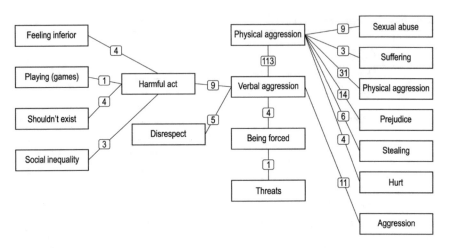

**Fig. 16.1** Tree of elements associated to "violence". Filter 1

Among the elements in the lower left quadrant one can highlight: *disrespect, hurt* (someone) and *unnecessary reasons* (for violence), the first one appearing more often. Such components refer to violence mostly from the perspective of its consequences for the victim: feelings of disrespect, of being hurt, and anger. They also present the attributed causes for violence, namely: *lack of communication, lack of character, lots of things, unnecessary reasons*, and *impose ideas*.

The lower right quadrant refers to the second or distant periphery. Table 16.1 shows that the second periphery is composed, on the one hand, of elements that seem to refer to the causes of violence, such as *social inequality, drugs*, and *playing games*. On the other hand, it is also constituted by elements associated with what violence causes (i.e., its consequences), from the victim's perspective: *feeling inferior, suffering, being forced*, and *shouldn't exist. Stealing* and *sexual abuse* are in this quadrant and seem to represent forms of manifestation of violence. In this quadrant there are also aspects associated with what is thought about the aggressor, i.e., *cowardice* and *bandit* (thug).

The words derived from the free evocations were grouped into categories with semantic proximity, which produced 17 distinct categories, encompassing 87% of the evocations. A similitude matrix analysis was done through IRaMuTeQ software, which generated a graphic representation (tree) of co-occurrences, showing the relation between the categories based on a minimum-co-occurrence filter (number of connections between two specific elements). The size of the vertices is proportionate to word frequency (categories), and the edges indicate word co-occurrence strength (Sá 1996).

Figure 16.1 presents the categories with one or more co-occurrence. The element *discriminatory act* is connected to the elements *feeling inferior, play games, shouldn't exist, social inequality*, and is more strongly connected to the *verbal aggression* element. The element *verbal aggression* appears linked to the elements *harmful act, disrespect, being forced, aggression*, and more strongly to the element physical aggression. The category *physical aggression,* besides being strongly connected to *verbal aggression,* seems to organise the elements *sexual abuse, suffering, psychological aggression, prejudice, stealing,* and *hurt.* The strong connection between the organising element *verbal aggression* and the category *physical aggression* indicates that the participants' social representation of violence is organised around such central elements.

## 16.4  Discussion

This study aimed to identify the social representations of violence among students from public schools that take part in the PSE in Florianópolis, Brazil. It was observed that violence is an integral part of the students' daily lives, pervading their spaces of conviviality, studies, and leisure; moreover, some other associated factors are also present in their everyday life, such as drug use and experiences of discrimination.

*Physical aggression, harmful act* and *aggression* constituted the probable core elements of the social representation of violence, which seems to indicate that violence is mostly associated to its visible manifestations. This perspective was also found in a study by Varela et al. (2012) on the social representations of domestic violence among women. The 30 women interviewed by the authors presented an ampler conceptualisation of what violence is, which was not reduced to physical aggression by the partner, encompassing also psychological, emotional, and symbolic aggression. However, the authors pointed that, in verbalising the experience of suffering violence, the female participants highlighted physical aggression. It is noteworthy that, in an analogous way, in this study 81.9% of our teenage participants affirmed having witnessed violent scenes. Such experience may have influenced the process of anchoring and objectification of the object, violence (Varela et al. 2012).

According to Moscovici (1985), to anchor means to classify. Facing an unknown object, the first step is to resort to classification through a word that already belongs to the individual's usual lingo or language. Thus the representation of violence appears anchored in its typologies, possibly observed by the teenagers themselves, for it is through referring to experiences and thought schemes already established that the object can be thought. Objectification refers to the process through which the elements of a representation are materialised into ideas and meanings and become reality expressions. In this sense, the objectification of violence occurs through aggression and more specifically through physical aggression, concepts which can be materialised because their consequences are more visible socially.

As elements that are close to the central core, the words *verbal aggression, psychological aggression,* and *prejudice* denote the less visible, more socially hidden expression of violence. Although violence is majorly characterized by explicit acts, there appears to be the acknowledgment of its manifestation through acts that do not imply bodily marks. The strong relationship between *verbal* and *physical aggression* indicates that the participants' social representation of violence is organised around these elements. In a similar way, in research developed by Ribolla and Fiamenghi Jr. (2007) with 46 adolescents, violence was characterised by both physical and moral aspects, underlining physical acts against the other (fights, murders), destructive acts towards the environment, manifestations of prejudice (age, race, social class), and also violence against oneself (self-inflicted). Bearing in mind that social practices are transformed and orientated by social representations, and vice versa (Doise 2001), it is possible to argue that, in pointing to psychological aggression and prejudice as characteristics of violence, the teenagers become better able to identify such conduct and act in a preventive way, denaturalizing acts of verbal and psychological violence.

In representing violence, the teenagers also denote the justificatory function of representations, for, according to Abric (2003), if the representations guide behaviours, they also allow one to justify them *a posteriori*. The peripheral elements identified in this study encompass the attributed causes regarding violence occurrence, namely: *social inequality* and *drugs*. Such results were also found in a study by Santos et al. (2010), in which the majority of the 109 adolescents interviewed cited as causes of violence elements such as markers of social class, illegal drugs, lack of dialogue and different values. According to Almeida and Santos (2013), such explanations or rationales of a more sociological nature point to the appropriation of an academic discourse, a reified knowledge, which sees the causes of violence in socio-economic, structural issues, such as drug use.

The teenage participants also pointed to playing games and games in general as associated with violence. It is worth noting that 44.2% of the males and 17.3% of the female participants reported playing violent videogames. The co-occurrence relation between the harmful act and playing games may point at the games as something that is damaging somehow. In such a context, games can be understood by the teenagers as causes of violent behaviour, or else as an expression of violence. However, as the data in Lemos et al. (2014) indicate, even if one considers a parallel between the increases in exposition to violent media and in criminal practices by youngsters, violence in the media is just one of the multiple factors that may contribute to violence; games may thus impact cognitively and psychologically, as both stimulus and elicitor of emotional catharsis.

There also appeared elements associated with what violence causes, from the perspective of its victims: *feeling inferior, suffering*, and *being forced*. It is important to highlight that 81.9% of the participants reported that they had witnessed violent scenes, and 32.9% affirmed having suffered some form of discrimination at school. Cavalcanti et al. (2006) found analogous results investigating the social representations of sexual violence against women, especially in the women's own perspective: social representations were associated with the ideas of suffering and forced sex. Our adolescent subjects, possibly because of their experiences with violent scenes, presented representational elements that signify violence based on the characterization of the feelings attributed to victims of violence, for, according to Doise (2002), positional standings can be anchored in collective symbolic realities, such as value hierarchies and the social experiences that are shared with other individuals.

There are also aspects associated with what is thought about the aggressor, namely: *lack of character, cowardice,* and *bandit* (thug). This seems to present violence as a monocausal phenomenon, which would be more easily explained through searching for one guilty party only. According to Rotter (1966), individuals interpret phenomena by attributing internal or external causes to them; that is, for the person to compose a locus of control (an expectation of internal or external control), he or she needs first to attribute causal relations to events.

## 16.5   Conclusion

This study aimed to investigate the social representations of violence among students from public schools that take part in the PSE in Florianópolis, Brazil. Results indicate that social representations of violence are elaborated and disseminated by the youngsters focusing on three main aspects: (a) physical violence; (b) verbal violence; and (c) internal causes (perpetrator characteristics) and external causes (social inequality, drugs, and games) of violence. Thus, in identifying the social representations and the main forms of violence experienced by the teenagers, this study brings forth new subsidies and practical information for educative practices in schools.

It is important to know the social representations of violence and their associated factors in the context of the adolescents' daily lives, in order to contribute to preventive works and to promote a culture of peace in schools – both being important aspects of PSE work in Brazilian public schools.

## References

Abramovay, Miriam; Rua, Maria das Graças, 2002: *Violência nas escolas* (Brasília, DF: UNESCO).
Abric, Jean-Claude, 2003: "Abordagem estrutural das representações sociais: desenvolvimentos recentes", in: Campos, Pedro; Loureiro, Marcos (Eds.): *Representações sociais e práticas educativas* (Goiânia: UCG): 37–57.
Almeida, Luciana; Santos, Maria de Fátima, 2013: "Representações sociais de violência urbana entre policiais civis", in: *Psicologia: teoria e prática*, 15,2: 76–91.
Boulding, Elise, 1981: "Las mujeres y la violencia social", in: UNESCO: *La violencia y sus causas* (Paris: UNESCO): 265–279.
Camargo, Brigido Vizeu; Justo, Ana Maria, 2013: "IRAMUTEQ: um software gratuito para análise de dados textuais", in: *Temas em psicologia*, 21,2: 513–518.
Campos, Pedro; Guimarães, Silvia, 2003: "Representações de violência na escola: Elementos de gestão simbólica da violência contra adolescentes", in: *III Jornada Internacional e I Conferência Brasileira sobre Representações Sociais* (Rio de Janeiro, RJ: UERJ): 2492–2509.

Campos, Pedro; Torres, Ana Raquel; Guimarães, Silvia, 2004: "Sistemas de representação e mediação simbólica da violência na escola", in: *Educação e Cultura Contemporânea*, 1,2: 109–132.

Cavalcanti, Ludmila; Gomes, Romeu; Minayo, Maria, 2006: "Representações sociais de profissionais de saúde sobre violência sexual contra a mulher: estudo em três maternidades públicas municipais do Rio de Janeiro, Brasil", in: *Cadernos de Saúde Pública*, 22,1: 31–39.

Doise, Willem, 2001: "Atitudes e representações sociais", in: Jodelet, Denise (Ed.): *As representações sociais* (Rio de Janeiro: UERJ): 187–203.

Doise, Willem, 2002: "Da psicologia social à psicologia societal", in: *Psicologia: Teoria e Pesquisa*, 18,1: 27–35.

Domenach, Jean-Marie, 1981: "La violência", in: UNESCO: *La violencia y sus causas* (Paris: UNESCO): 33–45.

Elias, Norbert, 1994: *O processo civilizador* (Rio de Janeiro: Zahar).

Flick, Uwe, 2009: *Introdução à pesquisa qualitativa* (São Paulo: Artmed).

Guimarães, Silvia; Campos, Pedro, 2007: "Norma social violenta: um estudo da representação social da violência em adolescentes", in: *Psicologia: Reflexão e Crítica*, 20,2: 188–196.

Jodelet, Denise, 1989: "Representações sociais: um domínio em expansão", in: Jodelet, Denise (Ed.): *As representações sociais* (Rio de Janeiro, EDUERJ): 17–44.

Lemos, Igor; Gouveia, Raimundo; Alves, Lynn, 2014: "The social representations of violence by electronic games users", in: *Gerais: Revista Interinstitucional de Psicologia*, 7,2: 199–207.

Melo, Monica; Barros, Érika; Almeida, Andréa, 2011: "A representação da violência em adolescentes de escolas da rede pública de ensino do Município do Jaboatão dos Guararapes", in: *Ciência & Saúde Coletiva*, 16,10: 4211–4221.

Minayo, Maria; Souza, Edinilsa, 1999: "É possível prevenir a violência? Reflexões a partir do campo da saúde pública", *Ciência Saúde Coletiva*, 4,1: 7–32.

Ministério da Sáude, 2011: *SIM-Sistema de informações de mortalidade* (Brasil: Ministério da Sáude).

Moscovici, Serge, 1985: "Social influence and conformity", in: Lindzey, Gardney; Aronson, Elliott (Eds.): *The handbook of social psychology* (New York: Random House).

Moscovici, Serge; Hewstone, Miles, 1984: "De la science au sens commun", in: *Psychologie sociale*: 539–566.

Moscovici, Serge, 2000: "Prefácio", in: Guareschi, Pedrinho; Jovchelovitch, Sandra (Eds.): *Textos em representações* (Porto Alegre: Vozes).

Moscovici, Serge, 1961: *A psicanálise, sua imagem e seu público* (Petrópolis: Vozes).

Moser, Gabriel, 1991: *A agressão* (São Paulo, SP: Ática).

Ratinaud, Pierre; Marchand, P., 2012: "Application de la méthode ALCESTE à de 'gros' corpus et stabilité des 'mondes lexicaux': analyse du 'Cable-Gate' avec IraMuTeQ", in: *Actes des 11eme journées internationales d'analyse statistique des données textuelles* (Liège, Belgium): 835–844.

Reichenhein, Michael; Werneck, Guilherme, 1994: "Anos potenciais de vida perdidos no Rio de Janeiro: as mortes violentas em questão", in: *Cad Saude Publica*, 10,1: 188–198.

Ribolla; Maria; Fiamenghi Jr., Geraldo, 2007: "Adolescentes na escola: representações sociais sobre violência", in: *Psicologia Escolar e Educacional*, 11,1: 111–121.

Rotter, Julian, 1966: "Generalized expectancies for internal versus external control of reinforcement", in: *Psychological monographs: General and applied*, 80,1: 1–28.

Sá, Celso, 1996: *Núcleo central das representações sociais* (Petrópolis: Vozes).

Santos, Maria de Fátima; Almeida, Angela; Mota, Vivian; Medeiros, Izabella, 2010: "Representação social de adolescentes sobre violência e suas práticas preventivas", in: *Temas em Psicologia*, 18,1: 191–203.

Souza, Edinilsa; Assis; Simone; Silva, Cosme, 1997: "Violência no município do Rio de Janeiro: áreas de risco e tendência da mortalidade entre adolescentes de 10 a 19 anos", in: *Rev Panam Salud Publica*, 1,5: 389–398.

Tavares dos Santos, José Vicente; Didonet, Beatriz; Simon, Carlos, 1999: "A palavra e o gesto emparedados: a violência na escola", in: *XXII Encontro anual da anpocs* (Porto Alegre: Secretaria Municipal de Educação).

Tavares dos Santos, José Vicente, 2004: "Violências e dilemas do controle social nas sociedades da 'modernidade tardia'", in: *São Paulo em perspectiva*, 18,1: 3–12.

Varela, Silmara; Oliveira, Neura; Freire, Janete; Ferreira, Pedro; Santos, Selma; Díaz-Bermúdez, Ximena; Shimizu, Helena, 2012: "Representações sociais acerca da violência doméstica das mulheres moradoras do da comunidade do Paranoá/Itapoã de Brasília-DF", in: *Tempus Actas de Saúde Coletiva*, 6,3: 277–293.

Velho, Gilberto, 2000: "Violência, reciprocidade e desigualdade: uma perspectiva antropológica", in: Velho, Gilberto; Alvito, Marcos (Eds.): *Cidadania e violência* (Rio de Janeiro: UERJ): 11–25.

# Chapter 17
# Quality School Education from the Perspective of Young Students: What Is the Future?

Sandra Lúcia Ferreira

## Introductory Comment

**Alda Judith Alves Mazzotti**

*While the importance of knowing[1] perceptions, expectations and values of students about the formative process offered by school has been increasingly emphasized as essential data for the construction of high quality public education, there is little research that investigates how students evaluate their learning and its possible contribution to their own future. Based on this statement and on the assumption that the formative processes are crucial for the construction of subjectivity, Professor Sandra Ferreira took on this challenge.*

*I knew Professor Sandra Ferreira from debates carried out at the International Center of Studies on Social Representations and Subjectivity in which we both*

---

Sandra Lúcia Ferreira is a teacher and researcher at the City University of São Paulo (UNICID) and an invited professor at the Paulista State University (UNESP). She is an associate researcher at the International Centre for Studies in Social Representations- Subjectivity and Education (CIERS-ed) at the Carlos Chagas Foundation (FCC) and the Centre for International Studies in Social Representations (NEARS). Email: 07sandraferreira@gmail.com.

---

[1]Alda Judith Alves Mazzotti is Emeritus Professor at the Federal University of Rio de Janeiro and at the Estácio de Sá University, where she is a full professor in the area of Educational Psychology. She completed her doctoral studies in educational psychology at New York University. Her academic production focuses on research methodology, where her work is an obligatory reference in the area of education. She develops with this methodological rigour countless studies in social representations, having implemented a research group that focuses on analysing teacher knowledge, teacher training and work, teacher identity, school failure, public school students, and child labour. Email: aldamazzotti@gmail.com.

© Springer Nature Switzerland AG 2021
C. Prado de Sousa and S. E. Serrano Oswald (eds.),
*Social Representations for the Anthropocene: Latin American Perspectives*,
The Anthropocene: Politik—Economics—Society—Science 32,
https://doi.org/10.1007/978-3-030-67778-7_17

*participated. The strength of the Center's knowledge about evaluation and techniques of data analysis, allied to its aim to apply them to the improvement of practices and policies focused on education, have accredited her to carry out this challenge.*

*The presented text is an excerpt from the research titled "Institutional Self-Assessment: Perceptions of Elementary School students about Quality of Education", which served as the foundation for a database about the subject.*

**Abstract** This article is the result of an information-centric survey of 227 pre-adolescents – thirteen to fourteen years old – from elementary school. Its objective is to reflect on the teaching of these young people, whose different faculties, potentials and skills equip them to be part of the decision-making process about teaching methods. The discussion is supported by Social Representations Theory because it reflects the interdisciplinary potential, combining students, teachers and managers in a one-dimensional perspective. For this purpose, it is a precondition to involve these educators in the process of continuing training, making them active researchers of the educational phenomenon, capable of improving daily practice by means of scientific methods. The methodology is defined by the application of questionnaires to focus groups in order to reveal their opinions about the daily routine at school. The results indicate that Quality in School Education is associated with respect and trust between students and educators, activities that make students feel valued, and the participation of students in processes which affect the school dynamics. There is also evidence that substantiates the view that Quality in School Education is based on a fictional social space, a school which does not actually exist. In this sense, Quality is designed for an imaginative future far removed from the present situation in which the students find themselves.

**Keywords** Social representations · Pre-adolescents · Elementary school · Quality of education · Future social spaces

## 17.1 Introduction

The intention of the research initiative "Institutional Self-Assessment: Perceptions of Elementary School students about Quality of Education", conducted by Fundação de Amparo à Pesquisa do Estado de São Paulo – FAPESP (São Paulo State Research Foundation), Process 15/11305-2, was to understand the articulation of what it is to be a young student finishing elementary school and, especially, a young student studying at a public school in the current social climate. In addition, the research sought to understand the thinking of these young Municipal and São Paulo state students about the Quality of School Education and their views on how schools could improve in the future.

Elementary education – split into Elementary I (children from six to ten years old) and Elementary II (from eleven to fourteen years old) – is suffering deep social and cultural changes – intolerance of diversity, exacerbated consumerism, socio-economic inequalities, among others. Consequently, this portion of the population no longer identifies the school space as a place of conviviality. Clearly, that presents a problem, considering the unquestionable importance of school education in public opinion. It is therefore urgent to review the way educational space is organized in schools, and to pay attention to the views of students from the sixth to the ninth grades, when pre-adolescents are in a period of autonomous construction marked by relational conflicts, and in transition to another phase of their education: high school. Thus, this work considers the importance of reflecting on ethical and moral issues, an approach justified by the weakening of standards and values (Touraine 2016). There is also justification for the non-participation of students who are uncertain and insecure (Bauman 2007).

It is assumed that students will recognise the importance of this period in their education. Yet Elementary School II often signals a phase of troubling issues, which has instigated studies and research on different aspects of the situation (Aguiar/Conceição 2009; Costa/Koslinski 2006; Resende/Pasian 2017).

Some field research on different perspectives of school management and education obtained results that reaffirm the main hypothesis, i.e. that pre-adolescent students have strong opinions, and their attitudes are not devoid of meaning when they are asked about their school environment and what it should be like in the future.

Committed to dialogue with young people, the report is based on their life stories, according to their scholastic past, combined with a discussion about social spaces associated with the respondents' understanding of the concept of Quality School Education. The questions which the pupils were asked focused on values, the role of the students, and their views on how to achieve quality in education in the future. The analysis of such questions is based on the publication *Social Representations: Explorations in Social Psychology* (Moscovici 1988), which asserts that any interpretation of reality should connect the objectivity of historical analysis to the subjectivity of life experience. The study therefore aims to understand the production of senses by the participants in spaces of intersubjective relations in a given social context, without the dichotomy between the internal and the external, conceiving the human being as a builder of and built by social reality (Lane 1984).

## 17.2  The Purpose of This Research

Pre-adolescence is a dynamic concept that is based on a historical representation of the young population. Unlike the concept of adolescence, established by the Child and Adolescent Charter (Estatuto da Criança e do Adolescente – ECA), according to Law No. 8069/90, which is defined by the age range of 12–18 incomplete years old, pre-adolescence, in this article, refers to a period not necessarily delimited by age, but which includes other aspects related to intense biological, social and cultural transformations, associated with the school environment. This implies that pre-adolescents can only be understood in the context of the society in which they are inserted – in this case, the school environment – as individuals and societies are inseparable (Salles 2005). This assumption is supplemented by understanding of the school environment, a place where initiations, systematizations and the establishment of structuring relationships and critical discussions will happen (Alarcão 2007).

So, the research focuses on the contribution of formative processes to the construction of autonomous and compromised subjectivities. It also recognizes the fact that young students are capable of reflection and have specific knowledge that can create and re-create practices, enabling them to act as producers and transformers of reality.

Therefore, this study seeks to provide advice on improving the teaching of pre-adolescents, recognizing them as capable of expressing themselves in the present and of projecting their expectations for the future, based on their continuation in education (high school), and aiming at the completion of their basic education.

> If the rights of the child to participate are to be carried out in practice, it is essential that they find sensitive educators/school staff: educators who recognize their skills and willingness to develop and learn and who, at the same time, are open to aspects of vulnerability and dependency. (Bae 2016: 13)

This research was developed in two distinct and complementary phases. The first focused on the implementation of a training process for the teachers and managers of the schools involved. Its aim was to contribute to the training of the educators in the schools – teachers and managers – reflecting and exploring concepts related to the theme of educational evaluation in order to make it feasible to train teachers to simultaneously conduct research into the reality of what happens where they live, expanding the perception of quality for pre-adolescents, and the educational practices developed every day in school. This phase is described in depth by Rosito (2018), who gives details of the training given to researcher-teachers. The second phase was supported by theory, in order to foster understanding of how the concept of Quality School Education is being constructed by young students, in a complex way referenced by symbolic and representational processes. Such concerns can be summed up in the following guiding questions for the investigative work:

a. How do young people describe Quality School Education and what social representations and values support their choices and perspectives?
b. What subjectivities are being built?
c. How do young people express their role in this context, considering their status as students who are graduating from elementary school?
d. How do they envisage the school developing in future?

## 17.3  Psychosocial Approach: A Choice

In accordance with the psychosocial paradigm of understanding the human phenomenon from the dialectic relationship between subjects and the environment into which they are inserted, this study was undertaken from the psychosocial perspective of Social Representations Theory:

> the theoretical-methodological tradition of this approach entails studying the social individuals' systems of shared knowledge, the symbolic processes of the groups involved, and the relationships established with a social object. (Ferreira et al. 2017: 888)

Opposed to separating the individual from its context, the psychosocial approach of SRT is formed by the conceptual matrix created by Moscovici (1978), with relevant interdisciplinary input, which has been refined by the contribution of scholars such as Jodelet (2001, 2005, 2009b, 2015, 2017), Doise (2002), Fabric (1994), Salvador and Monteiro (2013), Sá (1996), and Jovchelocitch (2000), among others.

Education has been one of the most privileged fields of knowledge from this perspective, since SRT offers an important tool for analysing signification in the educational space. Although research into education and social representations is a recognized field, there have so far been few studies which focus on the evaluation of pre-adolescents related to factors associated with Quality Education and the future.

The studies of Paredes/Pecora (2003) and Marcondes et al. (2018), as well as other investigations conducted by the International Research Centre on Social Representations (Núcleo de Pesquisa Internacional em Representações Sociais – NEARS, hosted at the Pontifical Catholic University of São Paulo – PUC-SP), are of relevance to the aspects associated with the future. Theoretically, integrated research into these issues – social representations, quality education and the future – should lead to improvements in the teaching conducted in schools. According to SRT, this is because the individual is a creator of symbolic and social systems, specifying how he demonstrates his inventive ability to take part of the world. The aim to use the theory as a means of emphasizing the role of the symbolic in the guidance of human conduct is inspired by this quotation: "I wanted to define social psychology problems and concepts from this phenomenon [social representations], emphasizing its symbolic function and its power to build reality" (Moscovici 1978: 14).

Based on this statement, this study recognizes the importance of analysing the objective or formal characteristics of a case and includes in its research approach an appreciation of the process of appropriation, construction and alteration of the reality by the subject. Thus, the importance of identifying the subject in social representations is highlighted, in addition to the contemporary reflection of the return to the subject in a political and historical perspective, treating subjectivity without losing the social dimension (Jodelet 2009a, 2015). The theory developed by Moscovici offers an alternative to the comprehension "of the subjectivities' constitution from the social conditionings, without disregarding the subjects as builders of their circumstances" (Novaes 2015: 332). Social subjectivity is used as a category for understanding socially constructed symbolization, and takes into account the dual mediation of social subjects with each other during their interactions. This makes it possible to analyse their views on what they consider to be *Quality School Education*, and also their perception of their role in the construction of their stories within school spaces, while at the same time outlining their expectations for the future.

When addressing the question of the future, there is another dimension of analysis that needs to be evaluated, given that the future is included in the notions of temporality and social times. The present is the relevant horizon, because it is during the present that the future is built (Araújo 2005, 2011). The future is designed by slight ruptures in the different action systems, such as education, family, politics, economics, experiences and their forms of regulation, and these shifts play a fundamental role in installing and perceiving changes and transformations in the present:

> The future is not concrete, but manipulable, that is, beyond the present; it is a process built every second; that's why we have responsibilities towards it. As an important contribution [...] there is the analysis that there are several temporal planning subsystems in society with specificities that include temporal expectations and standards, forcing individuals to develop behaviours and to think about their paths and choices in a specific way. Such configurations are structured according to the processes of organization of society and are influenced by different social classes and belonging groups, being associated with socialization and identity. (Marcondes et al. 2018: 75–76)

In such thinking, the combination of these three themes – social representations, quality education and the future – has faciliated analysis, in the places where elementary students are taught, of social representations that support the construction of subjectivity, values, attitudes, emotions and ways of dealing with the temporality of pre-adolescents and how these elements can impact on their understanding with regard to the schools they want.

## 17.4    The Route Taken: Empirical Research

The methodology chosen was listening to pre-adolescents. By giving voice to students to express what they experience in school, the research has shown that they want a high quality education. Listening to what they say puts them on a privileged level, because they become considered generators of relevant contributions, capable of reorientating teachers and the actions of the management. The assumption that supports this conclusion is that the school is a democratic place where everyone can express ideas and opinions. This listening was carried out by means of a questionnaire and a focus group – a type of research that enables interactions between people who are invited to reflect on a theme – in order to explore indicators associated with the topic of Quality School Education and the Future.

The research was developed by a partnership between the postgraduate education programme of the City University of São Paulo (UNICID) and four public schools of the state and municipal networks, located in the city of São Paulo: two on the east periphery and two in more centralized regions – one in Vila Carrão district (east region) and the other in the Jabaquara district (south region). The partnership was formed between eighteen teachers and managers of the schools involved, four students of the Master's degree programme and 227 students. Involvement was voluntary. Schools located in different regions also belong to different networks. However, it was possible to identify similar characteristics between them: most students live in social vulnerability situation and there are ongoing projects to improve learning and students' behaviour, indicating that the Public School is a place of conflict. Teachers and Master's degree students of the partner university (UNICID) were also involved.

## 17.4.1 Instruments

The research made use of the principle that the possibilities presented in a certain reality, or put better, the view of the reality, are always greater than those contained in the sample of any investigative process (Garcia 1981). The intention is not to use methods that paralyze the reality, capturing standardized situations and docile characters. The intention in this instance was to unravel processes and dynamics that allow students to think about the topic Quality School Education and the Future.

It is important to clarify that the present work is a clipping from the research project "Institutional Self-assessment: Perceptions of Elementary School Students about Quality Education", which organized and systematized a relevant database from the information collected from the participants. In order to collect and analyse data for the study, the manifestations of the participants were considered in relation to the questionnaire issues Q(5) and Q(6) and the documentation generated from the seven focus groups at the surveyed schools.

Questions 5 and 6 were:

a. Question 5 – Nowadays, there is a lot of discussion about the SCHOOL EDUCATION offered to children, young people and adults. It is also discussed that SCHOOL EDUCATION needs to be good QUALITY. What, in your opinion, constitutes good quality school education?
b. Question 6 – Considering what you wrote about QUALITY SCHOOL EDUCATION, answer: is the EDUCATION offered at your school GOOD QUALITY? Justify your answer.

The work was conducted according to methodological guidelines for the formation of focus groups (Morgan 1988; Wholey et al. 1994). Groups of five students were formed and the dialogues were recorded at the locations where the seven focus groups were conducted. At three schools the groups took place during the morning and the afternoon, and at one school during the morning only. In addition, written notes of the meetings were made. There were two motivating questions for the meetings: What changes would you like to see? What difficulties are you finding? Four categories were explored to guide the work:

a. The concept of quality: What is the concept of Quality School Education that is understood and experienced? What is the school's policy with regard to listening to students?
b. Student image: How do you interpret actions organized by the students themselves? What is the perception of the role of students in the school's decision-making process? Regarding the students' relationships inside the classroom: What do you think of this place, in particular? What do you think about the proposed projects?
c. Relational processes: What do you think about the link between family and school? What do you think about the relationship between students and school staff?
d. Future: What are the current dynamics of the school? What do we need to keep or change?

The reactions of the participants to the two questions, in addition to the information produced by the seven focus groups, were analysed, making it feasible to access the social representations, since the opinions expressed conceptions, values, feelings and ideologies, and therefore representations.

## 17.4.2 Procedures

Evaluating the results involved organizing the data analysing what the participants said. Two procedures were used to explore the themes of textual material, according to the principle of methodological diversity:

a. Analysis of content: In order to analyse the views expressed by students during the focus groups, assumptions of qualitative research were adopted, and the work was carried out using a method of content analysis adapted from Bardin (1977). According to this perspective, the analyses were organized in three stages:

1. Pre-analysis.
2. Analytical description.
3. Interpretation, inference or discussion of the results.
   During the pre-analysis, floating reading procedures were carried out, i.e., a type of intuitive reading, open to all ideas, reflections, hypotheses, etc., constituting a kind of individual *brainstorming*. During the analytical description, thematic analysis was chosen, a type of analysis of content that consists of observing the core parts of senses that compound the communication. Finally, during the inferential interpretation, interpretive analytical categories were organized to reveal what indicators were associated with the concept of Quality Education and the future.

b. Data processing by using the software *Analyse Lexicale par Contexte d'un Ensemble de Segments de Texte (Alceste)*. According to the software's assumption, when a text is produced by different individuals, it is possible to understand points of view that are shared by a social group at a given time (Kronberger/Wolfgang 2002). Therefore the classes of words produced by the software could be analysed as thematic categories of the senses attributed to the participants, making it possible to identify evidence of social representations which, associated with other results, contributed to the researchers' understanding of the symbolism in the evaluated context (Secchi et al. 2009). It is important to clarify that due to the technical conditions of the software Alceste, which requires calculations based on the laws of vocabulary distribution, separating the responses by school was not possible, nor was analysing the individual contents. This was because the Initial Context Units (ICU)[2] need a corpus with enough words to start the analysis procedures. The initial corpus was divided into four schools and generated the utilization of Elementary Context Units (ECU)[3] under the ideal consideration of 60%. Therefore, the written replies were carried out considering the total set of information of each group/school, as follows:

Suj1=School1(TUR1);
Suj2=School1(TUR2-late);
Suj3=School2(TUR1-morning);
Suj4=School2(TUR2-late);
Suj5=School3(TUR1-morning);
Suj6=School3(TUR2-late);
Suj7=School4(TUR1-morning).

Profile data – age, gender, school, race, school type, city where study and link/school – were generated through the software *Statistical Package for the Social Sciences (SPSS)*, collecting records of all the participants.

## 17.4.3 Who Are the Young People Who Participated in the Research?

A total of 227 students who were graduating from elementary school in municipal and state public institutions participated in the research. Approximately 48% were

---

[2]Initial Context Units (ICU) are defined according to the researcher and the nature of the textual data. In this study, as the data were provided from responses to a questionnaire, each item answered was considered an ICU. The ICU set was the corpus of analysis, processed by the software.

[3]Elementary Context Units (ECU) are small segments of participants' responses to a specific item, typically three lines in length, scaled by the software depending on the size of the corpus, and complying with the punctuation and the order of appearance in the text.

female and 51% male, and 87% of the participants had previously attended *edu-cação infantile* (pre-elementary school). Participants were stratified in three age groups, the majority in the range from 13 to 14 years old (73%), followed by the 15 years old range, corresponding to 14% of the respondents. When asked about their race, they recognized themselves as white (34%), black (22%) and brunette (38%) people. Almost 70% of the participants had studied at public (rather than private) schools for most of their lives. The data obtained from the seven focus groups were compiled and submitted for specific processing, according to the analysis guidelines. The results facilitated analysis of the implicit meanings of the questions posed in the discussion roadmap as well as the degree of importance attributed by the students to each category investigated, raising awareness of the challenges which need to be addressed with regard to the various values associated with the theme of Quality School Education and the Future.

## 17.5 Analyses: What Do Pre-adolescents Think About Good Quality and Future School Education?

In accordance with the methodological guidelines and the aims of the project, the data were processed cumulatively in order to organize and systematize the infor-mation. Analyses and synthesis were conducted to map the responses from the questionnaires and the focus groups. It was then possible to deduce what pre-ado-lescents think about Quality School Education and the Future.

### 17.5.1 What the Pre-adolescents Said in the Questionnaire

Analysis of the answers to the questions (a) What, in your opinion, constitutes good quality school education? (Q5); and (b) Is the education offered at your school good quality? (Q6), has already been undertaken and can be found in Ferreira (2018) under the title "Representações sociais e valores na perspectiva dos estudantes concluintes do ensino fundamental" ("Social representations and values from the perspective of students graduating from elementary school"). This study sought to access the categories of meanings, with the aim of identifying the design of cognitive representational elements, attitudes, emotions, values and conflicts and the main symbolic elements that emerged in the expectations about Quality School Education and the role of these pre-adolescents during the final phase of elementary school.

   Analysis of the textual content reveals consensual points of view among the pre-adolescents who participated in the study. The results indicate that the rep-resentational composition of Quality School Education and the Future is asso-ciated with the professional work of the teacher, revealing objects of articulated representation: school (structure) and teacher.

## 17.5.2  *What the Pre-adolescents Said in the Focus Group*

Data obtained from the application of the four focus groups were subjected to specific treatments in accordance with the analysis guidelines. The results made it possible to analyse the meanings that were implicit in the questions posed in the discussion roadmap, as well as the degree of importance attributed by the students to each category. This revealed the challenges which need to be considered with regard to the different indicators associated with the theme of Quality School Education.

It was found that there is distinct position/opinion in the investigated groups, because each of the four classes proposed by Alceste chose a predominant position, with specific content that personalized the concerns of the groups that were sampled.

Related to the software criteria, the percentage of accepted textual production of the four schools and their respective products of the focus groups was 68% of the ECUs, revealing a homogeneous opinion that allowed analyses and conclusions to be made. It is important to note that the ECUs do not correspond to the number of participants, because they consist only of excerpts from the written narratives.

The overall software analysis of the entire corpus identified four categories in the participants' responses. Based on the Descending Hierarchical Classification (DHC), these were: (a) Professional Competence: beyond the action of copying; (b) Coexistence: the school as a psychosocial space; (c) Dialogue: confidence in the partnership between pupils and educators; and d) Respect: the inclusion of everyone.

In order to contextualize the ECUs, here are some excerpts from the most significant written narratives to illustrate each category:

Category 1 – Professional Competence: beyond the action of copying. This category was composed of 121 ECUs, representing 21% of the total analysed in the Descending Hierarchical Classification set. This block sense was constituted by the following most significant words: "texto" [text] (30) textos [texts] (2)"; "desenh" - desenhando [drawing] (3) desenhar [draw] (3) desenho [draw] (23)"; "copi" – copia [copy] (6) copiando [copying] (8) copiar [copy] (16) copias [copies] (1)"; "lousa" [blackboard] (23)"; "pass" – passa [pass] (26) passado [past] (2) passam [pass] (4) passando [passing] (8) passar [pass] (10) passasse [passed] (1)"; and "livro" – livro [book] (18) livros [books] (1)".

The characteristics of this grouping are: studying at Municipal School (1), during the morning period. Sample of their opinions:

Unité Textuelle No. 181 KHI2[4] $= 29$ Individual No. 1 *School_1 *Municipal_2 *Morning_1

---

[4]Chi-square distribution is used in this research, because its values quantitatively reveal the relationship between the products of the research – via analysis of the focus group documentation – and the planned distribution of the phenomenon.

*She is going to show a picture of Cubism. She is going to tell us to do it. Then she is going to present a text about Cubism. She is going to tell us to copy it! Then she changes to another theme or presents more drawings.*

Unité Textuelle N° 217 KHI2 = 21 Individual No. 1 *School_1 *Municipal_2 *Morning_1

*There was a teacher, then one went out and another came in, and, I realized I'll take grade ten, nine. My mark was decreasing, so the grade in that subject was decreasing, then I don't even have a clue, because I don't do it anymore, I don't see it, I don't want to do what she proposes, because it doesn't suit me.*

Category 2 – Coexistence: the school as a psychosocial space. This category was composed of 46 ECUs, representing 8% of the total analysed. This block sense was constituted by the following most significant words: "usar" [use]; "barulho" [noise]; "multiuso" [multipurpose]; "pátio" [yard]; "precari" [precarious]; "apresentac" [presentation]; "fornec" [supply]; "informática" [informatics]; and "laboratório" [laboratory]. The characteristics of this grouping are: studying at Municipal School (2), during the afternoon. Sample of their opinions:

Unité Textuelle No. 346 KHI2 = 37 Individual No. 2 *Municipal School_2 *Afternoon:

*So, the organization of presentations is very precarious, because there is too much noise, we can't work on it. So, we needed, I don't know, a quiet place.*

Unité Textuelle No. 358 KHI2 = 26 Individual No. 2 *Municipal School_2 *Afternoon:

*Oh, that desk is dirty, that desk is broken. I don't like it. And the person itself makes a point of getting the school spaces dirty. At least here in the [...] we have ample rooms for research, such as the video room where we can do different activities, and I think this is very important because it helps to diversify knowledge.*

Category 3 – Dialogue: confidence in the partnership between pupils and educators. This category was constituted by 339 ECUs, representing 59% of the total analysed. This block sense was constituted by the following most significant words: "pai" [father]; "convers" [talk]; "aluno" [student]; and "reunião" [meeting]. The characteristics of this grouping are: studying at Public State School (5 and 6), during the morning. Sample of their opinions:

Unité Textuelle No. 627 Khi2 = 11 Individual No. 5 *State School_5 *Morning:

*Sometimes, it just interferes. That's it? I think it has constantly, but how can we choose the words? By both sides, the pupils and of the teachers, because many times someone may be wrong. The student is not always wrong. They must listen to the student and the teacher too.*

Unité Textuelle No. 814 Khi2 = 11 Individual No. 6 *State School_6 *Late:

*There are positive points and, also, negatives. Do you think that if both parents and students participate in the parent-teacher meetings it would be better, more democratic? Yes. So, the student could defend himself if he didn't do it, he can try to convince his parents.*

Category 4 – Respect: the inclusion of everyone. This category was composed of 67 ECUs, representing 12% of the total analysed. This block sense was constituted by the following most significant words: "limp" [clean]; "tia" [aunt]; "banheiro" [restroom]; "reunião" [meeting]; "suj" [dirty]; "consider" [consider]; "porta" [door]; "funcionário" [school staff]; "cozinha" [kitchen]; "interfer" [interfere]; "coordenador" [coordinator]; and "respeit" [respect]. The characteristics of this grouping are: studying at Municipal School 3 and 4, Public State School 7, during the morning. Sample of their opinions:

Unité Textuelle N.º 500 KHI2 = 14 Individual No. 3 *Municipal School *Afternoon:

> He's talking about everyone, teachers, directors, everyone. No exception. But there is a teacher who supports us, listens to us. He says that we should be heard and that our voice is important in the school as well.

Unité Textuelle Nº. 858 Khi2 = 28 Individual No. 6 *State School 7 *Afternoon_2:

> Sure! I think cameras should be installed to avoid it [graffiti] at school. During the morning, the students responsible should help to clean the school. And if it took an entire afternoon to clean the desks, would the students keep them clean? Many of them would not!

## 17.6 Looking Ahead: What Is the Future?

> Listening to the students was a challenge that educators proposed to face through the creation of a space so that the pupils' opinions could be expressed, liberated and perform their role. Pre-adolescents have opinions, defend their values, desire to conquer things, but often they do not have the opportunity to talk about them, even when what they say hurts and haunts us. (Gomes/David 2018: 25)

The results indicate that the pre-adolescents' experience of taking part in the study within the school space was recognized as evidence of their appreciation of the school. Such experience can be identified as an intentional, interactive and constant educational action based on motivation and the ability to strengthen the principle of *Freedom of expression*,[5] which recognize that the graduates of elementary school have their own ideas and points of view, capacities and potentials, and skills to be part of the decision-making process about their lives, according to their cognitive and moral development.

The participation of teachers and students, understood as a Right, was identified as a multidimensional process – conviviality, organization and learning – in which different practices and assumptions are recognized. The practices defined for the research design can be listed as:

---

[5]The term *Freedom of expression* is used in the United Nations Convention on the Rights of the Child (UNCRC) of 1989, an international treaty that is part of the global system of human rights. Among other things, the UNCRC states that children have the right to "express their views, feelings and wishes in all matters affecting them, and to have their views considered and taken seriously". https://www.unicef.org.uk/what-we-do/un-convention-child-rights.

a. Accessible and understandable information, made available by the university staff and teachers about the research and the reality of the participant schools.
b. Construction and valorization of individual and collective opinions, experienced through reflection, discussion, exchange, synthesis and questioning.
c. Listening and expression that should be careful, respectful and receptive.

Because the research involved the participation of both teachers and preadolescents, the following dynamics were necessary: participatory environment (guarantee that the views expressed would be considered without intimidation or punishment); recognition of subjectivities (construction of trust connections); alterity (commitment to the other); opportunity (change in the mechanisms of discrimination and exclusion); prominence (significant learning agents).

In the macro plan of school systems and the microcosm of the social conflicts involving pre-adolescents, the increasing deterioration in the way that the school is valued by students is perplexing, and generates several reflections. The results show that school is important. This is recognized in the speech of one of the participants: "Good quality education to me is learning what will really help in the future." To these young people, school is seen as a promise of a worthy future, but their day-to-day experience of school is characterised by unsatisfactory learning – it could be improved by more engaging content and better resources, such as a science lab to help in the learning process – disrespectful relationships – not all students are educated or sometimes they do not seem to have good qualities and teachers and school staff do not respect the student – nonsense classes – some teachers fail to explain clearly – and vacant lessons – yes, there are a lot of classes that do not happen because of the absence of teachers. This study confirms that, while valued by society, schools struggle to perform their role, according to the research participants.

The results also indicate that, for elementary school graduates, Quality School Education originates from two articulated representation objects interacting every day in restructuration: (a) the existence of components that show diverse informative, affective, attitudinal, identity and context elements with conflicting logics; and (b) the absence of indicators that are able to describe the actions of those young people who could demonstrate signs of prominence. This is a widely used term that is understood to equate to multiple powerful words, such as: attitude, proactivity; achievement; transformation, and reflection, among others. Such words were not identified in the speeches of the participants.

These findings are supported by the studies of Arruda (2014), who, when investigating the social representations of complex social objects, talks about multiple networks that connect and organize themselves through their historical and relational characteristics. In the research project discussed in this chapter, this concept of multiple networks is reflected in the young students and their school experiences, which connect elements of the representational field and construct the social thinking. Consequently, the results are an indication of the presence of different layers within the representational composition of young people, revealing psychosocial constructions and the production of subjectivities. Students' manifestations are marked by the sociopolitical environment that plays a part in undermining the democratic image of the school. The pre-adolescents say:

a.  There are many problems in school, and there are several people there who do not want them to succeed in life;
b.  Some students respect each other, but others use dirty words; and
c.  Nobody respects us; they do not care about our views or pay attention to what we say.

Taking into consideration the fact that social representations are always subject to change due to social conditions, mainly influenced by the public spaces of interaction, particularly in public schools, it could be expected that the representations of the elementary school graduates are formed by elements of great mutability, depending on their experiences at school as a *locus* of privileged relations for them. The results of the research, however, offer evidence of the presence of anchorage processes based on a fictional social space, i.e. a school which does not actually exist. A 'private' school.

This study aimed to examine the role of these pre-adolescents and analyse their expectations and choices in relation to the future. The data showed low-level evidence of a prominent role via an attitude identified during the students' voluntary participation in focus groups. This was manifested in the pre-adolescents' significant preoccupation with relationship issues between themselves and the different members of school staff – cooks, janitors, doorman, assistants, the Principal – who need to be respected by students, but also need to respect them. One student said: "I respect them, they respect me! It's a two-way process."

Although they have few opportunities to act, pre-adolescents find the support necessary for survival in their nearest concrete/structural elements. School environments can miraculously offer them the chance to transform their lives. In this scenario, social representations are being composed through objectification and anchorage processes on the basis of what the pre-adolescents experience and what they want to achieve.

Therefore, such results reflect a protective and subjective way of tackling the flaws that are present in the living spaces, because the construction of life projects requires guidance from a short-range perspective rather than a long-term one, and responding to opportunities requires people to exercise the virtue of flexibility (Bauman 2017).

In the face of what has been reported and knowing the difficulties involved in public education, it is timely to believe in the possibility of expanding dialogues between schools and universities; teachers and pupils; students and students, with the aim of changing representational compositions and objectifications of the quality of education to induce protagonists, producers and, consequently, reality transformers to take action in new directions.

Given this scenario and the difficulties involved, it is necessary to believe in the dynamic character of social representations that constitute a network of meanings with characteristics of fluidity and multidimensionality which facilitate their incessant mobility in societies where the communication processes are ever faster

354                                                                                               S.L. Ferreira

and multidirectional. This structuring dynamism allows social representations to be updated according to the experiences of the group. This characteristic of representations is proven, showing that they present an always existent ability, balanced by more structured elements, to integrate new content and hence to update the functionality and meaning of Quality School Education (Jodelet 2001). This research thus emphasizes the need to create spaces for dialogues, reflections and experiences in the pre-adolescents' teaching processes, aiming at the renewal of the future perspective, to generate actions that allow pre-adolescents to recognize themselves as reality producers and transformers in new directions.

# References

4567891011121314151617181920212223242526272829

Abric, Jean-Claude (Ed.), 1994: *Pratiques sociales et représentations* (Paris: Presses Universitaires de France).
Aguiar, Fernando Henrique Rezende; Conceição, Maria Inês Gandolfo, 2009: "Expectativas de futuro e escolha vocacional em estudantes na transição para o ensino médio", in: *Revista Brasileira de Orientação Profissional*, 10,2: 105–115.
Alarcão, Isabel, 2007: *Professores reflexivos em uma escola reflexiva* (São Paulo: Cortez).
Araújo, Emília Rorigues, 2011: "A política de tempos: elementos para uma abordagem sociológico", in: *Revista Política e Trabalho*: 19–40.
Araújo, Emília Rorigues, 2005: *O futuro não pode começar* (Braga: Universidade do Minho).
Arruda, Angela, 2014: "Representações sociais: dinâmicas e redes", in: Sousa, Clarilza Prado de; Ens, Romilda Teodora; Villas Bôas, Lúcia; Novaes, Adelina de Oliveira; Stanich, Kamila A. Biasoli (Eds.): *Angela Arruda e as representações sociais: estudos selecionados* (São Paulo/ Curitiba: Fundação Carlos Chagas/Champagnat Editora): 39–66.
Bae, Berit, 2016: "O direito das crianças a participar – desafios nas interações do quotidiano", in: *Da Investigação às Práticas*, 6,1: 7–30.
Bardin, Laurence, 1977: *Análise de conteúdo* (São Paulo: Martins Fontes).
Bauman, Zygmunt, 2007: *A vida fragmentada: ensaios sobre a moral pós-moderna* (Lisbon: Relógio d'Água).
Bauman, Zygmunt, 2017: *Tiempos líquidos: vivir en una época de incertidumbre* (Barcelona: Tusquets Editores).
Brasil, 1990: *Lei Federal nº 8.069: Dispõe sobre o Estatuto da Criança e do Adolescente* (Brasília: Congresso Nacional).
Costa, Marcio da; Koslinski, Mariane Campelo, 2006: "Entre o mérito e a sorte: escola, presente e futuro na visão de estudantes do ensino fundamental do Rio de Janeiro", in: *Revista Brasileira de Educação*, 11,31 (January/April): 133–201.
Doise, Willem, 2002: "Da psicologia social à psicologia societal", in: *Psicologia: Teoria e Pesquisa*, 18,1 (January–April): 27–35.
Ferreira, Sandra Lúcia, 2018: "Representações sociais e valores na perspectiva dos estudantes concluintes do ensino fundamental", in: Almeida, Julio Gomes; Diniz, Priscila M. (Eds.): *Avaliação institucional no ensino fundamental: O que pensam os alunos sobre educação de qualidade* (Curitiba: CRV): 215–228.
Ferreira, Sandra Lúcia; Marcondes, Anamérica Prado; Novaes, Adelina, 2017: "Indicadores psicossociais: um olhar ampliado para processos educativos", in: *Estudos em Avaliação Educacional*, 28,69 (September–December): 874–894.
Jodelet, Denise, 2015: "Problemáticas psicossociais da abordagem da noção de sujeito", in: *Cadernos de Pesquisa*, 45,156 (April–June): 314–327.

Jodelet, Denise (transl. Ulup, Lílian), 2009a: *Loucuras e representações sociais* (Petrópolis/RJ: Editora Vozes).

Jodelet, Denise, 2009b: "O movimento de retorno ao sujeito e a abordagem das representações sociais", in: *Sociedade e Estado*, 24,3 (September–December): 679–712.

Jodelet, Denise (Ed.), 2001: *As representações socia* (Rio de Janeiro: Editora da UERJ).

Jodelet, Denise, 2017: *Representações sociais e mundos de vida* (Paris/São Paulo/Curitiba: Éditions des archives contemporaines/Fundação Carlos Chagas/PUCPRess).

Jovchelocitch, Sandra, 2000: *Representações sociais e esfera pública: a construção simbólica dos espaços públicos no Brasil* (Petrópolis/RJ: Vozes).

Kronberger Nicole; Wolfgang, Wagner, 2002: "Palavras-chave em contexto: análise estatística de textos", in: Bauer, Martin; Gaskell, George (Eds.): *Pesquisa qualitativa com texto, imagem e som: um manual prático* (Petrópolis: Vozes): 416–442.

Lane, Sílvia Tatiana Mauer, 1984: "A psicologia social e uma nova concepção do homem para a psicologia", in: Lane, Sílvia Tatiana Maurer; Codo, Wanderley (Eds.): *Psicologia social: o homem em movimento* (São Paulo: Brasiliense): 10–19.

Marcondes, Anamélia Prado; Sousa, Clarilza Prado; Ferreira, Sandra Lúcia, 2018: "Formação docente nas licenciaturas: um olhar psicossocial com foco nos espaços sociais, tempos e valores", in: Sousa, Clarilza Prado; Marcondes, Anamérica Prado; Jardim, Anna Carolina Salgado; Coêlho, Vanusa dos Reis (Eds.): *Qual o futuro? Representações sociais de professores, jovens e crianças* (Campinas: Pontes): 67–96.

Morgan, David, 1988: *Focus group as qualitative research* (Newbury Park: Sage).

Moscovici, Serge, 1978: *A representação social da psicanálise* (Rio de Janeiro: Zahar).

Novaes, Adelina, 2015: "Subjetividade social docente: elementos para um debate sobre políticas de subjetividade", in: *Cadernos de Pesquisa*, 45,156 (April–June): 328–343.

Paredes, Coelho Eugênia; Pecora, Ana Rafaela, 2003: "Questionando o futuro: as representações sociais de jovens estudantes", in *Psicologia: Teoria e Prática*, 6 (December): 49–65.

Resende, Gisele Cristina; Pasian, Sonia Regina, 2017: "Inclinações motivacionais de adolescentes concluintes do ensino fundamental em Manaus a partir do BBT-Br", in: *Revista Brasileira de Orientação Profissional*, 18,2 (July/December): 233–247.

Rosito, Margarete May Berkenbrok, 2018: "Formação do professor pesquisador: um processo de desenvolvimento da autonomia e de escuta dos sujeitos", in: Almeida, Julio Gomes; Diniz, Priscila Martins (Eds.): *Avaliação institucional no ensino fundamental: o que pensam os alunos sobre educação de qualidade* (Curitiba: CRV): 179–186.

Sá, Celso Pereira de, 1996: *Núcleo central das representações sociais*. (Petrópolis/RJ: Vozes).

Salles, Leila Maria Ferreira, 2005: "Infância e adolescência na sociedade contemporânea: alguns apontamentos", in: *Estudos de Psicologia*, 22,1 (January/March): 33–41.

Salvador, J. M. V.; Monteiro, M. B. (Eds.), 2013: *Psicologia social*. (Lisbon: Fundação Calouste Gulbenkian).

Secchi, Kenny; Camargo, Brigido Vizeu; Bertoldo, Raquel Bohn, 2009: "Percepção da imagem corporal e representações sociais do corpo", in: *Psicologia: Teoria e Pesquisa*, 25,2 (April–June): 229–236.

Touraine, Alain, 2016: *El fin de las sociedades* (Ciudad de México: FCE).

Wholey, Joseph S.; Hatry, Harry P.; Newcomer, Kathryn E., 1994: *Handbook of practical program evaluation* (San Francisco: Jossey Bass Publishers).

# Chapter 18
# Social Representations in Motion: Concept Construction on Changing Subjects and Contexts

**Cristiene Adriana Silva Carvalho and Luiz Paulo Ribeiro**

## Introductory Comment

**Maria Isabel Antunes-Rocha**

*This text presents[1] the construction process and the initial delimitations of the concept of social representations in motion (SRM), formulated during the work of the Social Representations Study Group (acronym in Portuguese: GERES), which is associated with the postgraduate programme "Education: Knowledge and Social Inclusion", located within the Faculty of Education at the Federal University of Minas Gerais in the city of Belo Horizonte, Brazil.*

*In that context we worked with articles, dissertations, theses and professional research works. Priority was given to the context of the peasants' struggle for the right to schooling, especially access to higher learning, but in recent years the research has been expanded to include other subjects in different contexts (Ambrósio 2014; Angelina 2014; Telau 2015; Amorim-Silva 2015). Between 2004 and 2020 we conducted*

---

Cristiene Adriana da Silva Carvalho holds a PhD in Education from the Federal University of Minas Gerais. She is a pedagogue and teacher of Performing Arts. Email: cristienecarvalho@gmail.com.

---

Luiz Paulo Ribeiro is a full-time professor and researcher at the Federal University of Minas Gerais. He is a psychologist with a PhD in Education. Email: luizpr@ufmg.br.

---

[1]Maria Isabel Antunes-Rocha is a professor at the Federal University of Minas Gerais (UFMG). She coordinates the Social Representations Study Group (GERES), which develops research connected with the research and formation of professionals in basic education and education in the field. She gained her undergraduate and postgraduate degrees at the Federal University of Minas Gerais (UFMG) and undertook postdoctoral studies at the State University of São Paulo Júlio de Mesquita Filho. Her work in professional teacher training involves active minority groups, facilitating the development of research in social representations in movement, with important contributions to the Social Representations Theory. Email: isabelantunes@fae.ufmg.br.

nineteen studies, including dissertations, theses, postdoctoral reports and professional research developed under the coordination of Maria Isabel Antunes-Rocha. All the works contributed in some way to the construction of the concept of SRM.

The term evolved during research undertaken by the Social Representations Study Group (GERES), operating in the research areas of psychology, psychoanalysis and education during the postgraduate programme "Education: Knowledge and Social Inclusion" in the Faculty of Education at the Federal University of Minas Gerais. GERES is organized by professionals with different areas of knowledge who work in educational contexts. The group is committed to the production of knowledge about the social representations constructed by subjects who are experiencing situations that demand changes to their ways of thinking, feeling and acting. The research led us to question why we create representations. In Moscovici's opinion (2012), social representations are created to make the unfamiliar familiar. According to Moscovici (2012: 61), unfamiliar "things that are not classified and have no name are strange, non-existent and, at the same time, threatening". Thus, we saw that, when confronting something 'strange', there is the possibility of change – and a limit to the extent of change – in the representational universe of the subject. In the face of 'strange', subjects impelled to the inference can exhibit three reactions, or 'motions': (a) refusing to experience the new, rejecting the opportunity to change and strengthening already established knowledge; (b) adhering fully to the strange, promoting a break with the past; or (c) initiating a process of familiar re-elaboration, integrating the new progressively (Antunes-Rocha et al. 2015).

To build the concept of social representations in motion, we have elaborated its conceptual, analytical and methodological contours. In conceptual terms, we observed that the concept of motion accurately explains the process of forming social representations in contexts that push for change. We have discovered that subjects can shift their ways of thinking, feeling and acting to ensure that their core social representation remains unchanged.

In analytical terms, we have already managed to capture three motions: a motion that is intended to maintain social representations; another motion that is guided by the denial of previous social representations and the adoption of cognitive, affective and attitudinal maps that the subject considers appropriate for handling the new situation; and a third that is characterized by motions that move away from previous social representations in search of new information and attempts to adapt to the demands of the new context. The focus of current research is to try to understand the cognitive, affective and attitudinal strategies that characterize each motion.

In methodological terms, we have adopted methodological tools (data collection and analysis) to identify the phenomenon that is in motion. The two pieces of research presented here have inserted the methodological proposal into a descriptive qualitative approach from the analysis of categories: articulation of the research object within the context of the type of Social Representations Theory (SRT) approach used, and methodological procedures for capturing the motion of social representations. The analysis made it possible to observe the dialogue between the methodological approaches of these pieces of research from the perspective of improving the concept of "Social Representation in Motion", which has made it possible to verify such social representations in subjects inserted into *changing contexts. Social Representations Theory was regarded as a point of reference, furthering the analysis of social representations of subjects under pressure to change their formative processes.*

*This work presents a dialogue between two doctoral research works focusing on the study of social representations of violence and artistic practices in the formative process and practice of licentiate undergraduate students of rural education in the Faculty of Education at the Federal University of Minas Gerais. Through this discussion, we hope to demonstrate the paths taken in the construction of the concept of SRM.*

# References

Ambrósio, Ana Esperança Futi Bambi, 2014: "Um estudo sobre as representações sociais dos pais e encarregados de educação do colégio Padre Buílu em Cabinda/Angola – relação família-escola" (MSc dissertaton, Federal University of Minas Gerais, Faculty of Education).

Amorim-Silva, Karol Oliveira de, 2015: "Educar em prisões: um estudo na perspectiva das representações sociais" (MSc dissertation, Federal University of Minas Gerais, Faculty of Education).

Angelina, Casimiro Kâmbua, 2014: "A relaçao escola-família: um estudo sobre as representações sociais de pais e encarregados de educação sobre a escola do ensino primário do Chiwéca em Cabinda/Angola" (MSc dissertation, Federal University of Minas Gerais, Faculty of Education).

Antunes-Rocha, Maria Isabel; Amorim-Silva, Karol Oliveira de; Benfica, Welessandra Aparecida; Carvalho, Cristene Adriana da Silva; Ribeiro, Luiz Pablo, 2015: "Representações sociais em movimento: desafios para elaborar o estranho em familiar", Paper presented at the 12th National Conference of School and Educational Psychology and the 37th Annual Conference of the International School Psychology Association, São Paulo, Brasil: pp. 836–841.

Moscovici, Serge, 2012 [1961]: *A psicanálise, sua imagem e seu público* (Petrópolis/RJ: Vozes).

Telau, Roberto, 2015: "Ensinar – incentivar – mediar: dilemas nas formas de sentir, pensar e agir dos educadores dos CEFFAS sobre os processos de ensino/aprendizagem" (Federal University of Minas Gerais, Faculty of Education).

**Abstract** This text presents the construction process and first delimitations of the concept of social representations in motion (SRM), formulated from the work of the Social Representations Study Group (GERES – Federal University of Minas Gerais). GERES is organized by professionals with different fields of knowledge who work in educational contexts. The group is interested in the way in which social representations, as part of the production of knowledge, are constructed by subjects who are experiencing situations that demand changes to their ways of thinking, feeling and acting. According to Moscovici, social representations are created to make the unfamiliar familiar. We realised that there is the possibility of change – and a limit to the extent of change – in the representational universe of the subject when facing the unfamiliar. In the face of what is 'strange', subjects tend to have one of three reactions, or 'motions': (a) refusing to experience the new, rejecting the opportunity to change and strengthening knowledge that is already instituted; (b) fully adhering to what is strange, breaking with the past; or (c) initiating a process of familiar re-elaboration, integrating novelty progressively.

Between 2004 and 2020 we conducted nineteen studies, including dissertations, theses, postdoctoral reports and professional research developed under the coordination of Maria Isabel Antunes-Rocha. All the works contributed in some way to the construction of the concept of RSM. To build the concept of social representations in motion, we elaborated its conceptual, analytical and methodological contours. The results of this research answered our questions about why we create representations.

**Keywords** Social Representations in Motion (SRM) · Education · Novelty integration

## 18.1   Introduction

For the purposes of this text, priority was given to two research works (Ribeiro 2016; Carvalho 2017) prepared in conjunction with students of UFMG's licentiate undergraduate course in rural education (LeCampo/UFMG), since these works express a synthesis of what has been developed by the Social Representations Study Group (GERES) in terms of theoretical and methodological awareness.

The University's licentiate course (LeCampo/UFMG) was created in 2005 as an experimental course in response to the demand of peasant social movements for the University to construct a higher-learning formative process for educators which could meet their political, pedagogical, social and cultural expectations. The proposal formulated presented the following characteristics: formative locations and schedules are alternated, one period at the university and another at the community; contents are organized according to knowledge area; management is shared between university representatives and representatives of the social movements/labour unions; students undergo a special admission process and the political/pedagogic project clearly expresses the intention to consolidate the formative process of teachers with a school project, which in turn is consolidated with a rural project in connection with a society project (Antunes-Rocha/Martins 2009).

This course has been regularly offered by UFMG since 2008. It should be noted that this proposal was adopted by the Brazilian Ministry of Education as a template for the creation of the Rural Education Licentiateship Support Programme (PROCAMPO). Today, about forty-four universities offer that course (Molina/Hage 2016).

When the first class of students was created, it became clear that the entire experience would be challenging, from the initial formative stage to professional practice. Among other aspects, the social and cultural characteristics of the students, the way they organize their schedule, space and contents, the shared management model, and the explicit connection between the formative process, school and the life project helped teachers, students and representatives of the social movements and labour unions understand that something new was under construction and that, therefore, the problems faced must be seen from a procedural perspective (Antunes-Rocha et al. 2013). This is how research groups emerged, formed from the demands above. One of them was related to the challenges presented by the students with respect to the construction of a professional identity capable of appropriating academic knowledge and practices without losing their peasant roots.

Research which uses social representations as its theoretical reference starts from just such a perspective. From the outset, this approach's potential to expose the tension, from a procedural perspective, between different kinds of knowledge and practices became evident insofar as it generates the knowledge that enables the students' reality to be understood from many possible perspectives. Starting with all the reflections produced, we prepared some indicators that resulted in the formulation of the concept of SRM to indicate a process that shows subjects facing new situations in their lives, but does not necessarily involve changes in the way they think, feel and act. This is because, in each result, we saw that the subject, while retaining his or her existing knowledge about a certain object, kept moving in terms of knowledge, feelings and attitudes to ensure that representations would not change. With that indicator we conducted the inquiry through the comprehension of 'motion', that is, how possibilities and limits occurred when students were dealing with the new objects they faced during their learning and practical processes as rural educators.

During that journey, in search of a theoretical evolution, we attempted a dialogue in accordance with the Moscovician proposal, the contributions of Denise Jodelet, the works produced by the group from Aix-en-Provence under the direction of Pascal Moliner, and the works developed in the International Centre of Studies on Social Representations and Subjectivity – Education, under the direction of Clarilza Prado de Sousa. With respect to Denise Jodelet and Clarilza Prado de Souza, we have sought closer contact through the participation of both of them in our defence committees, as well as in debates and round-table discussions at academic events.

Ever since the first theoretical publications on social representations, it has been possible to identify curiosity about the reasons why we create a social representation and how that procedure occurs. According to Moscovici (2010), social representations are created to make us familiar with that which is unfamiliar. With respect to that which is unfamiliar, Moscovici (2010: 61) says that "things that are not classified and not given a name are strange, non-existent and, at the same time, threatening." Also according to Moscovici (2010: 55), when faced with that which is strange, there is the possibility of change – and a limit to the extent of change – in the subject's universe of representations.

Moscovici (2012: 59) says that "when we study a representation, we must always try to find out the unfamiliar characteristic that motivated it". In that task of making familiar that which is unfamiliar, of internalizing that which is strange, he identifies two processes: anchorage and objectification. Anchorage has to do with the representation's social roots, with the new object's cognitive and affective integration into the matrix of pre-existing affections and with the transformations which, on account of that, occur in the one and in the other. Objectification has to do with the process of making the new object concrete (Moscovici 1978: 289). Hence, a social representation may be understood as knowledge in motion, since it is also produced in a context of movement and, for that reason, demands changes.

Serge Moscovici (2012) indicated that there are two processes to which the process of constructing social representations are related: objectification and anchorage. These two processes show how a social representation is formed; one can see how something that is initially strange becomes familiar. Likewise, there is another relevant concept in that journey: pressure toward inference (Moscovici 2010: 228). This concerns a subject's need to give meaning to a certain object when faced with social and individual, past and present references. The object exerts a force on the subject(s), impelling it/them to produce a meaning, at times something new in the socio-cognitive constitution. Thus, we can say that the non-fixed character of social representations is not something recent, since this characteristic has been demonstrated by research works throughout SRT's theoretical trajectory. What does change with time is the way researchers refer to this motion, as well as the methodologies used in its observation/capture and the approaches taken. It should be noted that, according to Moliner (2001), despite the innovative character of those studies, for over thirty years (1960–1990) they remained on the margins of publications in this field.

Therefore, the terms 'genesis' and 'construction' (Moscovici 2012), 'dynamics' (Moliner 2001) and 'transformation' (Andriamifidisoa 1982) have been used to refer to the non-fixed characteristic of social representations. After considering a variety of names, the Social Representations Study Group (GERES) opted for the term 'motion', because the group identified that subjects, even those who tend to refuse change, take cognitive, affective or attitudinal actions in an effort to preserve their knowledge and practices.

Conversely, Moliner (2001) shows that there are some problems of a theoretical-methodological nature when studying the motion of social representations, implied in both the theoretical structuring about the state of a representation's stability, which the structural approach has been able to deal with, and the methodology necessary to longitudinally capture the representational phenomenon, something done due to the absence of techniques and, inevitably, to the issue of the time used by the researcher in such an enterprise. According to him, the solution to this problem lies in experimental studies, since they attempt to create the ideal conditions for producing motion in a social representation.

Social representations are believed to be inextricably linked to social practices; they mutually generate each other, with social representations guiding and determining social practices, and social practices acting on the creation or transformation of social representations in a spiral and continuous movement, back and forth (Tafani/Bellon 2005). According to Moliner's (2001) hypothesis about this, the individuals' movement within the social field translates into evolutions at the level of social representations. Based on those considerations, Flament (1989) and Tafani/Belon (2005) indicate two proposals for the change in social representations. In Flament (1989), the transformation of social representations may be abrupt or gradual. Depending on the level of distinction between social practices and social representations, the presentation of the 'new' may be a rupture with the past or, if they are similar, the past is slowly modified, bringing forth a new reality.

Tafani/Belon (2005) present the change of social representations in three stages: (a) progressive integration, where there are enough good arguments for the change; hence, it takes place progressively; (b) slow transformation, where there are not enough good arguments and reasons for the change of SR, and time for sharing (communication) is necessary for both subjects, making it weak in most cases; (c) abrupt transformation, where there is a good stock of good arguments and, since those contradict that which is familiar to the group, they end up bursting the "central system's core" of the belief at hand.

How do we deal with these issues in our work and in the construction of the concept of SRM? We start by saying that we understand the dimension of that which is new, strange and unfamiliar, which boosts the creation of a social representation in the subjects and the context under. It is possible to is possible because of our knowledge that the socio-cultural trajectory of young Brazilian peasantsis marked by their lack of rights. In other words, they belong to population groups without access to social benefits such as health, education, or leisure, among others, being also historically marked by disenfranchisement and prejudice (Antunes-Rocha et al. 2012). From the mid-1980s, when these populations began to act together by organizing social and labour union movements and starting to fight for land and rights (Caldart 2004), they established new forms of relationships with society and with themselves.

The experience of being a higher-learning student in a public university in itself introduces a new dimension. During the course of study, the contact with academic knowledge emphasizes that strangeness inasmuch as the students are exposed to scientific categories and analyses that, in some cases, refute part of the knowledge consolidated in rural life experience. Even the right of access to a higher-learning course is seen by the students as strange. School trajectory data show that they are the first in their family to have access to university. The formative process that is necessary for them to work as educators presents itself as a paradox. The students' former belief that they did not have the right to access schooling has led to them now being in the position of professionals within an institution that has always been absent in their lives and in the lives of their relatives. Ultimately, our research has been unveiling a series of new and strange situations for the students.

The research showed that those subjects presented at least three ways of organizing their social representations when faced with the innovative situation: (1) keeping their previous knowledge and refusing to enter into a dialogue with academic knowledge; (2) denying their previous knowledge and adopting academic knowledge as their only reference; (3) embracing the tension between the two kinds of knowledge and synthesizing them.

It is necessary to bear in mind, however, that the progressive scale of social representations should not be regarded as linear, with the tendency of one kind of 'knowledge' replacing the other in the construction of social representations; there is rather a coexistence of types of knowledge "that correspond to different needs and play different roles in social life" (Jovchelovitch 2011: 23).

Finally, in that apprehension of motion, the question remains as to what may cause it. Moliner (2001) and Moliner and Guimelli (2015) point out that when there are context and attitude alterations at the informational level, social representations are in motion. This can be verified by reading the research developed by Ribeiro (2016) and Carvalho (2017).

## 18.2 The Motion of Social Representations of Rural Violence

Ribeiro (2016) developed his research on the assumption that the violence experienced by the peasant population in Brazil is historical, marking the social imaginary and memory, as well as the identity and social representations of peasants. That research work considers that, since the Portuguese 'discovery' of Brazil in 1500, rural populations have been deprived of their rights as a result of economic development processes and projects that focused on domination and the exploitation of the land's natural resources, as well as the implementation of an agricultural export model.

Ribeiro (2016) remarks that since the earliest years of colonization, the expropriation of native Indians from their lands, the use of slave work, the hindrances to the possession of the land, the exploitation of immigrant peasant workers, the rural flight, the extinction of religious and confessional communities and resistance camps, agrarian policies focused on monoculture and the use of agrochemicals, the assassination of community leaders, and the educational public policies that disfavour the peasant population are expressions, among many others, of the violence suffered by peasants in the course of Brazil's history.

Such a thesis, however, is an attempt to go beyond the mere rescue of victims from a vicious cycle. It goes to show that, parallel to government investment and domination strategies, peasants developed actions and ways to get organized in the face of violence: escaping, finding refuge in *quilombos*, founding peasant Edens such as Canudos and Contestado, organizing groups to obtain justice, and creating communities, labour unions and social movements to resist the assaults of the capital owners and their destructive forms of production. Ribeiro (2016) believes that the struggle for rural education, as well as licentiate courses in rural education, is another way of producing something in the face of the absence of educational public policies for the benefit of peasant populations.

It was in the face of these individual and collective subjective actions which emerged as a response to violence that, inspired by the Rural Education Licentiateship's transformative potential, Ribeiro (2016) identified the type of teacher training that he thought would work in rural contexts. He reached this decision because teachers from a rural background act as agents of transformation in their communities, and also maintain contact with the countryside or consider themselves to be peasants, in addition to being engaged in the formative process of teacher training in a context of intense conflicts and tensions. Social Representations

Theory with a procedural approach (Jodelet 2001) was the methodology chosen for the research, with the aim of capturing the motion of such subjects during the development of their personal and social trajectory in the face of a life of suffering, experienced or discussed in the contexts of their material and subjective life story.

Data collection through questionnaires and semi-structured interviews took place in the summer of 2016 (Dec-Jan). Initially, an attempt was made to conduct narrative interviews, but the resultant information did not have the form of an actual narrative. Therefore, in order to analyse the data, the researchers decided to use the model proposed by Amorim-Silva (2015). This employs a procedure called 'trajectory analysis', which consists of analysing the contents and organizing points of familiarity between all the interviewees so as to find common experiences and/or similar forms of thinking, feeling and acting that may indicate social representations. In all, 108 questionnaires and seventeen interviews were completed by the licentiate undergraduate students of rural education in the Faculty of Education (FaE-UFMG).

The trajectory analysis revealed five social representations with respect to rural violence: submission, naturalization, abandonment, assistance (aid), and resistance. Each of those representations is connected with instructions to take action (Jodelet 2005), imbricated in the sociability and subjectivity of the individuals and in constant correlation with other representations, such as those of school, peasant, and, especially, the countryside. Namely: the *social representation of submission* says that, in the face of the violence experienced, there is a need to submit to survive; the *social representation of naturalization* speaks of the non-visualization of the violence experienced, since it is part of everyday life and therefore common, natural; the *social representation of abandonment*, laden with the historical scars of a people obliged to flee several times, is related to escaping from a territory where violence took place in order to reach other alternatives in other places; the *social representation of assistance* is related to the social representation of naturalization, since the subjects do not see the violence they suffer, but rather the violence suffered by others who are attempting to intervene and help; conversely, in the *social representation of resistance*, the individual is supported by both the social context and the awareness his or her involvement in creating this context, and hence notices rural tensions and realises that the only way to overpower violence is to fight against it by claiming rights, overcoming inequality of access, and altering the image of the countryside as a backwards place.

After identifying the social representations present in the participants' speeches with the help of IRaMuTeQ, a software interface for multidimensional text and questionnaire analysis, the groups that moved their social representations in similar directions were identified. Figure 18.1 highlights the fact that the subjects under research have, in their life trajectories, at least two comparison points, which will be called the initial point and the final point. The initial point is related to everyday experiences before admission to LECampo, i.e. childhood and relationship with the community and family, and the final point is related to social representations at the time of the interview, as a student of LECampo and, at the same time, a future educator in a rural school.

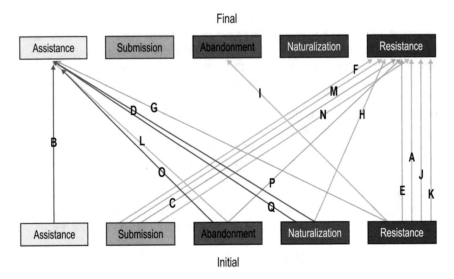

**Fig. 18.1** Trajectories of social representations of the subjects under research with respect to rural violence *Source*: Ribeiro (2017)

In the figure above (18.1), it can be seen that some groups of subjects go through the same social representation motions. Subjects 'A', 'E', 'J' and 'K' remained in the social representation of resistance; subject 'H', 'M' and 'N' started from a social representation of naturalization (not noticing violence), then moved towards a social representation of resistance; subjects 'D' and 'Q' started from a social representation of naturalization of violence toward a social representation of assistance, that is, believing that intervention is necessary to put an end to violence; subjects 'C' and 'F' moved from a social representation of submission to a social representation of resistance; 'O' and 'L' started from a social representation of abandonment in the countryside for survival toward a social representation of assistance; finally, subjects 'B', 'P', 'G' and 'I' moved in different directions with respect to the other subjects, each one in a different direction, and therefore cannot be classified in a group.

Subjects 'A', 'E', 'J' and 'K' have social representations of violence as something that takes place in the countryside and is inherent in political and power relations; they see resistance as a form of fighting back, a strategy used to struggle for the rights of all subjects in the collective. 'A', 'J' and 'K' are members of the same social movement that battles for the implementation of land reform settlements in the region of Paracatu, Minas Gerais. For those interviewees, the rural man is a strong subject who possesses important forms of culture and knowledge. For that group, the chief form of violence that affects rural populations is the withdrawal of rights, such as access to education and to the land.

For 'H', 'M' and 'N', the tone was "now I can see the violence". If, previously, they could not notice the existence of violence, from the moment they took part in social movements and in the Licentiate Course in Rural Education they began to realize how rooted violence was in the community's everyday life. Previously it was naturalized, but now the subjects are able to fight against it both for themselves and for others surrounding them. They now view the role of rural schools as that of "opening the subjects' eyes" so that they can see. There is also a motion toward the appreciation of rural subjects, of their social practices, culture and knowledge. All those three now live in the countryside with their families and their discourses expressly show their wish to remain in the countryside. Also noticeable in the reports of these three was the narrative according to which forestry is a means of production that worsens the life conditions of peasant populations, committing violence and changing the rural populations' lifestyle.

'C' and 'F' shared the motion from a social representation of submission (i.e. using silence and acceptance as a survival tactic), to a social representation of resistance, fight and permanence in the countryside. Both are women who are greatly concerned with the role of women in peasant social relations and in labour relations in the countryside. Resisting, for these two interviewees, reveals the possibility of rejecting their families and community, as if questioning or not accepting the rules represents some kind of defilement. Tensions causing the motion of social representations are mostly related to family issues, or access to study and gender issues in the countryside.

'D' and 'Q' are residents of urban areas and have social representations of rural violence mostly as sexual assault and violence against women in small communities. If they now witness such violence when they are at the rural school, they associate the school with the need for intervention.

'L' and 'O' presented the same motion, from a social representation of abandonment, escaping from the countryside, moving now to a social representation of assistance. Both are subjects who moved away from the countryside to proceed with their studies and now reside in the city's urban area. There are indications of an attempt to return to the countryside to aid other subjects represented as lowly, impoverished and uninformed.

In all the above cases, the essential role of the licentiate undergraduate course in rural education in the forms in which rural violence is represented became evident. Even with varied social representations and social representation motions, the subjects showed how much LECampo contributes in the quest to overcome the violence experienced by rural populations, especially with respect to access to higher learning in a federal institution, which is something regarded as way too distant for most peasants. Experience of various phenomena, discussions with other undergraduates and staff, and the emphasis on collectivity and experience-sharing, including narratives of various degrees of violence, all help the students to prepare for and cope with the different scenarios they might encounter in real life.

Discrimination and prejudice against rural citizens have been described in various contexts, including schools, banks, universities, streets, and city spaces. Sometimes that violence was not noticed or perceived, but now they can see how country people are seen as backwards and undeveloped. Experiencing such types of violence causes the rural citizens to avoid social spaces, according to their reports.

Violation of rights and the denial of access to public goods, services and policies were also listed as reasons for violence against rural citizens. Many times their rights are denied and rural populations naturalize this denial of access to public policies such as health and education, which is another form of violence against that population.

Violence against women appeared in almost all reports. Women were portrayed as submissive to men, sexually assaulted, both physically and morally, working in the field and at home, providing care for the children and the husband. The participants explained this thread of violence as the result of a traditional view of the roles of men and women in the countryside, which needs to be reviewed to boost equality.

The misunderstanding of rural social movements and their struggle for the land also appeared as motivators for physical and psychological violence suffered by rural citizens. The concept that they are 'troublemakers', 'idlers' and 'invaders' justifies the actions of great landowners, militias and police through truculent interventions, persecution, assassinations and threats, and, once again, gives rise to expropriation and the denial of rights.

Finally, the tensions that lead individuals to change is something that appears in all reports. Those tensions are referred to as "pressure toward inference", according to which the individual cannot fail to subjectively produce the legitimacy of ideals, perspectives and social representations. In that scope, the list includes the history of dispossessions, failure to have access to public policies, physical, sexual and psychological aggressions experienced at home, the anguish of witnessing violence against their own students, homophobia, discrimination for being peasants, devaluation of their production, horror in the face of violence against women and the elderly, and incestuous relations. All those reported situations, in addition to the participation, experience and connection with rural social movements and labour unions, form the context in which social representations of rural violence and their movements are produced by licentiate undergraduate students of rural education in the Faculty of Education of the Federal University of Minas Gerais (FAE/UFMG).

## 18.3   The Motion of Social Representations in the Artistic Practices of Rural Teachers

The work produced by Carvalho (2017), with the title "Social representations of artistic practices in the work of rural teachers" analysed social representations in motion from the practical experience of rural teachers who graduated from the licentiate undergraduate course in rural education in the Faculty of Education at the Federal University of Minas Gerais (FAE/UFMG). That study started from the understanding that artistic practices are forms of conception, understanding and fruition of artistic manifestations by the subjects. In that context, the analysis of social representations involves attempting to analyse artistic manifestations based on the values, beliefs and tensions present in the art in accordance with its social meaning. The starting point was the assumption that rural teachers' artistic practices are characterized by the attempt to establish a dialogue between the academic content of their training course and the affirmation of their identity as peasants, incorporating the elements of reflection, fruition and artistic construction as well. That dialogue is pervaded by a movement of construction and reformulation of their social representations, characterized by different orientations with respect to artistic practices, which can be analysed from the concepts of objectification and anchorage belonging to Social Representations Theory.

In the face of this change-provoking situation, Carvalho (2017) proposed this study with the intention of gaining awareness of the motion in the forms of the thinking, feeling and acting of rural education teachers, as far as artistic practices in their activity as teachers are concerned. Such a perspective is also an attempt to investigate how the subjects under research reformulated the social representations of artistic practices they had held during their undergraduate formative years, taking their debut as teachers as the change-provoking point. An attempt was also made to identify the concepts of objectification and anchorage in Social Representations Theory in the process of constructing the artistic practices of the subjects under research.

Two research subjects were analysed in order to obtain an understanding of this reformulation: Camila and Lucas, graduates from LECampo/FaE/UFMG, specifically from the class majoring in languages, arts and literature (LAL) which was admitted in 2010 and graduated in January 2015. The choice of subjects in that class was an attempt to continue the analysis of social representations undertaken during the Master's research of Carvalho (2015), seeking to realize the relationship that exists between the teachers' formative process and practice in the construction and alteration of their social representations with respect to artistic practices.

The study's methodology was constructed from the methodological theoretical reference of Social Representations Theory (SRT) proposed by Moscovici (2012), specifically from the procedural approach defined by Jodelet (2001). Data were accordingly collected through narrative interviews and participative observation.

The use of narrative interviews to collect data – utilising the work of Jovchelovitch and Bauer (2013) as a guide – was regarded as crucial for a deeper understanding of the information provided by the subjects. It represented the possibility of articulation of the subjects' experiences from the repertoire of important words and facts, seen as essential for the construction of the web of groups' social representations of artistic practices, since it enabled us to discover elements of subjects' identities and orientations based on their choices.

Participative observation, in its turn, enabled the researchers to monitor the subjects during their teaching activities, as well as their social, cultural and family interactions, exposing the various facets of the teacher's experience. It should be noted that, in dealing with the topic of rural education, being close to the subjects

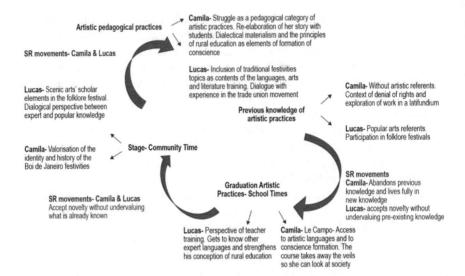

**Fig. 18.2** Motions of Camila's and Lucas's social representation constructions *Source*: Carvalho (2017: 264)

was crucial, as they commuted to the schools, enabling the researchers to understand their daily struggles, elements that helped in the contextualization of the reality of rural education.

Such procedures made it possible to reconstruct the history of the territory of the towns under research, and to survey the territorial, economic, educational and cultural characteristics experienced by the subjects in the course of their schooling, university years and insertion into the activity of teaching. The reconstruction of the subjects' trajectories was made with the intention of analysing the presence of artistic practices in the scope of artistic and cultural activities, professional trajectory and participation in social movements, labour unions and collective organizations, introducing a wide dimension that allows the analysis of artistic practices from the perspective of seven categories: (I) pedagogic practices, (II) artistic practices, (III) formative process, (IV) rural education, (V) mystique, (VI) transforming education and (VII) dialectical and historical materialism.

The analytical process was made up of six movements: (I) data transcription, (II) reconstruction of Camila's and Lucas's life trajectories, (III) analysis of the seven themed categories, (IV) analysis of the motion of change in the way Camila and Lucas thought, felt and acted during their course years, including the training process, internship and initiation into the activity of teaching, (V) apprehension of the motions of objectification and anchorage, and (VI) lexical analysis from the IRaMuTeQ software program. In this chapter, the focus is on the presentation of the fourth motion – analysis of the motion of change in the way of thinking, feeling and acting, as can be seen in Figure 18.2.

We can learn, by analysing Figure 18.2 (Motions of Camila and Lucas' social representation constructions), that the positioning made for each motion contains an accumulation of socio-cultural issues that touch the educators' social representations of their artistic and pedagogic practices. It can be noted that the construction process of Lucas's and Camila's social representations on artistic practices is marked by the third perspective of social representations in motion, which is characterized by change in the face of a new situation, where the subjects accept what is new without undervaluing what is already known (Antunes-Rocha et al. 2015).

Particularly remarkable is the fact that, although both have been allocated in the same type of motion, the trajectory of each one is marked by peculiarities related to collective dynamics within the historical, political and social contexts of the rural subjects. Such elements confer on these peculiarities the sense of sharing, which can be understood from the perception of Jodelet (2001) about the social dynamics of social representations.

> There are representations that fit us hand in glove, or that go all the way through the individuals: those imposed by the dominant ideology or those related to a condition defined in the social structure. However, even in those cases, sharing implies a social dynamic that explains the representation's specificity. This is what has been developed by the research that relates the representation's social character to the individual's social insertion. (Jodelet 2001: 32)

It should be noted that only in Camila's first motion, between her admission to the licentiate undergraduate course in rural education and her involvement in the formative process, did she totally reject the former state to embrace the new one. However, such a rejection was marked by her insertion into the context of constructing artistic and critical knowledge of rural education, a perspective that, in her own words, uncovered her eyes. From that point, the motions of both were distinguished by the dialogue with the new challenging context of changes and the reformulation, taking into account the formerly acquired knowledge.

The description and analysis of this fourth orientation, characterized by Camila's and Lucas's insertion into the activity of teaching, makes it possible to realize that the reality faced by teachers is the strange factor that destabilizes their pedagogic and artistic practices. The analysis took place in an effort to see what provocations caused destabilization and what the motions of those teachers were in order to stabilize such destabilizations.

It became evident that Camila and Lucas had to accept their new condition of school teachers, adapting the contents of the courses in languages, arts and literature, methods and theoretical formulations which they had taken during their undergraduate formative years. The task of inserting rural education in the school context is the chief challenge of that familiarity. Both educators are seen as pioneers from the moment they construct practices and share with other teachers and educators their reflections on the rural school's role and the contention that it has to become a reality for rural citizens.

It is also clear that the condition of pioneers of those subjects compels them to construct a rural school from the actions developed in the classroom. Camila has been observed to cling to her formative years at LeCampo, in connection with her life experiences and the way she has overcome oppression, as well as the new things that she has been learning through her Master's studies and discussions with the multidisciplinary group. Lucas, in turn, clings to what he learned in his formative years at LeCampo, in connection with his labour union experience. Both of them have demonstrated the influence of the formative process – an experience they have in common – on their way of handling challenging elements of their concrete tasks as teachers. Moscovici (2010), speaking about lifestyles and the senses constructed in representations, is helpful in the understanding the sharing and operationalization of knowledge and belief as a common interest.

> The moment when knowledge and technique are transformed into beliefs is when people come together and become a force that is able to transform individuals from passive members into active ones who take part in collective actions and in all that brings life to common existence. Societies break down when there is only power and various interests that unite people, when there is no gathering of ideas and values in which they believe, that may unite them through the common passion of transmitting them from one generation to another. (Moscovici 2010: 173)

From the perspective of Moscovici (2010), it is possible to notice that Camila and Lucas share a common knowledge and technique learned during their formative years at LeCampo, operationalizing this representation in the artistic and pedagogic practices developed by them as teachers. To that end, they also keep dialoguing with the historical and social elements, as well as the concrete challenges faced in their experience as teachers. This dialogue is pervaded with a motion of construction and reformulation of their social representations, characterized by different orientations with regard to artistic practices, which can be analysed from the concepts of objectification and anchorage inherent in Social Representations Theory.

Analysing the motion of Lucas's and Camila's social representation constructions provides information about the importance of the training process as a change-producing element. Camila and Lucas joined the course with different perspectives of political awareness and artistic experiences. For Camila, the course made access to both artistic practices and political engagement possible, while for Lucas, it enriched his reflections on popular arts, introducing erudite elements and his formative process to become a teacher into his reflections. Thinking about the practices of both of them as teachers brings the realization that both started from the perspective of knowledge development, integrating languages, arts and literature, and incorporating in their practices the perspective of the transformative power of education.

## 18.4   Final Considerations

By discussing the perspective of motion, this work attempts to advance analysis of the conceptual and morphological aspects of social representations. In conceptual terms, the motion has been discussed as a representational phenomenon in its dynamic characteristic. The attempt to put into record the creation of a discussion under the Social Representations Study Group (GERES) arises from an effort to systematize collaborative discussions, studies, findings and publications, giving due acknowledgements in the process of constructing academic knowledge and in the systematization of a concept that may lead to the discovery of the formulation processes of social representations.

In methodological terms, the journey is towards positioning the subject's narrative as a central element in the process of capturing social representations. In recently produced works by the group, practices of participative observation have been tested. The intention is to work through round-table discussions and/or focus groups in the near future.

Challenges do exist. Some are already part of the discussion and others have gone unnoticed so far. Affective aspects, for example, are an issue to be discussed in forthcoming works. The way that forms of thinking and feeling are related to practices is another topic already under study, as can be seen in the works of Carvalho (2017). The link between the different motions taken by the subjects and the aspects related to their peasant identity has appeared more often in more recent works, and is certainly another topic that requires a closer look. The dimension of the subjects as participants in collective processes of opposition, both in social movements and labour unions of bigger or smaller territorial extent, presents itself as an enormous challenge in terms of empirical research.

# References

Ambrósio, Ana Esperança Futi Bambi, 2014: "Um estudo sobre as representações sociais dos pais e encarregados de educação do colégio Padre Builu em Cabinda/Angola – relação família-escola" (MSc dissertaton, Federal University of Minas Gerais, Faculty of Education).

Amorim-Silva, Karol Oliveira de, 2015: "Educar em prisões: um estudo na perspectiva das representações sociais" (MSc dissertation, Federal University of Minas Gerais, Faculty of Education).

Andriamifidisoa Danichert, Irène, 1982: "La transformation d'une représentation sociale: exemple des relations sociales à Madagascar" (PhD dissertation, University of Provence).

Angelina, Casimiro Kâmbua, 2014: "A relaçao escola-família: um estudo sobre as representações sociais de pais e encarregados de educação sobre a escola do ensino primário do Chiwéca em Cabinda/Angola" (MSc dissertation, Federal University of Minas Gerais, Faculty of Education).

Antunes-Rocha, Maria Isabel; Martins, Aracy Alves, 2009: "Formar docentes para a educação do campo: desafios para os movimentos sociais e para a universidade", in Antunes-Rocha, Maria Isabel; Martins, Aracy Alves: *Educação do campo: desafios para a formação de professores* (Belo Horizonte: Autêntica): 25–39.

Antunes-Rocha, Maria Isabel; Leite, Maria Alcira; Nascimento, Margarete de Castro; Amorim-Silva, Karol Oliveira de, 2013: "Representações sociais em movimento: desafios para tornar o estranho em familiar", Paper for the *Anais da VIII Jornada Internacional* and the 6th Brazilian Conference on Social Representations, Recife, Brazil.

Antunes-Rocha, Maria Isabel; Martins, Maria de Fátima Almeida; Martins, Aracy Alves, 2012: *Territórios educativos na educação do campo: escola, comunidade e movimentos sociais* (Belo Horizonte: Autêntica).

Antunes-Rocha, Maria Isabel; Amorim-Silva, Karol Oliveira de; Benfica, Welessandra Aparecida; Carvalho, Cristene Adriana da Silva; Ribeiro, Luiz Pablo, 2015: "Representações sociais em movimento: desafios para elaborar o estranho em familiar", Paper presented at the 12th National Congress of School and Educational Psychology and the 37th Annual Conference of the International School of Psychology Association, São Paulo, Brasil: 836–841.

Caldart, Roseli Salete, 2004: *Pedagogia do movimento sem terra: escola é mais do que que escola* (São Paulo: Expressão Popular).

Carvalho, Cristene Adriana da Silva, 2015: "Práticas artísticas dos estudantes do curso de licenciatura em educação do campo" (Federal University of Minas Gerais, Faculty of Education).

Carvalho, Cristene Adriana da Silva, 2017: "Representações sociais das práticas artísticas na atuação de professores do campo" (Federal University of Minas Gerais, Faculty of Education).

Flament, Claude, 1989: "Structure et dynamique des représentations sociales", in: Jodelet, Denise (Ed.): *Les représentations sociales* (Paris: PUF).

Jodelet, Denise, 2001: *As representações sociais: um domínio em expansão* (Rio de Janeiro: EdUERJ).

Jodelet, Denise, 2005: "Experiência e representações sociais", in: Menin, Maria Suzana de Stefano; Shimizu, Alessandra de Morais (Eds.): *Experiência e representação social: questões teóricas e metodológicas* (São Paulo: Casa do Psicólogo): 23–56.

Jovchelovitch, Sandra, 2011: *Os contextos do saber: representações sociais, comunidade e cultura* (Petrópolis/RJ: Vozes).

Jovchelovitch, Sandra; Bauer, Martin, 2013: "A entrevista narrativa", in: Gaskell, George; Bauer, Martin (Eds.): *Pesquisa qualitativa com texto, imagem e som: um manual prático* (Petrópolis/RJ: Vozes).

Molina, Mônica Castagna; Hage, Salomão Mufarrej, 2016: "Riscos e potencialidades na expansão dos cursos de licenciatura em educação do campo", in: *Revista Brasileira de política e administração da educação*, 32,3: 805–828.

Moliner, Pascal, 2001: "Introduction", in: Moliner, Pascal (Ed.): *La dynamique des représentations sociales: pouquoi et comment les représentations se transforment-elles?* (Grenoble: Presses Universitaires de Grenoble – PUG): 7–14.

Moliner, Pascal; Guimelli, Christian, 2015: *Les représentations sociales: fondements théoriques et développements récents* (Grenoble: PUG).

Moscovici, Serge, 2012 [1961]: *A psicanálise, sua imagem e seu público* (Petrópolis/RJ: Vozes).

Moscovici, Serge, 2010: *Representações sociais: investigações em psicologia social* (Petrópolis/RJ: Vozes).

Ribeiro, Luiz Pablo, 2016: "Representações sociais de educandos do curso de licenciatura em educação do campo sobre a violência" (PhD dissertation, Federal University of Minas Gerais, Faculty of Education).

Ribeiro, Luiz Pablo, 2017: *O campo, a violência e a educação do campo: representações sociais de educandos do curso de licenciatura em educação do campo sobre a violência* (Rio de Janeiro: Gramma).

Tafani, Éric; Bellon, Sébastien, 2005: "Études expérimentales de la dynamique des représentations sociales", in: Abric, Jean Claude (Ed.): *Méthodes d'étude des représentations sociales* (Toulouse: ERES – Horscollection): 255–277.

Telau, Roberto, 2015: "Ensinar – incentivar – mediar: dilemas nas formas de sentir, pensar e agir dos educadores dos CEFFAS sobre os processos de ensino/aprendizagem" (Federal University of Minas Gerais, Faculty of Education).

# Chapter 19
# Social Representations: A Bet on Social Change

**Mireya Lozada and Adelina Novaes**

*My interest in social representations is firmly underpinned by their importance to processes of social change. And what do we mean by social change? That people are able to think in advance. That people do not only act according to their interests. People must be able to anticipate some things, and thus have a representation of the situation and of the future… and have a language to be able to talk about the future, and that is where the space of social representations is found* (Moscovici 1999: 303).

**Abstract** This chapter shows exactly what we would like to highlight as a working model of present and future psycho-social perspectives for Latin America in the Anthropocene based on social representations. It is the result of shared reflections between two researchers from different generations and national contexts regarding the potential of Social Representations Theory to promote ethico-political change.

**Keywords** Latin America · Ethico-political change · Psycho-social perspective · Social Representations Theory

Mireya Lozada is Professor and Researcher at the Universidad Central de Venezuela (Venezuela). Email: mireyaloza@gmail.com.

Adelina Novaes is Researcher at the Fundação Carlos Chagas (Brazil) and Professor and Researcher at the Universidade Cidade de São Paulo (Brazil). Email: adelnovaes@gmail.com.

## 19.1  Introduction

This book brings together studies that are part of a new generation of research and seek to contribute to Social Representations Theory (SRT) in the context of the Anthropocene from a Latin American perspective. The proposal guiding the book was for chapters written by emerging researchers to be commented upon by academics with established trajectories in Latin America. The project thus seeks to show how SRT, developed forty years ago by researchers trained in Europe (Banchs/Arruda 2014), has gained new potential in the tropics, developing strongly and helping the theory to continue through the training of young researchers. It is, therefore, a book that deals with the time, the space and the embodied subjects that have seen SRT grow in Latin America and project itself into the future.

This chapter, however, has a different structure. When we were invited to write it, we realized that the partnership in which we have been involved, to establish the Latin American research network in movement: psycho-social perspectives,[1] shows exactly what we would like to highlight as a working model and the investment we make for the future. Thus, to conclude the book, we proposed to the editors a single text, that would be the result of shared reflections between two researchers from different generations and national contexts regarding the potential of Social Representations Theory to promote ethico-political change.

SRT's interest in change is not new, finding expression in Serge Moscovici's classic studies of minority influence. While this theme attracted collaborators (like Doise, Doms, Gonzales Montes, Ibáñez, Mugny, Papastamou and Pérez, among others), the studies on minority influence were traditionally focused on relations between groups and the possibility of groups bringing information into a debate that was new and unsettling enough to disrupt the *status quo* and result in changes to symbolism and social meanings.

Specifically with regard to Latin America, since 1992, when the first International Conference of Social Representations was held in Ravello, Italy, various studies have examined forms of social thought among diverse cultures across the continent, constituting a contextual turn that has brought to light our reality and its complexity in the construction and transformation of social representations. Fluctuations between logics of resistance and social change in the region offer important heuristic possibilities for the contextual direction, when articulated with the hermeneutical, discursive and affective directions posited by the human sciences in recent decades, which privilege interpretation, language and emotion in their approach to analysing social dynamics (Lozada 2002).

---

[1]In Spanish: Latinoamerica en movimiento: miradas psisociales. The Network, currently in formation, aims to promote exchange, reflection and the construction of knowledge around psychosocial dynamics in Latin America, from an ethico-political perspective.

In this sense, the adoption of Social Representations Theory in our studies is linked to its potential to promote creativity and change, since SRT is radically different from both social determinism, which explains human beings as products of society, and pure voluntarism, which regards the human being as a free agent. Thus, the studies that bring us together share the same dissatisfaction that feeds Latin American social psychology, namely: its constant preoccupation with immediate social reality, its emancipatory vocation and its political-reflective character.

In this regard, in this chapter, we both seek to offer a personal form of communication and presentation of the objects which have held our attention and the research trajectories that have crossed thanks to our common attachments and interests, seeing as, in our work together, hopeful horizons beckon with great force.

We certainly celebrate this with enthusiasm as, with great interest, we watch the movements of our people. We congregate and join forces to identify, through the study of subjectivities and social thought, hubs of resistance despite the hardships faced in our Latin American societies. Between hope and uncertainty, we advance in a collective project of research as we face the task presented to us.

## 19.2  Latin America: Between Uncertainty and Hope

Numerous dichotomies are frequently invoked when thinking about Latin America. In contemporary times, we highlight two: uncertainty and hope. For Bengoechea,[2] Latin America swings between one and the other, as even if it has seen a weakening of populism and progress in Colombia's peace process in recent years, from the socio-economic point of view, the Latin American subcontinent is trapped between two major powers: the United States and China, between the expected expansion of the US economy and the slowdown of the Asian giant. In addition to Latin American countries' increasing debts, from a political point of view and within the framework of the region's democratic weakness, corruption is the main threat, tending to grow in the minds of voters and weaken the 'social glue'.

According to Santos (2016), uncertainty is generated between fear and hope, which are not equally distributed across social groups or historical epochs. For the author, the uncertainties of our time concern knowledge, democracy, nature and dignity. "This means that the epistemological and experiential distribution of fear and hope is defined by parameters that tend to benefit social groups with greater access to scientific knowledge and technology" (Santos 2016: 50). According to him, democratic processes became uncertain as they were manipulated by the holders of social and economic power, while human rights became a hegemonic

---

[2]Bengoechea, Gonzalo Gómez, 2017: *Latinoamérica, entre la esperanza y la incertidumbre*, at: www.cambio16.com/actualidad/latinoamerica-entre-la-esperanza-y-la-incertidumbre (14 May 2017).

narrative referring to the dignity of human beings, in a context where 'naturally' inferior groups were never completely defeated (Santos 2016).

Our interest lies between these fears and hopes, and amongst these uncertainties. Rather than digging deeper into the worries and proposals of Latin American researchers regarding the impacts of global economic processes and the crisis of the democratic state in the region, the return of militarism or the emergence of neo-authoritarian models, we are interested in reflecting on subjectivities, knowledge and social thought in Latin American studies on social representations in this context. More than a defence of culture, privileged in Latin America during the transition from "theories of dependency to theories of resistance", according to Morris/Schlesinger (2000), it is about the dialogue between society and culture:

> Undoubtedly, the relationship between society and culture is the axis of social representations, at the very intersection of society and culture, because nothing becomes a social reality if it does not have a certain cultural inscription, which means inscription in the beliefs of the people (…). We can say then that when we speak of this intersection of society and culture, we can understand and speak of "common sense". Even though I am not the first to observe it, we have to recognize that if something is not passed to common sense, that is to action, in the way people position themselves in relation to others within categories of thought, nothing can become a living reality, that is: a reality of relationships and a reality of action. (Moscovici 1999: 302)

We are participants and disseminators of this commitment to the future, which is based on a dialogue between researchers from Latin American countries, which addresses certain parallels, juxtapositions and contradictions between biographical trajectories and social itineraries. In writing this chapter, our interests are united with those of other colleagues[3] from different countries within our continent, from different generations, and from different traditions within SRT, but who share with us an investment in such social exchange.

This convergence of interests stems from the advances made by SRT on a continental scale. Catalysed by networking, it has permitted many of us to engage with questions related to the pressing challenges and social developments in Latin America, but above all allowed young researchers to be exposed to the diversity of approaches, knowledge and themes among researchers already consolidated in the field of SRT.

This is the case of the Latin American research network in movement: psychosocial perspectives, which is the product of previous studies, developed within the framework of the Latin American Imaginaries Project (Arruda/de Alba 2008;

---

[3]Those who have participated in the creation of the Network: Alfredo Guerrero Tapia (Mexico), Angela Arruda (Brazil), Clarilza Prado de Sousa (Brazil), Eduardo Aguirre (Colombia); Eréndira Serrano (Mexico); Erika Souza (Brazil); Franscisco Portugal (Brazil), Jorgelina Di Iorio (Argentina), Josefina Parra Toledo (Mexico); Lúcia Pintor Santiso Villas Bôas (Brazil); María de Fátima Flores Palacios (Mexico); Martha De Alba (Mexico); Paulo Afrânio Sant'Anna (Brazil); Silvia Vidrio Gutierrez (Mexico); Susana Seidmann (Argentina); Juana Juarez (Mexico).

Arruda/Sousa 2013). The seed of our partnership lay in this Project,[4] which was composed of researchers from Latin America and Europe, coordinated internationally by Angela Arruda and in Brazil by Clarilza Prado de Sousa. The "Imaginaries", as it became known among the participating students and researchers, resulted in courses, meetings and publications.

In the Network's current configuration, researchers with different experiences seek to highlight voices, symbols and narratives that share an ethico-political dimension (Sawaia 1999a), claiming an inclusive alterity (Arruda 1998) that allows us to re-signify the democratic imaginary as a participatory project that is felt and shared by different social and political sectors across Latin America, taking into account the advance of processes of social and political polarization in different regions and countries, which define an "Other as enemy" (Lozada 2014).

As Touraine points out (2016: 13), in a "communication society", power seeks to dominate representations, opinions, decisions, lifestyles and everything that relates to personality, without renouncing control of material goods and capital. In light of this, and of the meanings and narratives in dispute in our context, the Network pursues a set of objectives and problematic axes:

- To problematize the triad: stability, resistance, innovation, so as to understand the construction, transformation and re-signification of social representations in Latin America, in democratic and neo-totalitarian contexts, and in the context of rising global geopolitical tensions.
- To recognize the transits of the social subject in Latin America and its impact on representations of identity, in global contexts of communication and "cultural hybridization" (García Canclini 1989).
- To discuss the consensual nature of social thought and the notion of common sense in the current Latin American context, where processes of fragmentation are intensifying and different forms of struggle and social resistance between majorities and minorities are being staged.
- To develop joint research projects between different countries in Latin America, enhancing possibilities for change through articulations between social representations and social influence by minorities.

Recognizing the dialogical nature of knowledge development, it is also urgent to address the complexity of social thought claimed by SRT and its complementarity with concepts such as moral thinking and critical thinking, formulated by Kantian traditions and the critical theory of the Frankfurt School, respectively. Similarly, there is the great challenge of producing comprehensive knowledge about social dynamics in Latin America and the construction of social subjectivity (Novaes 2015; Passeggi/Novaes 2018) within the logic of exclusion-perverse inclusion (Sawaia 1999b), or exclusive inclusion (García-Guadilla 2003).

---

[4]Mireya Lozada, was part of the body of researchers who developed the Latin American Imaginaries Project. Within the scope of this Project, Adelina Novaes undertook her undergraduate degree under the supervision of Angela Arruda and her Master's and PhD under the supervision of Clarilza Prado de Sousa.

## 19.3   A Look at Our Work

Our adoption of Social Representations Theory in our studies is based on its potential to promote change. In identifying symbolic production as the construction of singular subjects in interaction, SRT recognizes the determinants of such productions, but does not rule out the possibility of innovation and social change, preserving the creative capacity of subjectivity within its analysis.

Marková (2006) has already shown us that there is considerable symmetry between the concepts of stability and change in terms of their theoretical conditions. Her examination of psychological theories of social knowledge indicates that, in general, they assume stability, rather than change. This was not the case with the theories developed by Moscovici. In particular, in elaborating his genetic and innovation theories, Moscovici assumed that change was possible.

For him, social transformation is made possible by innovation or, in other words, by the mutation of representations that occurs through the incorporation of new information by social subjects fully engaged in interaction and immersed in their social, cultural and historical context. According to Moscovici,

> Innovation is imperative in society, with the same right as conformity. From this point of view, innovation should not be considered a secondary phenomenon, a form of deviation or non-conformity, but must be taken for what it is: a fundamental process of social existence. Innovation presupposes a conflict whose solution depends on both the forces of change and the forces of control. The tension between those who must defend certain norms, opinions or values, so as to change existing ones, is the result on which the evolution of a society rests. (Moscovici 2011: 238)

By adopting SRT, we seek to contribute to the construction of socio-historically situated and collective knowledge of innovation and resistance. On the other hand, following the emancipatory pedagogical research of Paulo Freire (1967) and Ignacio Martín Baró (1986), we position ourselves in a permanent demand for comprehension, intervention and psychosocial support for the population of Latin America, in their everyday lives beset by serious structural problems. However, an inescapable responsibility and ethico-political challenge is to look critically at the task ahead, even in the midst of urgent circumstances.

Doise (1986: 254) offers support in conceiving of "social representations as generative principles of position-taking linked to specific insertions in a set of social relations that organize the symbolic processes that intervene in those relations".

Both in defining our professional identity and characterizing our technical interventions, we, as psychologists, privilege some aspects at the expense of others, through the arrangement of the priorities of mental health and social changes that justify and legitimize our levels of commitment in public or private spheres. The notions of 'compromise' derived from these two positions accounts in part for the link between the discourse and action of the individual. On the other hand, in agreement with a vision that is widespread in militant circles in Latin America, it is also linked to a historical commitment in which the theory-praxis link appears to be associated with the sturdiness and permanence of principles that drive social transformation, along Marxist or Gramscian lines. Thus, the work of the psychologist,[5] or professional social scientist, is assimilated in the first case as a 'socially sensitive expert' and in the second as a 'committed professional'.

Even if both tendencies are underpinned by the "promise" of carrying out and sustaining action (Arendt 1968) in some way, and each retains an essential core that sustains its professional practice and known symbolic universe, the often unforeseen violent and dramatic character of many of the problems faced by different populations in Latin America redefines the intentionality, the temporality and the situational moorings of these commitments and how they are represented (Lozada/Rangel 2001).

The ethical challenge leads us to assume increasing degrees of commitment and allows for the emergence of consensus, favouring spaces of coexistence in extremely vulnerable contexts and the development of action by different groups who, during other irreconcilable moments, are committed to different theoretical and methodological principles. These "consensual universes", continuous and visible creations permeated by meanings and purposes with a human voice, can, in fact, achieve the goal of all representation: to make the unfamiliar familiar, to make it easier to understand of society, and to guide behaviour (Moscovici 1984: 24).

As shown by Palmonari and Zani (1989), when we situate our work as psychologists in the opposition social-individual it acquires social significance by providing a technical scientific solution to the individual; or by responding with specific tools to the needs of the subject that allow him or her to confront reality and transform it. However, in both cases the psychosocial dimension appears blurred. The psychological is understood as referring to the individual and the internal, while the social refers to the conditions of the 'context'. The approach to this problematic from the psychosocial perspective, which in many cases is defined as the combination of the sociological and the psychological, goes beyond the individual-society dichotomy towards an integrated vision in which subject and society are understood as parts of the same process, as integrated and mutually constitutive realities.

Thus, we understand that psychosocial intervention does not undermine people's knowledge, their history or their possibility of facing different and critical situations. Hence the importance of dialogue with affected communities in order to define the objectives and processes of intervention, in order to rebuild the social fabric (Beristain 1999). Therefore, we emphasize the need to recover ethical, affective, individual and collective references that foster commitment to the task of reconstruction in the context of social fragmentation in our region, while at the same time strengthening processes that construct citizenship. This means facilitating and promoting the inclusion of different social groups and enabling

---

[5]According to Martín-Baró (1984: 504), "In more direct terms, mental health is a dimension of the relationships between people and groups rather than an individual state, although that dimension takes root in a different way in the organism of each of the individuals involved in these relationships, producing diverse manifestations ('symptoms') or states ('syndromes')."

the required participatory processes. In this context, the ethico-political challenge involves assuming the transformations that each group has a right and duty to contemplate. This responsibility places us in a position of ethical obligation to the other, according to Levinas's (1995) sense of a responsibility for the other through dialogue.

Our psychosocial work expands when it demands the crossing of geographical boundaries, and is strengthened through joint action. Thus, we join our efforts in working along different paths, in different countries, but with the same purpose of change, in favour of the social development of our people. In many ways, we are united by our investment in SRT as an instrument of change. For us, this investment is more than an academic-professional effort; it is a political commitment.

# References

Arendt, Hannah, 1968: *Between past and future* (New York: Penguin Books USA).

Arruda, Angela, 1998: "O ambiente natural e seus habitantes no imaginário brasileiro-negociando a diferencia", in: Arruda, Angela (Ed.) *Representando a alteridade* (Petrópolis: Vozes): 17–46.

Arruda, Angela, 2002: "¿Hacia una epistemología de representaciones sociales propia a América latina?", Paper for the 6th International Conference on Social Representations – Thinking Societies: Common Sense and Communication, Stirling, Scotland, 27 August–1 September.

Arruda, Angela; de Alba, Martha (Eds.), 2008: *Representaciones y espacios imaginarios: aportes desde latinoamerica* (Mexico City/Paris: Universidad Autónoma Metropolitana/Maison des Sciences de l'Homme, Anthropos).

Arruda, Angela; Sousa, Clarilza Prado, 2013: *Imaginário e representação social de universitários sobre o Brasil e a escola brasileira: um estudo construído com múltiplas possibilidades* (São Paulo: Annablume).

Banchs, María Auxiliadora; Arruda, Angela, 2014, "Entrevista concedida à María Aurixiadora Banchs", in: Sousa, Clarilza Prado; Ens, Romilda Teodora; Villas Bôas, Lúcia; Novaes, Adelina; Stanich, Karina (Eds.), *Angela Arruda e as representações sociais: estudos selecionados* (Curitiba-São Paulo: Champagnat/Fundação Carlos Chagas).

Baró, Ignácio Martín, 1984: "Guerra y salud mental", in: *Estudios Centroamericanos*, 429, 430 (El Salvador): 503–514.

Baró, Ignácio Martín, 1986: "Hacia una psicología de la liberación", in: *Boletín de Psicología*, 22 (El Salvador): 219–231.

Beristain, Carlos Martín, 1999: *Reconstruir el tejido social: un enfoque crítico de la ayuda humanitaria* (Barcelona: Icaria).

Doise, Wilheim, 1986: "Les représentations sociales: définition d'un concept", in: Doise, Wilheim; Palmonari, Augusto (Eds.): *Les représentations sociales: un nouveau champ d'étude* (Genève: Delachaux & Niestlé): 81–94.

Freire, Paulo, 1967: *Educação como práctica da liberdade* (Rio de Janeiro: Paz e Terra).

García Canclini, Néstor, 1989: *Culturas híbridas: estrategias para entrar y salir de la modernidad* (Mexico City: Grijalbo).

García-Guadilla, Maria del Pilar, 2003: "Politization and polarization of Venezuelan civil society: Facing democracy with two faces", Paper for the 24th International Congress of the Latin American Studies Association, Dallas, USA, 27–29 March.

Levinas, Enmanuel, 1995: *Difficile liberté* (Paris: Albin Michel).

Lozada, Mireya, 2002: "Representaciones sociales en Latinoamérica: el giro contextual", Paper for the 6th International Conference on Social Representations – Thinking Societies: Common Sense and Communication", Stirling, Scotland, 27 August–1 September.

Lozada, Mireya, 2014: "Us or them? Social representations and imaginaries of the other in Venezuela", in: *Papers on Social Representations*, 23: 21.1–21.16.

Lozada, Mireya; Rangel, Ana, 2001: "Reconstruir desde la gente: la cuestión psicosocial", in: San Juan, Cesar (Ed.): *Catástrofes y ayuda en emergencia* (País Vasco: Icaria): 217–233.

Marková, Ivana, 2006: *Dialogicidade e representações sociais: as dinâmicas da mente* (Petrópolis, RJ: Vozes).

Morris, Nancy; Schlesinger, Philip, 2000: "Des theories de la dépendance aux theories de la résistance", in: *Hermes*, 28 (Paris): 19–33.

Moscovici, Serge, 2011: *Psicologia das minorias ativas* (Petrópolis, RJ: Vozes).

Moscovici, Serge, 1999: "Lo social en tiempos de transición: Lozada, Mireya en diálogo con Serge Moscovici", *Revista SIC*, 617 (Caracas): 302–305.

Moscovici, Serge, 1984: "The phenomenon of social representations", in: Farr, Robert; Moscovici, Serge (Eds.): *Social representations* (Cambridge: Cambridge University Press): 3–96.

Novaes, Adelina, 2015: "Subjetividade social docente: elementos para um debate sobre 'políticas de subjetividade'", in: *Cadernos de Pesquisa*, 45,156: 328–343; at: https://doi.org/10.1590/198053143205.

Palmonari, Augusto; Zani, Bruna, 1989: "Les représentations sociales dans le champ des professions psychologiques", in: Jodelet, Denise (Ed.): *Les représentations sociales* (Paris: Presses Universitaires de France): 319–339.

Passeggi, Maria da Conceição; Novaes, Adelina, 2018: "Núcleo central e narrativa: entre permanência e mudança das representações sociais do fazer docente", in: Abrahão, Maria Helena Menna Barreto; Cunha, Jorge Luiz; Villas Bôas, Lúcia. (Eds.): *Pesquisa (auto) biográfica: diálogos epistêmico-metodológicos* (Curitiba: CRV): 223–234.

Santos, Boaventura de Sousa, 2016: "La incertidumbre, entre el miedo y la esperanza", in: *El viejo Topo*, 346 (Barcelona): 49–53.

Sawaia, Bader, 1999a: "Introdução: Exclusão ou inclusão perversa?", in: Sawaia, Bader (Ed.): *As artimanhas da exclusão. Analise psicossocial e ética da desigualdade social* (Petrópolis: Editora Vozes): 7–13.

Sawaia, Bader, 1999b: "O sofrimento ético-político como categoria de análise da dialéctica exclusao/inclusão", in: Sawaia, Bader (Ed.): *As artimanhas da exclusao. Analise psicosocial e ética da desigualdade social* (Petrópolis: Editora Vozes): 99–119.

Touraine, Alain, 2016: *Le nouveau siécle politique* (Paris: Editions du Seuil).

## Other Literature

Bengoechea, Gonzalo Gómez, 2017: *Latinoamérica, entre la esperanza y la incertidumbre*; at: www.cambio16.com/actualidad/latinoamerica-entre-la-esperanza-y-la-incertidumbre (14 May 2017).

# About the Editors

**Clarilza Prado de Souza** (Brazil) is a full professor at the Pontifical Catholic University of São Paulo (PUC-SP). She graduated in psychology and obtained an MSc and a PhD in education from the Pontifical Catholic University of São Paulo. She undertook postgraduate specializations at the University of Marseilles and Harvard University and post-doctoral studies at the École des Hautes Études and at the Foundations des Sciences de L'Homme de Paris. She has complementary training in England, Sweden, France and Brazil.

She created and led the "UNESCO Chair for Teacher Professionalization", which assembles thirty-four research groups. At the time, this made it possible for the International Centre for Studies on Social Representations and Subjectivity – Education (CIERS-ed) at the Carlos Chagas Foundation to expand its work nationally and internationally. She currently coordinates the Work Group of Social Representations of the National Association of Research and Graduate Studies in Psychology – ANPEPP. She is the author of fifteen books, fifty scientific articles and fifty-three chapters published on educational evaluation, education, and social representations in Spanish, Portuguese and French.

Her recent authored and co-authored publications include: "Educação para a paz: análise das condições culturais de desenvolvimento no Brasil", in *Práxis Educativa*, 14 (2019): 1–21; "Estudo da alfabetização científica de alunos do 9° ano do ensino fundamental de um colégio particular da cidade de São Paulo", in *Debates Em Educação*, 11 (2019); "Large-scale assessment classroom and school: a necessary dialogue", in *Psicologia Da Educação*, 1 (2019): 13–23; "As crianças face a continuidade e a descontinuidade da mente: notas em psicologia social", in *Educação em Foco: Revista de Educação*, 24 (2019): 925–952; "Estudo da

C. Prado de Sousa and S. E. Serrano Oswald (eds.),
*Social Representations for the Anthropocene: Latin American Perspectives*,
The Anthropocene: Politik—Economics—Society—Science 32,
https://doi.org/10.1007/978-3-030-67778-7

alfabetização científica de alunos do ensino médio de um colégio de São Paulo", in *Revista Eletrônica Científica Ensino Interdisciplinar* (RECEI), 5 (2019): 537–544.

*Address*: Dr Clarilza Prado de Souza, Pontifícia Universidade Católica de São Paulo, Programa de Pós-Graduação em Psicologia de Educação e Programa de Pós-Graduação em Educação Profissional da PUC/SP, Programa de Pós-Graduação em Educação: Psicologia da Educação, Rua Monte Alegre, 984, Perdizes, 05014901 – São Paulo, SP – Brasil.

*Email*: clarilza.prado@gmail.com.

 **Serena Eréndira Serrano Oswald** (Mexico) is full professor at the Regional Multidisciplinary Research Centre, National Autonomous University of Mexico (CRIM-UNAM). In addition to a PhD in social anthropology (UNAM) and doctoral studies in psychology (CRISOL), she holds an MSc in social psychology (LSE), an MFT in systemic family therapy (CRISOL), and a BA Hons in political studies and history (SOAS). She has undertaken postdoctorate studies in sociology and gender (UNAM), and postgraduate specializations in teaching and supervision, infant and adolescent psychotherapy, Gestalt psychotherapy, person-centred therapy, and gender justice and public politics (FLACSO), and has eight professional diplomas in the social sciences, gender and psychotherapy.

She is certified by the National Council of Researchers (SNI) and by UNAM (PRIDE). She is also President of the Latin American and Caribbean Regional Science Association (LARSA-RSAI), was General Secretary of the Latin-American Council for Peace Research (2017–2019), and President of the Mexican Association of Regional Development (AMECIDER; 2013–2016). She has edited thirteen books and written more than sixty scientific articles and chapters on gender, social representations, resilience, migration, regional development, and environment in English, Spanish, Portuguese and Chinese.

Selected authored and co-authored publications include: "Social representations, gender and identity: Interactions and practices in a context of vulnerability", in *Papers on Social Representations*, 28,2 (2019): 4.1–4.19; *Estudio para el seguimiento de las recomendaciones generales, informes y pronunciamientos, emitidos por la CNDH. Tema: indígenas* (2019; CNDH-UNAM); *Regiones, desplazamientos y geopolítica: agenda pública para el desarrollo territorial* (e-collection, 2019; AMECIDER-IIEc-LARSA); *Migración, cultura y estudios de género desde la perspectiva regional* (e-book, 2019; AMECIDER-IIEc-UNAM);

*Perspectivas regionales de los estudios culturales y de género* (e-book, 2019; AMECIDER-IIEc-UNAM); *Climate change, disasters, sustainability transition and peace in the Anthropocene* (2019; Springer); *Riesgos socio-ambientales para la paz y los derechos humanos en América Latina* (2018; CRIM-UNAM); *Risks, violence, security and peace in Latin America* (2018; Springer).

*Address*: Dr Serena Eréndira Serrano Oswald, Priv. Rio Bravo No. 1, Col. Vista Hermosa, Cuernavaca, Morelos, Mexico CP 62290.

*Email*: sesohi@gmail.com.

# About the Contributors

**Adelina Novaes** is a researcher in the Educational Research Department of the Carlos Chagas Foundation (FCC), where she works as coordinator of the International Centre for Studies on Social Representations and Subjectivity – Education (CIERS-ed). She is a permanent researcher of the UNESCO Chair on Teaching Professionalization, a member of the scientific council of the Franco-Brazilian Serge Moscovici Chair and executive editor of the journal *Periodical Studies in Educational Evaluation*. She is *stricto sensu* lecturer at the Post-Graduation Programme in Education and the Professional Master's Programme in Educational Management Training, both at the City University of São Paulo (UNICID). She undertook her postdoctoral degree at the Pontifical Catholic University of São Paulo (PUC-SP) and the Department of Social Psychology at the London School of Economics and Political Science (LSE).
*Email*: adelnovaes@gmail.com.

**Adriane Roso** is Associate Professor at the Federal University of Santa Maria – UFSM (Graduate/Masters). She was a postdoctoral fellow at Harvard University (March 2018 to June 2019), and undertook postdoctoral studies in communication at UFSM. She gained her PhD in psychology at the Pontifical Catholic University (PUC-RGS), with a Fulbright Fellowship at Columbia University. She has a Master's degree in social and personality psychology (PUCRS). She is a public health specialist (UFRGS), a health management specialist (UFRGS), a CNPq Research Productivity Fellow, and conducts research at *Amparo Research State Foundation in Rio de Janeiro (FAPERJ)*.
*Blog*: www.psicologiasocialbrasileira.blogspot.com.
*Email*: adriane.roso@ufsm.br.

© Springer Nature Switzerland AG 2021
C. Prado de Sousa and S. E. Serrano Oswald (eds.),
*Social Representations for the Anthropocene: Latin American Perspectives*,
The Anthropocene: Politik—Economics—Society—Science 32,
https://doi.org/10.1007/978-3-030-67778-7

**Alcina Maria Testa Braz da Silva** is a professor and researcher on the science, technology and education postgraduate programme (PPCTE) of the Federal Centre for Technological Education Celso Suckow da Fonseca (CEFET) in Rio de Janeiro. She gained her BSc in physics at the Federal University of Rio de Janeiro (UFRJ; 1980), her MSc in material and metallurgical engineering at the Federal University of Rio de Janeiro (UFRJ; 1989), and her PhD in education at the Federal University of Rio de Janeiro (UFRJ; 1998). She has experience in education, and specializes in the following subjects: children's education; information and communication technologies; social representations; teaching and curricula in physics.
*Email*: alcina.silva@cefet-rj.br.

**Alda Judith Alves Mazzotti** is an Emeritus Professor at the Federal University of Rio de Janeiro (UFRJ) and at the Estácio de Sá University, where she is a full professor in the area of Educational Psychology. She completed her doctoral studies in Educational Psychology at New York University. Her academic production focuses on research methodology, and her work is an obligatory reference in the area of education. With this methodological rigour she has undertaken, and continues to develop, countless studies on social representations, having implemented a research group that focuses on analysing teacher knowledge, teacher training and work, teacher identity, school failure, public school students, and child labour.
*Email*: aldamazzotti@gmail.com.

**Alfredo Guerrero Tapia** is a senior researcher in the Faculty of Psychology at the National Autonomous University of Mexico (UNAM). He has a BSc in psychology, a Master's in educational psychology, and PhD in social-environmental psychology, all from UNAM. He has been working in the field of social representations, theoretically and empirically, since 1989. He is co-editor, with Denise Jodelet, of the book *Develando la cultura* (México: Facultad de Psicología/ UNAM, 2000). He belonged to the International Research Group on Latin American Imaginaries and Social Representations, supported by the Laboratoire Européen de Psychologie Sociale, of the Fondation Maison des Sciences de l'Homme, whose research output, "Images of Latin America and Mexico through mental maps", was published in the book *Espacios imaginarios y representaciones sociales* (Barcelona, Anthropos/UAMI, 2007).
*Email*: alfredog@unam.mx.

**Alicia Barreiro** has a PhD in educational sciences, a Master's in educational psychology and a psychology degree with a postdoctoral research stay. She is a researcher at the University of Buenos Aires (UBA) and the National Council for Research and Technology (CONICET). She is also academic coordinator of the Master's in teacher training at the Pedagogical University (UNIPE), and she teaches at the Chair of Psychology and Genetic Epistemology, Faculty of Psychology, University of Buenos Aires (UBA) and on the Master's course in cognitive psychology and learning (FLACSO). She directs various research projects and is the author of numerous publications in national and international scientific

journals on social representations and the effects of collective memory on inter-group relationships, moral development and the way children and adolescents understand the social world.
*Email*: avbarreiro@gmail.com.

**Aline Reis Calvo Hernandez** is an associate professor in the Faculty of Education at the Federal University of Rio Grande do Sul (UFRGS), and a guest professor on the Master's course in environment and sustainability at the State University of Rio Grande do Sul (UERGS). She is the leader of the research group "Political Psychology, Education and Present History" (National Council of Scientific and Technological Development – CNPq). She undertook postdoctoral research in social psychology at the Pontifical Catholic University of Rio Grande do Sul (2009; PUC-RGS), and gained her PhD in social psychology and methodology at the Autonomous University of Madrid (2005; UAM), and her MSc in education (2000) and BSc in psychology at the Pontifical Catholic University of Rio Grande do Sul (1998; PUC-RGS).
*Email*: alinehernandez@hotmail.com.

**Amanda Castro** is a professor in group training at the University of the Far South of Santa Catarina (UNESC), a supervisor in psychodrama at the Viver Psicologia Psicodrama School and the Amparense University Centre (UNIFIA), and a specialist in Developmental Psychology at the University of Araraquara (UNIARA). She gained her Master's in psychology at the Federal University of Santa Catarina (UFSC), and her PhD in social psychology and culture on the graduate programme of the Federal University of Santa Catarina (UFSC), where her research topic was representations and social practices. She conducts research connected with Social Representations Theory.
*Email*: amandacastrops@gmail.com.

**Andrea Barbará da Silva Bousfield** is a full professor in the Department of Psychology at the Federal University of Santa Catarina (UFSC). She gained her BSc in psychology at the Catholic University of Pelotas (UCPel; 2000), her MSc in psychology at the Federal University of Santa Catarina (UFSC; 2004), and her PhD in psychology at the Federal University of Santa Catarina (UFSC; 2007), with postdoctoral research at the University Institute of Lisbon (ISCTE-IUL) in Portugal and the University of Padua (UNIPD) in Italy. She coordinated the Laboratory of Communication and Cognition (LACCOS) and the psychology graduate programme (PPGP) at the Federal University of Santa Catarina (UFSC). Her areas of expertise are social psychology and social representations, knowledge, adolescence, and chronic disease.
*Email*: andreabs@gmail.com.

**Andréia Isabel Giacomozzi** is a professor in the Department of Psychology at the Federal University of Santa Catarina (UFSC). She gained her BSc, MSc and PhD in psychology at the Federal University of Santa Catarina (UFSC) with doctoral research stays in France and at the University of Padua (UNIPD) in Italy.

She is a member of the Laboratory of Communication and Cognition (LACCOS). Her areas of expertise are social representations, AIDS, women and gender, and sexuality.
*Email*: agiacomozzi@hotmail.com.

**Angela Arruda** is a pioneer in the study of social representations in Latin America, and a senior researcher at the Federal University of Rio de Janeiro and the University of Évora, with extensive experience in social psychology, with the emphasis on social representations and qualitative methodologies. She gained her BSc in psychology at the Federal University of Rio de Janeiro (UFRJ), her MSc in social psychology at the École des Hautes Études en Sciences Sociales (EHESS) in France, and her PhD in social psychology at the University of São Paulo (USP) with postdoctoral studies at the University Institute of Lisbon (ISCTE-IUL) in Portugal.
*Email*: arrudaa@centroin.net.br.

**Antonio Marcos Tosoli Gomes** is a full professor in the Department of Medical-Surgical Nursing and on the graduate programme in nursing in the Faculty of Nursing at the State University of Rio de Janeiro (UERJ). He is certified by the National Council for Scientific and Technological Development (CNPq) and is a young researcher at the Foundation Carlos Chagas Filho Research Support of the State of Rio de Janeiro.
*Email*: mtosoli@gmail.com.

**Brigido Vizeu Camargo** is a pioneer in the study of social representations in Brazil, and a professor at the Federal University of Santa Catarina (UFSC), where he founded the Laboratory of Communication and Cognition (LACCOS). He trained in psychology and undertook doctoral and postdoctoral studies in France. His contributions to the Brazilian Society of Psychology (SBP) and the Ibero-American Federation of Associations of Psychology (FIAP) reflect his interests in bridging social psychology in Brazil, France and Italy. He is currently interested in analysing the diffusion of Social Representations Theory in Brazil, as well as the construction of ideas and ideologies as representational systems.
*Email*: brigido.camargo@yahoo.com.br.

**Claudomilson Fernandes Braga** is a professor at the Federal University of Goiás (UFG) and a guest professor at the Universidad Estácio de Sá, the Lisbon Polytechnic Institute in Portugal, and the University of Buenos Aires in Argentina. He obtained his BA in public relations at the Fernando Pessoa University in Porto (Portugal) and his BA in social communications and public relations at the Federal University of Goiás, before taking his MSc in religion and his PhD in psychology at the Pontifical Catholic University of Goiás (PUC-Goiás), and conducting postdoctoral research in social psychology at the Rio de Janeiro State University (UERJ) and the Pontifical Catholic University of Goiás (PUC-Goiás). His research areas are social psychology, communication, group processes and identities.
*Email*: claudomilson_braga@ufg.br.

**Cristiene Adriana Silva Carvalho** obtained her BA in scenic arts at the Federal University of Ouro Preto (UFOP; 2008), and her MSc and PhD in education at the Federal University of Minas Gerais (UFMG). She specialised in educational research methods at the Federal University of Ouro Preto (UFOP; 2011) and in e-learning at the Federal Fluminense University (UFF; 2012). She is researcher at the Social Representations Study Group (GERES), University of Minas Gerais. Her areas of interest are social representations, education, theatre, music and culture.
*Email*: cristienecarvalho@gmail.com.

**Daniela B. S. Freire Andrade** is an associate researcher at the Federal University of Mato Grosso (UFMT), campus Cuiabá, and the coordinator of the Research Group in Child Psychology. She obtained both her BSc in psychology (1989) and her MSc in psychology (1992) at the Gama Filho University, and her PhD in education at the Pontifical Catholic University of São Paulo (2006; PUC-SP). She is a guest researcher at the International Centre for Studies on Social Representations and Subjectivity – Education (CIERS-ed) and a vice-coordinator at the National Association in Postgraduate Programmes and Research in Psychology (ANPEPP). Her research topics are education, social representations, child education, and child psychology.
*Email*: freire.d02@gmail.com.

**Denize Cristina de Oliveira** is a full professor at the Rio de Janeiro State University (UERJ). She gained her degree in nursing and her PhD in public health at the University of São Paulo (USP) and undertook postdoctoral research in social psychology at the École des Hautes Etudes en Sciences Sociales (EHESS) in Paris, France. She is an associate editor of various journals certified by the National Council for Scientific and Technological Development (CNPq). Her research areas are public health, nursing and social representations, focusing on topics such as social images and public health practices, symbolic incorporation of health systems, adolescence and health promotion, children's health, and HIV-AIDS.
*Email*: dcouerj@gmail.com.

**Deysi Ofelmina Jerez Ramírez** is an associate professor at the Research Institute for Risk Management and Climate Change at the University of Science and Arts of Chiapas (UNICACH), where she leads a training and research group and is the oordinator of the Master's in risk management and climate change. She gained her PhD in social and political science at the National Autonomous University of Mexico (UNAM), her MSc in social work at UNAM, and her undergraduate degree in social work at the Industrial University of Santander (Colombia). Her PhD thesis won the Serge Moscovici Prize in 2019. She is the research director in the risk management area of the company Engineering Geo-solutions and Systems (GISSA), and a member of the Mexican Association of Sciences for Regional Development.
*Email*: deysi.jerez@unicach.mx.

**Edna Maria Querido de Oliveira Chamon** is a full professor at the University Estácio de Sá and the University of Taubaté (UNITAU), in the Department of Architecture and Construction at the College of Civil Engineering, Architecture and Urbanism. She also collaborates at the University of Campinas (UNICAMP). She undertook her undergraduate and Master's degrees in education and her PhD in psychology at the University of Toulouse II (Le Mirail), and conducted post-doctoral research in education at the University of Campinas (UNICAMP). Her research on Social Representations Theory and education is a required reference in the field.
*Email*: edna.chamon@gmail.com.

**Elizabeth Fernandes de Macedo** was Professor of Curriculum Studies at the State University of Rio de Janeiro (UERJ), and Associate Provost for Graduate Programmes (2008–2015). She was a Visiting Scholar at the University of British Columbia (UBC), Canada, in 2007, and at Columbia University, USA, between 2013 and 2015. She obtained her PhD in education at the State University of Campinas (1996), and has served as President of the International Association for the Advancement of Curriculum Studies. From 2010 to 2013 she edited the journal *Transnational Curriculum Inquiry*, sponsored by that Association. She currently serves as an Associate Editor of the *Journal of Curriculum Studies* and as a member of the editorial board of *Curriculum Inquiry*, and coordinates an inter-institutional group of researchers who are working collaboratively with state-level bureaucracies, schools and teachers to produce situated public policies in response to the compulsory national curriculum.
*Email*: bethmacedo@pobox.com.

**Jorge Ramón Serrano Moreno** is a senior professor, retired, with five decades of academic experience. He served as a researcher at the Multidisciplinary Regional Research Centre (UNAM), the Centre for Economic and Social Studies of the Third World (CEESTEM), the College of Mexico (ColMex), and the Centre of Higher Research of the National Institute of Anthropology and History (CIS-INAH) in Mexico. He gained his PhD in social science – sociology at the University of Chicago, and has MSc degrees in sociology and philosophy, and undergraduate degrees in sociology, philosophy, theology and Letters. He has studied in Mexico, Italy, Germany, the USA and India. He is the founder of diverse international and national research associations and the author of twenty books, thirty book chapters and twenty-five research articles. He has directed, from a multidisciplinary perspective, over forty research groups and research projects, and was the tutor of almost fifty undergraduate and graduate students.
*Email*: jrsmhi@gmail.com.

**José Antonio Castorina** is a full professor at the Chair of Psychology and Genetic Epistemology at the University of Buenos Aires (UBA). He is Director of the Institute of Research in Educational Science (IICE-UBA), and a professor at the Pedagogical University of Argentina (UNIPE). He is certified by the National Council for Research and Technology (CONICET). He was awarded an honorary

doctorate by the National University of Rosario (Argentina), and is an honorary professor at the University of San Marcos (Peru). He gained his undergraduate degree in psychology at UBA, his Master's in philosophy at the Argentinian Society for Philosophical Analysis (SADAF), and his PhD in education at the Federal University of Rio Grande do Sul (UFRGS). He is currently researching genetic epistemology and the notions of school authority in children, inserting Piagetian theses into the contemporary debate on the relationships between knowledge, conceptual change, social representations and school learning.
*Email*: ctono@fibertel.com.ar.

**Lúcia Villas Bôas** is Director-Vice-President of the Carlos Chagas Foundation, Coordinator of the UNESCO Chair on Teaching Professionalization, and scientific director of the Franco-Brazilian Chair Serge Moscovici as well as a researcher and teacher at the Universidad de San Pablo (USP). She obtained her undergraduate degree in history from the University of São Paulo (1996), her Master's in education and PhD in psychology at the Pontifical Catholic University of São Paulo (PUC-SP), and undertook postdoctoral research at the École des Hautes Études en Sciences Sociales in France. She is member of the Scientific Council of the Réseau Mondial Serge Moscovici (REMOSCO/EHESS, France), the UNESCO Chair on Training and Professional Practices (France). She is executive director of the journal *Cadernos de Pesquisa* at the Carlos Chagas Foundation and a member of the social representations research group at the National Association in Postgraduate Programmes and Research in Psychology (ANPEPP).
*Email*: lboas@fcc.org.br.

**Luiz Paulo Ribeiro** is an associate professor in the Faculty of Education, Federal University of Minas Gerais (UFMG) in Brazil. He gained his undergraduate degree in psychology at the Pontifical Catholic University of Minas (PUC-Minas), his Master's degree in health promotion and violence prevention at the Faculty of Medicine of the Federal University of Minas Gerais (FM-UFMG), and his PhD in education knowledge and social inclusion at the Federal University of Minas Gerais (UFMG) with a postdoctoral residency studying the correlations between identity and social representations. Currently, he coordinates the activities of the Social Representations Study Group (GERES) and participates in the collegiate course in education in the countryside.
*Email*: luizpr@ufmg.br.

**Maria de Fátima de Souza Santos** is a full researcher in the Department of Psychology at the Federal University of Pernambuco (UFPE). She completed her undergraduate studies in psychology at the Federal University of Pernambuco and her PhD in psychology at the Université Toulouse le Mirail in France. She is certified by and a consultant to the National Council for Scientific and Technological Development (CNPq), and an evaluator for the Coordination for the Improvement of Higher Education Personnel (CAPES). She currently coordinates the Centre for Studies in Social Representations Serge Moscovici. She conducts research in the areas of social psychology and developmental psychology, with an emphasis on

social representations, working mainly on the following topics: violence, adolescence, old age, health, and social practices.
*Email*: mfsantos@ufpe.br.

**María de Fátima Flores Palacios** is a full researcher at the Peninsular Centre in Humanities and Social Sciences (CEPHCIS) at the National Autonomous University of Mexico (UNAM). She did her undergraduate degree in psychology at the National Autonomous University of Mexico (UNAM), with a specialization in mental health at the National Institute of Psychiatry. She undertook psychoanalytic training at the Comillas Pontifical University in Madrid and obtained her PhD in psychology at the Autonomous University of Madrid (UAM) in Spain. Certified by the National Council of Researchers (SNI) and a member of the Mexican Academy of Science, she was awarded the Sor Juana Inés de la Cruz Prize (2014). She is an invited professor at the University of Evora in Portugal, the Social Psychology Laboratory in France and the National University of Costa Rica. Her current areas of research are gender and social representations, vulnerability, HIV-AIDs, and public health.
*Email*: fatimaflor@hotmail.com.

**María de Lourdes Soares Ornellas** is a full professor at the State University of Bahía (UNEB), where she leads the Study Group in Education Research and Social Representations (GEPPE-rs) and coordinates the International Research Centre in Social Representations (NEARS). She obtained her BA in pedagogy and psychology at the State University of Bahía (UNEB), her MSc and PhD in educational psychology at the Pontifical Catholic University of San Pablo (PUC-SP), and undertook postdoctoral research at the University of San Pablo (USP). She is a psychoanalyst and a member of the Psychoanalytic Association of Bahía. She collaborates in the International Centre for Studies in Social Representations and Subjectivity – Education (CIERS-ed) and the study group in discourse analysis at the Catholic University of Salvador (UCSAL). Her research focuses on articulating education, psychoanalysis and SRT.
*Email*: ornellas1@terra.com.br.

**Maria Isabel Antunes-Rocha** is a professor at the Federal University of Minas Gerais (UFMG). She coordinates the Social Representations Study Group (GERES) that develops research underpinned with the research and training of professionals in basic education and education in the field. She has undergraduate and postgraduate degrees from the Federal University of Minas Gerais (UFMG) and undertook postdoctoral studies at the State University of São Paulo Júlio de Mesquita Filho. Her work in professional training for education constitutes a group of active minorities, allowing the development of research in social representations in movement, with important contributions to Social Representations Theory.
*Email*: isabelantunes@fae.ufmg.br.

**Mariana Bonomo** is a professor at the Espírito Santo Federal University (UFES) and collaborates at the University of Bologna (UNIBO), Italy. She obtained an

undergraduate degree at the Espírito Santo Federal University (UFES) and a PhD in psychology at the Espírito Santo Federal University (2010) with a sandwich PhD at the University of Bologna, Italy. She currently coordinates the exchange programme between the Espírito Santo Federal University (UFES) and the University of Bologna (UNIBO). Certified by the National Council for Scientific and Technological Development (CNPq), she is a member of the social representations research group at the National Association in Postgraduate Programmes and Research in Psychology (ANPEPP). Her research areas are social identity, social psychology, gender, mental health and hospital psychology.
*Email*: marianadalbo@gmail.com.

**Marlon Xavier** is currently a substitute professor in the Department of Psychology at the Federal University of Santa Catarina (UFSC), and an associate professor (on sabbatical) at the University Caxias do Sul. He obtained his PhD in social psychology at the Autonomous University of Barcelona (UAB; 2010) in Spain, his MSc in social psychology at the Pontifical Catholic University of Rio Grande do Sul (2001; PUC-RGS), and his BSc in psychology at the Federal University of Rio Grande do Sul (UFRGS; 1996). He has been a Visiting Research Fellow at the Laboratory of Communication and Cognition (LACCOS) and at the Bauman Institute/School of Sociology and Social Policy at the University of Leeds in England (2011). He is a Chartered Member of the British Psychological Society. His research areas are social representations, subjectivity, addiction, religiosity and consumerism.
*Email*: marlonx73@gmail.com.

**Mireya Lozada** is a full professor at the Institute of Psychology, Central University of Venezuela (UCV), with over three decades of research experience in social representations. She gained her PhD in psychology at the Université de Toulouse Le Mirail in France, and her MSc in social psychology at the Central University of Venezuela. She is the coordinator of the research project "Venezuela: Scenarios for the Future. In search of an inclusive and sustainable peace" And the coordinator of the Network Latin America in Movement: Psychosocial Outlooks ('La Red Latinoamérica en Movimiento: Miradas Psicosociales'). Her research areas are: social violence, peace, polarization of public space, damage repair, democracy, and everyday life.
*Email*: mireyaloza@gmail.com.

**Pedrinho Guareschi** is a professor and researcher at the Federal University of Rio Grande do Sul (UFRGS), an international lecturer with many decades of experience in Social Representations Theory, and an associate researcher at the Carlos Chagas Foundation (CCF). He was a professor at the Pontifical Catholic University of Rio Grande do Sul (PUC-RGS), a lecturer in *stricto sensu* at the Pontifical Catholic University of Paraná (PUC-PR), and a visiting professor at the Federal University of Health Sciences in Porto Alegre (UFCSPA). He has a BA in philosophy (Faculdade de Filosofia Imaculada Conceição), a BA in theology (Instituto Redentorista de Estudios Superiores de San Pablo), an MSc

in sociology (PUC-RGS), a BA in Letters (Passo Fundo University), an MSc in social psychology (Marquette University Milwaudee) and a PhD in social psychology (University of Wisconsin at Madison, USA). He has undertaken postdoctoral research at the University of Wisconsin, USA; the University of Cambridge, UK; and the University La Sapienza, Italy. He participates in the Latin American Network of Epistemological Studies in Educational Policy (ReLePe) and the Centre for International Studies in Social Representations (NEARS). His main research areas are mass media, ideology, social representations, ethics, communication and education.
*Email*: pedrinho.guareschi@ufrgs.br.

**Pedro Humberto Farias Campos** is a full professor at the Salgado Filho University (UNIVERSO) in Rio de Janeiro. He completed his undergraduate degree in psychology at the Catholic University of Goiás (1988), his Master's in Education at the Federal University of Goiás (1994), his Master's in social psychology at the Université de Provence (1995), and his PhD in psychology at the University of Provence (1998) in France. His areas of interest are social representations, the structural school of social representations, social exclusion, social practices, and violence.
*Email*: pedrohumbertosbp@terra.com.br.

**Priscila Pereira Nunes** is an undergraduate psychology student at the Federal University of Santa Catarina (UFSC), and a psychology assistant at the Júnior Autojun Company. She is involved in the Scientific Initiation Scholarship Programme (PIBIC) and is a member of the Laboratory of Communication and Cognition (LACCOS).
*Email*: priscila.ppnunes@gmail.com.

**Renata Lira dos Santos Aléssio** is an associate professor in the Department of Psychology at the Federal University of Pernambuco (UFPE). She has undergraduate degrees in psychology (UFPE, 2003) and in psychology training (2004), and undertook her Master's and PhD in social psychology at the Université D'Aix-Marseille I (Université de Provence, 2007) in France. She develops research projects in the areas of adult and adolescent health, bioethics, human embryos, ageing and intergroup relations. Her main research areas are social representations, violence, human development, educational practices and adolescence.
*Email*: renatalir@gmail.com.

**Rita de Cássia Pereira Lima** is a professor at the University Estácio de Sá (UNESA), and an associate researcher at the International Centre of Studies on Social Representations and Subjectivity – Education (CIERS-ed) at the Carlos Chagas Foundation. She gained her undergraduate degree in occupational therapy at the Pontifical Catholic University of Campinas (1982) and in educational sciences at the Université René Descartes Paris V (1987). Her PhD in educational sciences was obtained at the Université René Descartes Paris V (1994) in France. She is a member of the working group in social representations at the National Association in Postgraduate Programmes and Research in Psychology (ANPEPP), and a reviewer for various academic journals. Her areas of expertise are social representations and violence, teaching, school and professional education, and fashion. *Email*: rita.lima@estacio.br.

**Romilda Teodora Ens** is currently a research associate at the Carlos Chagas Foundation (FCC), where she participates in the International Centre of Studies on Social Representations and Subjectivity – Education (CIERS-ed) and the UNESCO Chair on Teacher Professionalization. She is *stricto sensu* lecturer in the programme of graduate studies in education at the Pontifical Catholic University of Paraná (PUC-PR). She graduated in education from the Federal University of Paraná (1971), and holds a degree in law from the Federal University of Paraná (1974), a Master's in education from the Federal University of Paraná (1981) and a PhD in educational psychology from the Pontifical Catholic University of São Paulo (PUC-SP, 2006). She undertook postdoctoral research at the University of Porto in Portugal, and participates in the Latin American Network of Epistemological Studies in Educational Policy (ReLePe) and the Centre for International Studies in Social Representations (NEARS). Her main research areas are educational policies, teacher professionalization and social representations. *Email*: romilda.ens@gmail.com.

**Sabrine Mantuan dos Santos Coutinho** is a researcher at the Fluminense Federal University (UFF) and an invited professor at the Espírito Santo Federal University (UFES). She obtained her graduate, postgraduate, doctoral and postdoctoral degrees in psychology at the Espírito Santo Federal University. She is a member of the research groups "Memory, Identity and Social Representations" at the National Association in Postgraduate Programmes and Research in Psychology (ANPEPP) and "Sociocultural Representations Identities and Practices" at the National Council for Scientific and Technological Development (CNPq). Her main areas of research are: gender, social representations, family and marital relations, motherhood and fatherhood, infants and adolescents in conditions of vulnerability, and psychological interventions. *Email*: sabrinems@hotmail.com.

**Sandra Lucia Ferreira** is a teacher and researcher at the City University of São Paulo (UNICID) and an invited professor at the Paulista State University (UNESP). She is an associate researcher at the International Centre of Studies on Social Representations and Subjectivity – Education (CIERS-ed) at the Carlos Chagas Foundation (FCC) and the Centre for International Studies in Social Representations (NEARS). She has an undergraduate degree in pedagogy (1987), a Master's in education (1992), and a PhD in education (2005), all awarded by the Pontifical Catholic University of São Paulo (PUC-SP). Her areas of expertise are social representations, educational evaluation, image, and institutional evaluation. *Email*: 07sandraferreira@gmail.com.

**Susana Seidmann** is a full professor of social psychology at the University of Buenos Aires (UBA), Dean of the Faculty of Humanities, and Director of the Master's in community social psychology at the University of Belgrano.

She is an associate researcher at the International Centre of Studies on Social Representations and Subjectivity – Education (CIERS-ed), Carlos Chagas Foundation, and the coordinator of the research group of the International Centre for Studies in Social Representations – Subjectivity and Education in Buenos Aires.
*Email*: susiseidmann@yahoo.com.ar.

**Suzzana Alicia Lima Almeida** is a professor at the State University of Bahía (UNEB). She has an undergraduate degree in pedagogy from the Catholic University of Salvador, a specialization in methodology and higher education teaching, a Master's in education from the Université du Québec à Chicoutimi in Canada, and a PhD in Education at the State University of Bahía (UNEB). She is a member of the Study Group in Education Research and Social Representations (GEPPE-rs) and of the National Council for Scientific and Technological Development (CNPq), linked to the International Centre of Studies on Social Representations and Subjectivity – Education (CIERS-ed). She is also on the executive committee of the State Symposium in Social Representations and Education (SIERS). Her areas of expertise are education, school, epistemology, social representations, teaching-learning, and reflexive teachers.
*Email*: suzzanaalice@hotmail.com.

**Tarso Mazzotti** is a full professor at the Federal University of Rio de Janeiro (UFRJ), and an associate researcher at the Carlos Chagas Foundation and the Estácio de Sá University. He has a degree in pedagogy from the State University Paulista Júlio de Mesquita Filho (1972), a Master's degree in education from the Federal Universidade of São Carlos (1978), and a PhD in education from the University of São Paulo (USP, 1987). He has extensive experience in the field of education, philosophy of education and social representations, and has made significant contributions to the study of the subject. His main areas of research are social representations theory, philosophy of education, epistemology, and social representations.
*Email*: tmazzotti@gmail.com.

**Zeidi Araujo Trindade** is full professor of social psychology at the Espírito Santo Federal University (UFES). She received her PhD in psychology at the University of São Paulo (USP), where she also undertook postdoctoral research. She is a member of the directory of the National Association in Postgraduate Programmes and Research in Psychology (ANPEPP), an evaluator for the Coordination for the Improvement of Higher Education Personnel (CAPES), and coordinator of the Social Psychology Network (RedePso). Her areas of expertise are: social representations, social practices, culture, gender, fatherhood and motherhood, reproductive health, and youth.
*Email*: zeidi.trindade@gmail.com.

# Index

© Springer Nature Switzerland AG 2021
C. Prado de Sousa and S. E. Serrano Oswald (eds.),
*Social Representations for the Anthropocene: Latin American Perspectives*,
The Anthropocene: Politik—Economics—Society—Science 32,
https://doi.org/10.1007/978-3-030-67778-7

Printed in the United States
by Baker & Taylor Publisher Services